POLYMER MODIFIERS AND ADDITIVES

PLASTICS ENGINEERING

Founding Editor

Donald E. Hudgin

Professor
Clemson University
Clemson, South Carolina

1. Plastics Waste: Recovery of Economic Value, *Jacob Leidner*
2. Polyester Molding Compounds, *Robert Burns*
3. Carbon Black-Polymer Composites: The Physics of Electrically Conducting Composites, *edited by Enid Keil Sichel*
4. The Strength and Stiffness of Polymers, *edited by Anagnostis E. Zachariades and Roger S. Porter*
5. Selecting Thermoplastics for Engineering Applications, *Charles P. Mac-Dermott*
6. Engineering with Rigid PVC: Processability and Applications, *edited by I. Luis Gomez*
7. Computer-Aided Design of Polymers and Composites, *D. H. Kaelble*
8. Engineering Thermoplastics: Properties and Applications, *edited by James M. Margolis*
9. Structural Foam: A Purchasing and Design Guide, *Bruce C. Wendle*
10. Plastics in Architecture: A Guide to Acrylic and Polycarbonate, *Ralph Montella*
11. Metal-Filled Polymers: Properties and Applications*, edited by Swapan K. Bhattacharya*
12. Plastics Technology Handbook, *Manas Chanda and Salil K. Roy*
13. Reaction Injection Molding Machinery and Processes, *F. Melvin Sweeney*
14. Practical Thermoforming: Principles and Applications, *John Florian*
15. Injection and Compression Molding Fundamentals, *edited by Avraam I. Isayev*
16. Polymer Mixing and Extrusion Technology, *Nicholas P. Cheremisinoff*
17. High Modulus Polymers: Approaches to Design and Development, *edited by Anagnostis E. Zachariades and Roger S. Porter*
18. Corrosion-Resistant Plastic Composites in Chemical Plant Design, *John H. Mallinson*
19. Handbook of Elastomers: New Developments and Technology, *edited by Anil K. Bhowmick and Howard L. Stephens*
20. Rubber Compounding: Principles, Materials, and Techniques, *Fred W. Barlow*

Additional Volumes in Preparation

POLYMER MODIFIERS AND ADDITIVES

edited by

John T. Lutz, Jr.

JL Enterprises
Bensalem, Pennsylvania

Richard F. Grossman

The Hammond Group
Hammond, Indiana

MARCEL DEKKER, INC.

NEW YORK • BASEL

Portions reprinted from *Thermoplastic Polymer Additives: Theory and Practice* (J. T. Lutz, Jr., ed.), Marcel Dekker, Inc., NY, 1989, and *Encyclopedia of PVC: Second Edition, Revised and Expanded, Volume 2* (L. I. Nass and C. A. Heiberger, eds.), Marcel Dekker, Inc., NY, 1988.

ISBN: 0-8247-9949-6

This book is printed on acid-free paper.

Headquarters
Marcel Dekker, Inc.
270 Madison Avenue, New York, NY 10016
tel: 212-696-9000; fax: 212-685-4540

Eastern Hemisphere Distribution
Marcel Dekker AG
Hutgasse 4, Postfach 812, CH-4001 Basel, Switzerland
tel: 41-61-261-8482; fax: 41-61-261-8896

World Wide Web
http://www.dekker.com

The publisher offers discounts on this book when ordered in bulk quantities. For more information, write to Special Sales/Professional Marketing at the headquarters address above.

Current printing (last digit):
10 9 8 7 6 5 4 3 2 1

PRINTED IN THE UNITED STATES OF AMERICA

Preface

The modification of the properties and processability of polymers through addition of other substances is a field huge in scope and fast moving in character. Many of the technologists who are ''skilled in the art'' could be described as having encyclopedic knowledge of some subset of this field. Is this a prerequisite for successful practice? Not necessarily. In common with other areas of applied science, what is necessary is an understanding of how additives function. The combination of this understanding and some familiarity with past work, applied with scientific discipline to a particular problem, gives the best chance of

1. Coming up with a solution rapidly enough for the experimental work to provide commercial success and personal satisfaction
2. Reaching a major breakthrough or invention

What do we mean by an ''understanding'' of how polymer additives function? Nothing more than a picture, an artificial construct that appears to account for experimental observations. The picture provides rationalization of why one additive appears more effective than another, why one polymer is more or less receptive, and so forth. The technologist, realizing that the ''mechanism'' is merely a picture, asks persistently: Is the picture distorted? Is it too simple? Do the observations at hand suggest a somewhat different picture, whose details will lead to a rapid solution/major breakthrough?

It is our purpose to bring up to date the art and science of modifying polymers through the use of additives. To that end, chapters on the most significant additives have been contributed by leading experts. The reader will find diverse approaches to understanding. This is entirely intentional. The ability to incorporate the rationalizations of others into one's own understanding is vital to education. It is, above all, our purpose to be educational, that is, to leave the reader somewhat closer to devising solutions.

John T. Lutz, Jr.
Richard F. Grossman

Contents

Contributors

Albin P. Berzinis   General Electric Company, Pittsfield, Massachusetts

Melvin M. Gerson†   Podell Industries, Inc., Clifton, New Jersey

Saul Gobstein   Sa-Go Associates, Inc., Shaker Heights, Ohio

John R. Graff†   Bayer Corporation, Haledon, New Jersey

Joseph Green   Consultant, Joseph Green Associates, Monroe Township, New Jersey

Richard F. Grossman   Halstab Division, The Hammond Group, Hammond, Indiana

Dale Krausnick   NSF Enterprises, Toledo, Ohio

John T. Lutz, Jr.   JL Enterprises, Bensalem, Pennsylvania

Thomas Martini   Hoechst AG, Frankfurt, Germany

K. K. Mathur†   Specialty Minerals, Easton, Pennsylvania

Robert P. Petrich   Rohm and Haas Company, Philadelphia, Pennsylvania

Robert A. Schweitzer   Owens Corning, Granville, Ohio

D. B. Vanderheiden   Specialty Minerals, Easton, Pennsylvania

Lewis B. Weisfeld   Consultant, Philadelphia, Pennsylvania

Albert W. Winterman   BASF, Wyandotte, Michigan

---

†Deceased.

Introduction to Plastics Additives

Plastics are chemicals that can be formed into useful shapes. Stated another way, plastics are organic materials that have a visual appearance appropriate to the intended function and that perform this function safely for the life of the product at a competitive cost. This book discusses ways of modifying the pure individual materials by means of additives to achieve these results.

Polymer additives fall into three general functional classes:

> Those which are essential for fabrication of parts
> Those which improve properties
> Those which correct problems caused by other additives

I. SOME FUNDAMENTALS ABOUT POLYMERS AND ADDITIVES

Polymers differ from all other chemicals because of their relatively long chains and high molecular weights. Plastics are polymers that have been modified for commercial use by the use of additives.

It is safe to assume that there is an additive to improve the best polymer for any application and that your job is to find it and use it properly. If you manufacture additives, assume that better additives are needed and can be found or, at least, that the materials handling of the additive can be modified to make it easier for customers to use or inventory.

II. FUNCTIONS OF ADDITIVES

Plastics additives serve three different and important functions. First, additives are required in order to process or fabricate many polymers. Polyvinyl chloride (PVC), polypropylene (PP), poly(ethylene-carbon monoxide) (E/CO), and polyoxymethylene (POM) have thermal stability limitations that force the producers to find and utilize stabilizers that allow melt processing without loss of properties. Even room-temperature storage requires that some polymers be stabilized from the effects of the oxygen in air.

Some plastics and monomers will not wet reinforcing glass fibers, and wetting agents must be used in order to process the mixture into a useful plastic. These additives are usually put onto the glass by the glass manufacturer. Sometimes, the plastic compounder adds his own wetting agent or coupling agent (there is a big difference) to gain a competitive advantage. This illustrates the second type of additive, those used for property enhancement. In the above example, a wetting agent may be necessary to process the plastic, but a coupling agent can be added as an enhancement to provide long-term durability.

Fillers improve the flexural modulus and DTUL (deflection temperature under load). Color, objectionable odor, surface gloss, and other properties can be improved by additives of this second category. Polymers having elastomeric or "rubberylike" properties improve impact resistance. Fibers can do a bit of both. Flame retardants enhance the safety properties of a plastic and make it more valuable. Additives of this second type pay for themselves by giving more performance value than the cost of the additive.

The third type or class of additive corrects the flaws of the first two classes. Plasticizers are required to fabricate flexible PVC; lubricants are required for unplasticized PVC. Some of these additives provide food for the growth of microorganisms such as bacteria and fungi. A further additive is needed to correct this flaw, created by required additives.

Mineral fillers reduce impact while increasing modulus. Impact modifiers are used to correct these flaws. Coupling agents along with decoupling agents (often wetting agents are used for this purpose) are used to correct these flaws also.

III. HOW THIS BOOK HELPS

This book is designed to give even the casual reader a better understanding of additives and how they relate to real-world situations. The various chapters were carefully written by experts who have worked for many years with the additives they discuss. All the fundamental types of additives are included.

IV. POLYMERS AND PLASTICS: WHY ADDITIVES ARE ESSENTIAL

Polymers are chemicals that have very high molecular weights. Molecular weight (MW) is a measure of the length of the chain (or chains) making up the polymer molecule. When the chains are linear (single chains) and not bonded (cross-linked) to other chains, the polymer is a thermoplastic. Thermoplastic polymers can be melted and remelted many times, unlike thermoset polymers.

If the chain is bonded to many other chains, then the plastic is crosslinked and it does not melt before it decomposes. This is the characteristic of thermoset polymers. If the chemical is very highly crosslinked and each atom is bonded to several other atoms, then the material is a ceramic. Diamond is carbon bonded to itself in a three-dimensional fashion; thus, it is the simplest ceramic. Other ceramics are alumina, silica, silicon nitride, boron carbide, and silicon oxynitride. Ceramics are not covered in this book. Metals composed of mostly single atoms are also not covered in this book.

Thermoset polymers do not melt and flow. For this reason, this class of polymer must be compounded and modified before crosslinking. Thermoplastics, on the other hand, which do melt and flow are usually compounded with additives after the linear polymer is made.

When thermoplastics melt, they flow with different degrees of difficulty. Viscosity is a measure of resistance to flow. High-MW polymers have high viscosities and they do not flow easily. Thermoset polymers have infinite or near-infinite viscosity because their MW is so high and they do not flow. Sometimes, a given part consists of a single molecule of thermoset plastic.

Thermoplastic polymer viscosity increases with molecular weight. As the chain grows, the viscosity increases forever. To summarize to this point: polymers differ from other molecules because they consist of long chains of monomer units. The process of polymerization adds value to monomers, so polymers are more expensive than monomers. In other words, the value of polymers is enhanced by their molecular weight. The higher the molecular weight (chain length) the higher the viscosity and the greater the cost of manufacturing.

Unlike viscosity, many mechanical properties reach a maximum and do not increase beyond that point. The specific point of maximum property attainment is different for each monomer.

Because polymers are made from monomers, they are more expensive than monomers. We pay a lot for molecular weight, so we want to be sure that we do not lose it during processing. MW is what makes polymers different from other chemicals and gives these chemicals mechanical properties that are commercially valuable. Because you have paid for molecular weight, you do not want to lose it. Stabilizers and antioxidants are additives that protect MW during processing and, consequently, save or maintain mechanical properties. This is an essential function of additives.

It is extremely difficult to melt process polymers like PP, PVC, E/CO, and acetal without additives because they will lose MW quite rapidly in the melt. Many other polymers such as polyethylene (PE), polystyrene (PS), poly(acrylonitrile-butadiene-styrene) (ABS), and many thermoplastic elastomers (TPEs) also benefit from stabilizers by avoiding excessive MW changes.

Plastics can be regarded as polymers that have been improved with additives and certain chemical modifications to make them commercially important materials. Plastics are important because they are easy to fabricate into commercially important parts.

To summarize to this point, additives and modifiers have three main functions:

Allow fabrication of useful and sometimes very complex parts
Enhance properties
Correct problems caused by other additives and modifiers

V. CHAIN LENGTH AND POLYMER PROCESSING

As noted earlier, there is a relationship between chain length and mechanical properties that follows directly from the nonbonded interaction theory of polymer strength. A polymer chain with strong electrostatic attractions to itself will reach its maximum strength at a relatively low molecular weight. On the other hand, a polymer chain with low attractive forces will require the summation of many more "mer group" forces to reach high levels of attraction. This point is illustrated with three important industrial polymers; polyethylene, Nylon 6,6, and polytetrafluoroethylene.

Nylon 6,6 has a backbone polymer chain that contains polar amide groups. These groups have great electrostatic attraction for each other, thus adjacent chains of Nylon 6,6 attract each other very strongly. Nylon 6,6 reaches a high strength at low chain lengths because of this. In fact, Nylon 6,6 (and other nylons as well) reach maximum strength at a chain length of about 100 monomer units.

Linear polyethylene, which has very low polarity, requires long chains to reach maximum tensile strength and other mechanical properties. Maximum properties of polyethylene usually occur after 1000 monomer units make a chain. This is over 10 times the corresponding value for nylon.

In the case of polytetrafluoroethylene (PTFE), which is one of the least polar polymeric materials known, the chain must be 5000–10,000 monomer units long before maximum properties are approached. This is due to both the low attraction of one "mer" unit for another and the size of the fluorine atom, which keeps the carbon atoms on adjacent chains further away from each other than in the case of polyethylene. This leads to extremely long chain lengths, and because long chains lead to high viscosity, the processing of PTFE (and UHMWPE) is extremely difficult by conventional methods of screw extrusion and injection molding.

All three of the above molecules are close to linear, with few side chains. When side groups are added, there is a reduction in viscosity and a requirement for slightly higher molecular weights in order to reach maximum properties. In general, the trade-off is acceptable because processing ease increases faster than properties fall. It is not a satisfactory answer if maximum properties are required for a given type of polymer. In those cases, external lubricants can be utilized to facilitate processing.

Often copolymers are used to enhance processability. This is accomplished entirely by altering the polymer chain structure and is a technique that has made polymers extremely versatile and useful materials. Copolymerization tends to tie up a production line; therefore, it can not be practiced unless there is a large and continuing need for the product. When demand is low, additives and blends must be used to obtain the benefits of copolymerization. Of course, additives can be beneficial to copolymers, just as they help homopolymers.

When homopolymers and copolymers can no longer meet the demands of customers, grafting, crosslinking, and additional selective modification by additives such as impact modifiers, fibers, fillers, colorants, and so forth must be utilized.

VI. TYPES OF PROPERTIES

There are many different types of properties of plastics that are measured routinely, and other properties that are only measured occasionally. Properties can be placed into several categories:

> Properties that change with processing conditions
> Properties that change with testing conditions
> Properties that change with pretesting conditioning
> Properties that are affected by impurities or flaws
> Properties that change with additives
> Properties that do not change

Although this book is about additives, the reader must remember that many other factors can change the properties of plastics. Cost is the driving force in undertaking all technical problems, and plastics formulation is no exception. Even when medical, military, or aerospace applications are involved, cost-effectiveness must be considered, and there are no long-term exceptions to economics.

Technical factors to consider when reviewing a new formulation undertaking include the following:

1. Ease of fabrication
2. Ease of assembly of parts made from the composite material

3. Part appearance
4. Environmental extremes of the application
5. The specifications that are set by your customer

Whenever possible, the formulator should learn about the applications for the plastic material that is being manufactured. This learning must include a wide range of secondary operations like painting, decorating, gluing, and heat welding. Many formulations that were considered successful from the standpoint of meeting all technical specifications have failed in commerce because of unspecified factors such as the vital issues of assembly, shipping, secondary processing, machining, and similar considerations.

The objective of the formulation problem is to meet all specifications (and extended specifications as defined earlier) in a single formulation in the most cost-effective manner. In the act of plastics formulation, there usually will be 4–10 specification criteria that are considered vital to the success of the material being developed. The dependent variables of the formulation procedure are physical properties, ease of processing of the composite, and attributes such as mentioned earlier. The independent variables are the additives and modifiers and the processes by which they are combined to make the required plastic material which meets or exceeds all specifications in a cost-effective way.

The proper choice of resin or resins may lead to automatic conforming to many or even a majority of the specifications. Use temperature, solvent resistance, color, and transparency are a few examples.

System constraints such as cost of materials, machines available for the compounding and fabrication work, and limits of concentration of certain additives should now be considered. In thermoplastic resins, there is a limit to the amount of solid fillers and reinforcements that can be used. More than about 60–70% by weight of fiber glass (this is around or over 30% by volume for many polymers) is extremely difficult to process by conventional extrusion in a resin like nylon or polyethylene terephthalate (PET). On the other hand, thermoset resins, with their special fabrication techniques that start with low-viscosity monomers, might hold up to and over 80% solids in processes such as filament winding, casting, pultrusion, or hand layups.

At this point, the formulator must select which few dependent variables or properties will be used to judge progress of the formulation. After some formulation work has been done and some examples are beginning to approach the final targets, more or all of the dependent variables should be measured. Of course, there are exceptions to this, but work will proceed much faster with fewer, better selected targets early in the process.

Some key points in the formulation process are as follows:

1. Can this plastic be fabricated into a useful shape?
2. Is the shrinkage correct?

3. Are all the property enhancements made (color, impact, load bearing, durability)?
4. Did any of the additives create a problem for which another additive is required?
5. Will the plastic function in its intended application?
6. Will the plastic survive unintended but possible end-use abuses?
7. Does the plastic meet all specifications, especially cost?

Of course, you will have to run all the specification tests and possibly more to answer these questions. If you suspect a problem before you get to the end of your work, you should add a quick test for the problem property to your short list of in-process tests. This book will aid you in addressing all of the those concerns about additives.

VII. SOME HIGHLIGHTS OF THIS BOOK

After reading this chapter, a brief review of all the chapters will provide a good overview of what to expect. There will be an in-depth review of antioxidants and ultraviolet (UV) stabilizers. Some of the main points are as follows:

1. There are primary and secondary antioxidants.
2. Some antioxidants turn color as they function.
3. Some antioxidants work during processing and others are more effective at room temperature for longer times.
4. There is a chemical/physical difference between antioxidants and UV stabilizers.

Filler basics encompass mass and volume pricing, size, shape, color, hardness, and so forth. These materials reduce price per kilogram, but they increase volume pricing because of their higher density, so they must contribute value to justify the increased part cost. Benefits of fillers include increased load bearing, hardness, and density.

The differences among coupling agents, surface modifiers, and wetting agents has to do with the chemical bonding versus physical interactions of each with the organic or inorganic material involved.

The chapter on flame retardants discusses test methods, nonhalogen types, antimony–halogen systems, smoke and toxic gases, and recent developments including phosphorus–bromine synergy.

Impact modifiers must balance load bearing and impact resistance.

Reinforcement technology includes natural and synthetic materials, specific strength, how fibers are used, and roving, chopped, woven, and the new grades of glass fibers.

The very important area of thermal stabilizers reviews the structure and degradation of PVC—heavy metals and alternative stabilizers for PVC and other polymers. This technology applies to many halogen-containing polymers such as polychloroprene, copolymers containing vinyl chloride monomer (VCM), and chlorinated polyethylene (CPE).

Plasticizers and lubricants for PVC and polymers other than PVC will be useful for many different formulations and processing situations. There are many interesting facts to know in this area of polymer modification:

> Color theory teaches the problems of making white whites and why base resin color is so critical. The cost of color is related to thermal stability and end-use durability.
>
> Cross-linking is an important way of increasing durability at elevated temperature while maintaining the benefits of thermoplastics during the shape generation stage of processing.
>
> Mechanochemistry deals with the necessary interaction of mechanical devices and chemistry. Mixing of viscous materials and maintaining temperature control are essential for the chemistry to occur in a controlled fashion.
>
> Foams and blowing agents have undergone many changes recently with the acceptance of exothermic chemical blowing agents and the replacement of chlorofluorocarbons (CFCs) with supercritical CO_2, cyclopentane, and isobutane.

Several miscellaneous additives to reduce electrostatic charge, interfere with biological growth, and perform the other necessary functions involved with compounding are covered in this book.

Ernest A. Coleman
CP Technology, Inc.
Stamford, Connecticut

POLYMER MODIFIERS
AND ADDITIVES

1

Antioxidants

Richard F. Grossman
Halstab Division, The Hammond Group, Hammond, Indiana

I. ROLE OF ANTIOXIDANTS

Plastics are generally considered durable. It is, in fact, complained that plastic articles are too durable, that they tend to persist forever in garbage disposal facilities; this has even led to the design of additives to counter this "defect." Without the incorporation of antioxidants, few plastics would have more than a brief service life. Most, actually, would have none, because articles of commerce could not be fabricated without irreversible destruction. The same need exists in a broad range of other organic materials. The sales volume of antioxidants in the United States has been projected to approach $1 billion in 1995 [1].

The reasons for this need are quite simple. Most polymers in practical use contain carbon–hydrogen bonds. The thermodynamic drive to convert a C—H bond to C=O and O—H is very favorable. Approximate bond energies are taken from Ref. 2:

C—H (primary)	95 kcal/mol
C—H (secondary)	90 kcal/mol
C—H (tertiary)	85 kcal/mol
C—H (allylic)	77 kcal/mol
O—H	110 kcal/mol
C=O	150 kcal/mol

The fuel value of C—H containing substrates is obviously substantial. Polymers devoid of C—H bonds (e.g., PTFE) are highly oxidation resistant; those with low levels (e.g., silicones) are close behind. Another factor is the tendency to

1

eliminate small molecules at elevated temperatures [e.g., HCl from poly(vinyl chloride) (PVC)]. In these cases, the polymer becomes more subject to oxidation with such elimination; in some cases, the reaction product functions to aid further degradative reactions. Additives that interfere with oxidation are considered anti-oxidants; those that scavenge elimination products or interfere with small-mole-cule elimination are most often called stabilizers.

The barriers to polymer degradation lie in three categories: activation energy, activation entropy, and competing reactions. To use the barrier of activation energy is to contemplate processing at the lowest possible temperature with the minimum shear rate. This, of course, is also consistent with the lowest output. Entropy of activation plays a role through the availability (or lack) of weak points on the polymer for oxidation. Thus, degradation can be approached through polymer design; this has, in fact, been a fruitful area of research for many years. Demand for the maximum in cost-effective processing has, nevertheless, mainly stimulated ingenuity in the use of additives to interfere with degradative processes.

A. Mechanism of Oxidation

The worst feature of oxidation is that, uninhibited, it is a chain reaction; that is, it is self-propagating, fast, and has a high yield [3]. In the first stage, initiation, polymer radicals are formed. This can occur through main chain or sidechain cleavage from shear [Reaction (1)] or from loss of hydrogen from a C—H bond [Reaction (2)]:

$$—C—C— \rightarrow —C\cdot + —C\cdot \tag{1}$$

$$—C—H \rightarrow —C\cdot \tag{2}$$

Reaction (2) may result from thermal instability, attack of atmospheric oxygen, or, most likely, loss of a hydrogen to another radical. In the absence of oxygen, the above radicals tend to recombine [reverse of Reaction (1)] with minimal degradative effects. Use of this feature (i.e., processing under a blanket of inert gas) is occasionally practical (e.g., bulk polymer grafting in an internal mixer). The simple propagation steps [Reactions (3) and (4)] are, of course, fast compared to Reaction (1) or (2):

$$—C\cdot + O_2 \rightarrow —COO\cdot \tag{3}$$

$$—COO\cdot + —C—H \rightarrow —COOH + —C\cdot \tag{4}$$

If these were the only reactions involved, antioxidant chemistry would be rela-tively simple. Reaction (4), however, not only leads to propagation but also gener-

ation of the hydroperoxide, —COOH. This is a source of further radicals; heat
or ultraviolet light can initiate Reaction (5):

$$-COOH \rightarrow -CO\cdot + \cdot OH \tag{5}$$

Both of the new radicals have sufficient activity to abstract hydrogen from the
polymer:

$$-CO\cdot + -C-H \rightarrow -COH + -C\cdot \tag{6}$$

$$HO\cdot + -C-H \rightarrow H_2O + -C\cdot \tag{7}$$

Thus, if Reaction (5) becomes prominent, the chain reaction develops a very high
yield through Reactions (6) and (7). Another unfortunate feature is the generation
of water, a prodegradant in many polymers. Most polymers, as received in industry, have a low level of hydroperoxide content. Reaction (5) is, therefore, commonly the most significant contributor to the initiation stated generally in Reaction
(2).

Hydroperoxide decomposition is known to be catalyzed by a number of
transition metals [4]. The path is given in Reactions (8) and (9) and summarized
in Reaction (10):

$$-COOH + M^n \rightarrow -COO\cdot + H^+ + M^{n-1} \tag{8}$$

$$-COOH + M^{n-1} \rightarrow -CO\cdot + OH^- + M^n \tag{9}$$

$$2 -COOH \rightarrow -COO\cdot + -CO\cdot + H_2O \tag{10}$$

For the reaction to proceed, the metal in question must have two transition states,
separated by one electron, that are relatively close in energy level. This is the
case with Fe, Cu, Ni, Mn, Co, V, Ti, and others. With electron-withdrawing
ligands, such as halogen, on the metal, coordination to hydroperoxide oxygen
becomes favorable and catalyst efficiency in Reactions (8)–(10) very high. This
has the effect of greatly lowering the activation energy for propagation of oxidation. The compounds in question are often contributed by polymerization catalyst
residues. In addition, fillers containing transition metal impurities or based on
such a metal (e.g., titanium dioxide) can produce suitable ions if polymer degradation generates acidic fragments (e.g., HCl from PVC).

B. Consequences of Oxidation

Termination of the sequence of reactions through recombination, unless simply
reversing the original initiation of Reaction (1), leads to crosslink formation:

$$-C\cdot + -C\cdot \rightarrow -C-C- \tag{11}$$

Reaction (11) describes the combination of two different polymer molecules, not
the reversal of Reaction (1), under anaerobic conditions. Crosslinking during

processing increases viscosity, reduces flow, and can lead to catastrophic degradation. During service, crosslinking leads to reduction in flexibility, chalking of filler, and eventual embrittlement.

Another possibility is that the polymer radical may fragment:

$$—C—C—C· \rightarrow —C{=}C + —C— \tag{12}$$

The hydrogen transfer needed to complete the olefinic structure generated may come internally or, by chain transfer, from another polymer molecule. An alternating structure where every other carbon can form a secondary radical (polypropylene) or, especially, a tertiary radical [poly(methyl methacrylate) (PMMA)] favors Reaction (12) over Reaction (11). A predominance of primary carbon atoms, as in polyethylene, favors cross-linking over fragmentation. The chain scission process leads to rapid lowering of molecular weight with consequent loss of physical properties. Chain scission is also favored as oxidation generates oxygen-containing radicals:

$$R—CO· \rightarrow R· + C{=}O \tag{13}$$

Thus, certain polymers (e.g., a ethylene propylene copolymer) may exhibit cross-linking followed by fragmentation as oxidation proceeds, or a balance of the two (which balance may be altered by additives).

An additional consequence is that the oxidized site will serve as weak point, a locus of further oxidation:

$$H—C—C{=}O \rightarrow ·C—C{=}O \tag{14}$$

Not only the radical intermediates but also the reaction products favor increased oxidation. The eventual products are typically unsaturated ketones and acids, diketones, ketoacids, and so forth. These are chromophores in themselves, and powerful chromophores if coordinated to trace metals. Thus, the polymeric composition commonly darkens and discolors. If the polymer was produced by a condensation reaction with water elimination, water generated by oxidation may cause hydrolytic depolymerization. For example, the combination of water and decomposing hydroperoxide will rapidly hydrolyze polyamides.

C. Primary Antioxidants

The plastics technologist's wish is that an antioxidant (A/O) will interfere with propagation as proposed in Reaction (15):

$$—C· + A/O \rightarrow [—C{\rightarrow}A/O]· \text{ complex} \tag{15}$$

that is, that the antioxidant complex the radical before it is oxidized. The complex will be insufficiently mobile for the subsequent propagation steps to continue efficiently. It will lose energy in harmless, frustrated vibration, then decompose to regenerate the unharmed antioxidant and a slower polymer radical. The latter will likely still oxidize, but it will be a poor participant in a chain reaction. Thus, the yield will be limited and overall degradation slowed.

As a second choice, it is hoped that an oxidized radical will be trapped:

$$—COO· + A/O → [—COO→A/O]· \qquad (16)$$

When the complex breaks up, the regenerated oxidized radical will still go on as above, but much more slowly. What has been accomplished is the conversion of high-speed, high-yield chain reaction oxidation into a slower, lower-yield free-radical oxidation, one that uses the antioxidant as a reverse catalyst (an additive that lowers a reaction rate and is itself not consumed).

If Reaction (15) or (16) does not reverse, eventually even these unwieldy complexes will abstract hydrogen (from water, —OH, or even from the polymer):

$$[—C→A/O]· → —C—H + A/O· \qquad (17)$$

that is, eventually the antioxidant will be oxidized. This is an inevitable consequence, not a desired event. Many descriptions in the literature are concerned only with the path of consumption of the antioxidant and the number of harmful free radicals it can consume on its way to more or less harmless oblivion. The practical technologist, familiar with the price structure of antioxidants, is concerned instead with how far the protective activities of such additives may be prolonged.

Radicals are thought of as electron-deficient species [5]. The traps used to complex them, generally referred to as primary antioxidants, are typically aromatic compounds containing an activated benzene ring, but which lack substituents that would react irreversibly with free radicals. The orbital that is electron deficient, because of the presence of one, instead of a pair of, electrons, can be visualized as overlapping the pi cloud of the aromatic ring. Electron-donating groups, such as alkyl substituents, are employed to boost the activating power of a phenolic or amino group. In many cases, they also shield the activating group sterically.

D. Hindered Phenolic Antioxidants

Simple phenols are readily oxidized to highly colored quinones. Their presence would interfere with chain reaction oxidation of a polymer, but at the cost of strong discoloration. More importantly, oxidation would remove the antidegradative effect of repeated radical trapping. Most commercial phenolic antioxidants are hindered phenols of the structure A. In the complex shown as structure B, the steric effects of the t-butyl groups inhibit oxidation (removal of hydrogen from the phenol by the radical). The blocking group at the para position inhibits oxidation at that site. Consequently, the complex is long lived in comparison to the chain reaction intermediates. This is what is needed to defeat the chain reaction mechanism.

The starting material for antioxidants of type A is di-*t*-butylphenol, readily and cheaply available through direct alkylation of phenol with aluminum as the catalyst, the effective reagent being the metal phenolate [6]. Di-*t*-butylphenol is readily alkylated in the para position by a number of reaction schemes (e.g., addition of its anion to an activated double bond).

(A)   (B)

The simplest member of the series is di-*t*-butyl-4-methylphenol (BHT), where R— is methyl. BHT is, in addition, the most commonly used (and lowest cost) representative. It is a highly effective antioxidant with a low order of toxicity, permitting use in a variety of substrates. Its effectiveness derives from its high mobility; this enables repeated trapping of radicals before it is itself oxidized. The path of the latter has been well studied [7]. In the initial step, the complex of structure B collapses, transferring the phenolic hydrogen to the radical, as shown in Reaction (18) (here the symbol A/O is used to represent the di-*t*-butylphenyl nucleus):

$$CH_3—[A/O]—OH + R· \rightarrow RH + CH_3—[A/O]—O· \qquad (18)$$

Step (18) is not catastrophic if another additive intervenes to regenerate the antioxidant. If this does not occur, the phenolic radical will tend to dimerize per Reactions (19) and (20):

$$CH_3—[A/O]—O· \leftrightarrow ·CH_2—[A/O]—OH \qquad (19)$$

$$2 ·CH_2—[A/O]—OH \rightarrow HO—[A/O]—CH_2—CH_2—[A/O]—OH \qquad (20)$$

Bisphenols such as the product of Reaction (20) are extremely prone to oxidation to quinoid structures such as given below because of the extended conjugation that develops:

$$O=[A/O]=CH—CH=[A/O]=O$$

The quinoid dimer of BHT is bright yellow (shifts in color and increase in intensity accompany complexing with trace metals, e.g., titanium).

If the group at the para position is not methyl, but a higher analog, for example, as in Irganox® 1076,

$$HO—[A/O]—CH_2—CH_2—CO_2—C_{18}H_{37}$$

dimerization does not lead to extended conjugation and the product has only slight color:

$$O=[A/O]=CH—CH—CO_2—C_{18}H_{37} \}_2$$

Eventually, however, further oxidation of such antioxidants will lead to extended conjugation and increase in color:

$$O=[A/O]=CH—C—CO_2—R \}_2$$

Antioxidant oxidation is particularly a problem with polymers that generate strong acids during degradation (e.g., PVC). If carbocations are formed, rearrangement of alkyl (particularly t-butyl) groups on the phenolic ring proceeds readily at processing temperatures, increasing the likelihood of forming colored quinoid oxidation products. Migration of alkyl groups can also occur under radical conditions (in this case, migration of methyl groups is favored), leading to bisphenols of the structure:

$$HO—[A/O]—CH_2—[A/O]—OH$$

These oxidize readily to semiquinones:

$$O=[A/O]=CH—[A/O]—O\cdot$$

that, unlike those encountered in the foregoing, tend to combine with polymer radicals, especially at chain ends, at the phenolic oxygen. These break down yielding unsaturation [8]:

$$—CH_2—CH_2—O—[A/O]—CH=[A/O]=O \rightarrow —CH=CH_2 \qquad (21)$$

with transfer of hydrogen regenerating bisphenol.

There is widespread belief that combinations of phenolic antioxidants that are primarily electron donors (e.g., di-t-butylated phenols) and methylene bisphenols (which can function above as electron acceptors) lead to diminished effectiveness and chances of increased discoloration. This may be true, but there does not seem to be direct data on the subject.

E. Antioxidant Mobility

No additives, other than those intended to bond chemically (e.g., coupling agents) are totally localized to a given polymer site. In static service, it is usually anticipated that most additives will stay put. In dynamic service, some migration may be anticipated (e.g., it is intended that protective waxes will migrate to the surface of the tire). During processing, there are additives that must have sufficient mobility to enable required flow of the composition. Additives used at substantial levels (fillers, plasticizers) are in this category. It is anticipated that the filled, plasticized polymer can be extruded, calendered, molded, and so forth into a finished article having (at least) the same homogeneity as the mixed compound. During processing, some rearrangement of proportions may occur, but it is only during applied strain. Substantial fractionation of a major component can occur, particularly from turbulent flow, but this is rarely desirable.

Additives of the third kind (the first, bonded, immobile; the second, reluctant movers) are those where repeated function is anticipated. These include antidegradants and lubricants. Such additives are used at low levels. They are successful because of the following factors:

1. The solubility parameter of the additive is close enough to that of the polymer (or polymer/plasticizer blend) to ensure good compatibility.
2. Despite the above, the additive contains a group (usually a nonpolar "tail") that has very limited solubility in the polymer. (If not, it becomes a plasticizer or filler.)
3. The active site of the additive has a strong attraction to polarized sites on the polymer.

Polymer degradation occurs most readily in areas that are under applied strain. Absent strain, radicals tend to recombine as rapidly as they are formed. The effect of strain is to increase the polarization of bonds between dissimilar elements. Consider the competition between Reactions (3) and (15):

$$—C· + O_2 → —COO· \qquad (3)$$
$$—C· + A/O → [—C → A/O]· \qquad (15)$$

How can the unwieldy antioxidant possibly win? To begin with, the oxygen molecule is not traveling alone. Its transport will invariably involve a complex with a small, mobile molecule (indeed, quite possibly the antioxidant!) An additive whose complex with atmospheric oxygen facilitates degradation would be classed as a phase transfer catalyst for oxidation. Antioxidants are reverse phase transfer catalysts; they (efficiently) bring the reactants into a complex that is unreactive. Thus, a more accurate model of antioxidation would be that of Reaction (22):

$$—C· + O_2 + A/O → [—C → A/O ←O_2]· \qquad (22)$$

An important factor in choosing an antioxidant from among the variety of those available is efficiency in the above process. This is intimately tied to the mobile behavior described earlier. As the polymeric composition is varied in polarity and viscosity, the relative merits of particular candidates can change greatly. This will be discussed in patterns of specific use.

A characteristic common to all primary antioxidants is that if the reaction were not impeded by bulky groups providing steric hindrance, they would be oxidized much more rapidly than a hydrocarbon polymer. Simple phenols and amines, for example, must be protected by inert gas to prevent atmospheric conversion to dark tarry substances. Consider the model implied in Reaction (22): atmospheric oxygen coordinates with the most (if it were not hindered) readily oxidized species, the antioxidant. The complex may or may not become associated with a polymer (or other) free radical. If it does, the overall complex may still break down to regenerate the reactants. In the initial stages, the time period in which no degradation is observed, this equilibrium prevails.

The factors favoring the competitive reactions of polymer and antioxidant oxidation are the concentration of polymer radicals (temperature, rate of shear, structure of the polymer) and of oxygen (aerobic exposure during processing) and the temperature (mobility of oxygen). Opposed to these are the concentration of antioxidant and its mobility in the system. The goals of the formulator are to prolong the initial stages, deferring the onset of measurable degradation, and to minimize the rate of the latter.

F. Secondary Antioxidants

The worst feature of free radical chain reaction oxidation of the C—H bond lies in hydroperoxide decomposition [Reaction (5)]. This step produces two additional active radicals from the initial antioxidants, often referred to as antioxidant synergists, which decompose hydroperoxides, yielding more or less harmless products. The major categories are phosphite esters and divalent sulfur compounds. The former abstract hydrogen from hydroperoxides through a 4-center, Wittig-type reaction:

$$—C—OOH \ + \ (R—O—)_3P: \ \rightarrow \ —C—OH \ + \ (R—O—)_3P = O \qquad (23)$$

The alcohol formed is usually more difficult to oxidize than the starting —C—H bond; furthermore, its oxidation tends, under ordinary conditions, not to follow a chain reaction path. With many phosphite esters, Reaction (23) is very fast. Such additives are highly useful in preserving color and providing process safety. Similar reactions are found with activated (e.g., beta to a carbonyl group) divalent sulfur compounds:

$$—C—OOH \ + \ (—COCH_2CH_2)_2S \ \rightarrow \ C—OH$$
$$+ \ (—COCH_2CH_2)_2S = O \qquad (24)$$

In this case, there is almost certainly participation by the activating group, because esters of thiodipropionic acid are more effective than simple organic sulfides. In the above reactions, the additives serve as special-purpose reducing agents.

Further along in the degradative process, activated divalent sulfur compounds and phosphite esters react with unsaturated carbonyl groups, neutralizing potential chromophores:

$$-C{=}C-C{=}O \ + \ (RO)_3P \rightarrow \ -C-C-C-O-R$$
$$| \qquad\qquad\qquad (25)$$
$$(RO)_2P{=}O$$

$$-C{=}C-C{=}O \ + \ -S- \ \rightarrow \ -S-C-C-C{=}O \qquad\qquad (26)$$

Many of the observations of degradative coloration fading during service, after initial intensification, are traceable to Reactions (25) and (26). These classes of additives are unique in their ability partly to reverse the oxidation process.

Although our best choice is for Reaction (22) to reverse, conversion to hydroperoxide will eventually become an increasingly competitive route:

$$[-C \rightarrow A/O \leftarrow O_2]\cdot \ \rightarrow \ [-COO \rightarrow A/O]\cdot \qquad\qquad (27)$$

This is still at a stage where overall polymer degradation may not be observable. Secondary antioxidants, in their destruction of hydroperoxides, here become regenerators of the primary antioxidant:

$$[-COO\rightarrow A/O]\cdot \ + \ (RO)_3P \rightarrow A/O \ + \ (RO)_3P{=}O \ + \ -CO\cdot \qquad (28)$$
$$[-COO \rightarrow A/O]\cdot \ + \ -S- \ \rightarrow A/O \ + \ -S{=}O \ + \ -CO\cdot \qquad (29)$$

At the sacrifice of the secondary, the primary antioxidant has been regenerated and the chain reaction broken. The released radical will no doubt be oxidized, but not autocatalytically and not at a highly destructive rate. It is usually worth trading one oxidized site to recover the primary antioxidant. The above model explains why it is customary to use two to four times the level (in equivalent weight) of secondary to primary antioxidant in practical formulations.

For this last mode of action to be significant, there must be primary–secondary antioxidant interaction; that is, the secondary must be located nearby. In fact, the infrared spectra of blends of hindered phenolic antioxidants with phosphites or thiodipropionate esters indicate coordination of phenolic oxygen to phosphorus or sulfur [9]. This is advantageous for the function of the primary antioxidant also. Both trivalent phosphorus and divalent sulfur are easily oxidized, are more readily oxidizable than —C—H, were it not for the special construction of the secondary antioxidants, where steric groups and internal coordination block an easy direct reaction. Thus, both primary and secondary antioxidants share the status of hindered oxidative substrates. Representing the secondary as P: (although just as easily S:), the combined antioxidant should be thought of as the complex [P: → A/O] or [S: → A/O]. The various competing reactions are at a locus that includes the antioxidant combination, atmospheric oxygen, and a polymer radical. In this way, the existence of a large number of commercial primary and secondary antioxidants can be understood; it is necessary to formulate combinations that can form mobile complexes readily in a particular polymeric composition. As the latter is diverse, the combinations become varied. Specific combinations of primary and secondary antioxidants will be discussed in Section IV.

Organic phosphites are subject to hydrolysis, as shown in Reaction (30):

$$(RO)_3P: + H_2O \rightarrow (RO)_2P—OH + ROH \tag{30}$$

The resultant product is isomeric and exists mainly in the P=O form:

$$(RO)_2P—OH \leftrightarrow (RO)_2PH=O \tag{31}$$

The initial hydrolysis products are good reducing agents and can react directly with oxygen or with hydroperoxides (primary and secondary functions). Keeping in mind that oxidation of C—H bonds invariably produces water at the oxidized site [Reactions (7) and (10)], phosphite hydrolysis will always be a factor. During processing, this is a favorable factor. Partial hydrolysis consumes water that would otherwise play a part in the generation of colored oxidation products, and it generates useful PH=O antioxidants. This behavior probably plays a part in the ability of phosphites to add to antioxidant protection in almost every polymer.

Simple phenolate anions are excellent leaving groups in the hydrolysis of organic phosphites, particularly in the presence of acids. Hydrolysis itself generates acidic P—OH and PH=O protons. The danger lies in continued hydrolysis past the first stage, yielding first RO—P(OH)H=O, then (HO—)₂PH=O, much stronger acids that may greatly accelerate formation of oxidized products. Trialkyl phosphites are more resistant to hydrolysis, but less useful as synergists for hindered phenols (as would be anticipated by considering the probability of forming a complex with the aromatic primary antioxidant, $(RO)_3P: \rightarrow A/O$ being favored if at least one RO— is also aromatic).

Triaryl phosphites are particularly subject to hydrolysis, as steric relief tends to promote loss of the third bulky group. This is a problem with tris(nonylphenyl) phosphite (TNPP), an additive with broad Food and Drug Administration (FDA) sanction that is, overall, the most widely used phosphite. Commercial grades are available containing up to 1% triethanolamine or tris(isopropanol) amine. The amines are complexed with the phosphite and suppress formation of acid products, inhibiting hydrolysis. Their use is beneficial in many cases where the antioxidant–phosphite combination is being used to preserve the polymer during storage, rather than in processing, such as inhibition of gel formation in synthetic elastomers. These grades should not be used in amine-sensitive polymers such as PVC. Hindered phenols having ortho and para positions blocked, such as tris(2,4-di-*t*-

butylphenyl) phosphite, are much more resistant to hydrolysis, but also much more costly. Aryl alkyl phosphites used in PVC are more resistant, in themselves, to hydrolysis than triaryl phosphites. However, during degradation of PVC and other halogenated polymers, HCl evolution at the reaction site invariably leads to the generation of R—OH when phosphites are present.

Thiodipropionate esters, on the other hand, are relatively resistant to hydrolysis, as are the initial oxidation products $(ROCOCH_2CH_2)_2S{=}O$. These fragment thermally, again yielding products that can serve both as primary and secondary antioxidants: $RO{-}COCH_2CH_2S{-}OH$ [10]. Eventually, they may be oxidized to strong acids, $-SO_2H$ and $-SO_3H$, but these processes are relatively slow. The general result is that divalent sulfur synergists are slower acting than phosphites, but often more persistent. Why not, therefore, use a combination of both types: organic phosphite for protection during processing and thiodipropionate ester for added service life? Unfortunately, Reaction (32) intervenes, potentially destroying both additives:

$$(RO)_3P{:} \ + \ -CH_2-S-CH_2- \ \rightarrow \ (RO)_3P{=}S$$
$$+ \ -CH_2-CH_2- \quad (32)$$

Both thiophosphates and their reaction products are, in addition, characterized by unattractive odors. Does this mean that such combinations can never be used? No. Reaction (32) is much slower than the useful reactions of phosphites and thiodiesters [Reactions (28) and (29)]. In certain circumstances, low levels of phosphite synergists can be used (and consumed) for protection during processing, with divalent sulfur compounds in the background for extended service life. Only occasionally is this practical.

G. Amine Antioxidants

A variety of aromatic secondary amines, Ar—NH—, function in an analogous manner to phenols, Ar—OH. The initial, and most useful, reaction is that of Reaction (22):

$$-C{\cdot} \ + \ O_2 \ + \ A/O \rightarrow [-C \rightarrow A/O{\leftarrow}O_2]{\cdot} \quad (22)$$

Amines are, in general, even more readily oxidized than phenols. This drive ensures that if the additive has sufficient mobility in the polymeric composition, it will be highly effective. They are used most frequently with substrates that are easily oxidized, such as gasoline and unsaturated elastomers. Secondary amine antioxidants are typically diaryl, Ar—NH—Ar. Their oxidation, and inevitable consequence, easily generates strong chromophores:

$$-C{\cdot} \ + \ Ar-NH-Ar \rightarrow -CH \ + \ (Ar)_2N{\cdot} \quad (33)$$
$$(Ar)_2N{\cdot} \rightarrow Ar{=}N-Ar{\cdot} \quad (34)$$

These easily condense to form very dark colors (comparable to aniline black). Many commercial products are diaryl phenylene diamines, Ar—NH—Ar—N-H—Ar. Their oxidation yields colored azo products, Ar—N=Ar=N—Ar. These, however, are useful secondary antioxidants:

$$Ar—N=Ar=N—Ar + 2\ —COOH \rightarrow Ar—NO=Ar=NO—Ar$$
$$+ 2—COH \quad (35)$$

The conjugated amine oxides formed are also strong chromophores. Amine antioxidants of this type can be used only in dark compositions, most often elastomers filled with carbon black. Even in these cases, the dark oxidation products often migrate to the surface of the finished article. Thus, their traditional designation: staining antioxidants (hindered phenols are similarly referred to as nonstaining antioxidants). The extent of conjugation in oxidized products can be limited by chemical structure. For example, in Ar—NH—Ar—NH—Ar, if the end aryl groups are naphthyl (di-β-naphthyl p-phenylene diamine), chromophore formation is reduced to the point where the product is referred to as "slightly staining." Similar considerations apply to the blockage of the para position instead by alkyl groups, as in Ar—C(CH$_3$)$_2$Ar—NH— structures (e.g., di-bisphenyl A amine). Amine antioxidants are rarely used in halogenated polymers other than elastomers (e.g., polychloroprene, chlorosulfonated polyethylene).

Combinations of phenolic and amine antioxidants are widely regarded as likely to reduce effectiveness and to produce highly colored products, seemingly without relevant data in the literature. Similarly, blends of different phenolic antioxidants are suspect (although blends of amine antioxidants are used routinely). Most of these anecdotal observations have (in the author's view) no basis in fact, but result from accidental formation of antioxidant–antioxidant complexes in polymeric compositions in which such interaction products have poor mobility. The object of the formulator is to match the mobility of the additive package to the polymeric composition. In this context, blends of antioxidants of radically different polarity can hardly seem a promising approach.

Recently a class of secondary antioxidants has been developed that functions efficiently with amines. These are 2-mercapto-benzimidazole (MB), 2-mercaptotoluimidazole (MTI), and the zinc salts of these, respectively (ZMB and ZMTI). The toluimidazoles are

MTI   ZMTI

Although the details are as yet unreported, it seems likely that these synergists behave in a manner analogous to that of phosphites and organic sulfur compounds with hindered phenolic antioxidants (i.e., to regenerate the primary antioxidant). It is also possible that this may involve a cyclic mechanism similar to those found with hindered amine stabilizers (see *Section* I.K).

H. Metal Deactivators

Metal catalysis [Reactions (8)–(10)] of hydroperoxide decomposition is such an important factor in maintaining chain-reaction oxidation that almost every polymer system that uses primary and secondary antioxidants also includes at least one metal deactivator. For metal catalysis to be effective, in addition to two available transition states separated by one electron, the specific metal compound must be able to coordinate closely with the hydroperoxide; that is, it must be favorable for peroxy oxygen to donate electron density to the metal. This requires an electron-withdrawing substituent, usually chloride. Metal deactivators are, therefore, of two general types: those which chelate the metal, making it less available, and those which capture chloride, replacing this ligand with one that is electron donating, again making the metal less available.

The most common chloride scavengers are metal stearates, primarily calcium stearate. In nonpolar and semipolar media, metal carboxylates exist in equilibrium between structures having one carboxylate oxygen linked to the metal, having both oxygen atoms coordinated to the same metal cation, and with carboxylate groups bridging two cations. These structures have, in many cases, distinct C—O symmetric and antisymmetric stretching frequencies in the infrared [11]. Alkaline earth stearates, in particular, complex M—X fragments, forming carboxylate bridges between the alkaline earth and the transition metal. Halide bridges may also form. The complexes may be approximated as

```
      O        Cl
      ||       / \
  R-C-O-Ca       M-
           \   /
           O   O
            \ /
            C-R
```

If analogous interactions are carried out in a test tube, the shifts in C—O stretching frequencies suggest the development of bridging between unlike metals [12]. The resultant complexes become relatively poor catalysts for hydroperoxide decomposition. Complex formation is an equilibrium; the success of metal stearate additives depends on sufficient concentration to drive the reaction to the right. This use is independent of lubrication effects that may also be provided by the same additive. Nevertheless, it should be recognized that if overlubrication leads to lubricant exudation, complexed catalyst residues may accompany it.

Chloride may also be scavenged by the addition of hydrotalcite, a synthetic clay having a platy magnesium–aluminum hydroxide structure, containing anions, typically carbonate, between the flat metal hydroxide plates [13]. This anion is exchanged for chloride. Again, this is concentration driven; levels of two to three times the expected trace chloride concentration are typical (the same levels would be used with a metal stearate). In this case, chloride is more strongly immobilized, leading to improved corrosion resistance as compared to metal carboxylates [14]. In nonionizing media, ion pairs prevail. Therefore, trapping the chloride anion has the effect of also capturing the transition metal cation, probably through adsorption to the porous hydrotalcite structure.

The lack of mobility of the final product is advantageous; however, the same factor limits its efficiency in attracting M—Cl. Thus, a phase transfer catalyst is useful to transport the transition. The most commonly used is, of course, calcium stearate. The two may be added separately, or the hydrotalcite stearate coated by the supplier. An interesting case is in the stabilization of polyolefins containing halogenated flame retardants. Here, a residual Ziegler–Natta catalyst is amplified by traces of chloride or bromide from the flame retardant. Organotin maleates, at similar trace levels, are effective as phase transfer catalysts for the conveyance of halide in such compositions.

Wire insulation forms a special case. The effect of the passage of electric current is to inject trace levels of copper ions into the insulation. Copper is an excellent catalyst for hydroperoxide decomposition. In the case of unsaturated elastomers, the effect is so severe as to require tin coating the conductor. A second factor is that in wire used for electronic and communications applications, metal impurities in the insulation reduce the data-carrying capacity of the overall construction. As a result, in many applications, additives are used that form highly stable coordination complexes with metals (in particular, copper). At one time, citric acid and ethylene diamine tetraacetic acid (EDTA) were in common use. They have been largely replaced by hydrazides, hydrazines, and oxamides having the following structures:

$$Ar—CH{=}N—NH—CO—CO—NH—N{=}CH—Ar \qquad \text{(hydrazide)}$$
$$Ar—CH_2CH_2CO—NH—NH—CO—CH_2CH_2—Ar \qquad \text{(hydrazine)}$$
$$Ar—CH_2CH_2CO—OCH_2CH_2NH—CO—CO—NH—R \qquad \text{(oxamide)}$$

The idea is to surround the metal completely with ligands. It can be seen that these nitrogenous compounds will also form salts with associated halide anions. The above types of compounds are expensive but extremely effective.

A problem with additives such as the above is that they are, in themselves, easily oxidized species. Furthermore, they cannot be protected by the introduction of steric blocking groups, because that would interfere with chelation. An interesting approach, now used by almost all suppliers, is to construct the end aryl groups as hindered phenolic antioxidants. An incoming hydroperoxide encounters not only a completely shielded (potential) metal catalyst, but a primary antioxidant (probably with a nearby associated secondary synergist) ready to destroy it. (It should be noted that considerable thought must be given to the exact spatial disposition needed for bound antioxidants to protect other parts of the same molecule.)

I. Ultraviolet-Light Absorbers

Consider the complex previously discussed, considered key to the useful behavior of antioxidants: [—C → A/O ← O_2], keeping in mind that A/O may represent some association of primary and secondary antioxidants. Let us postulate that such complexing may occur, in some fraction of the total cases, before actual radical formation. Consider, therefore, [—C—Y → A/O ← O_2], where C—Y represents a weak C—H or C—C bond, perhaps on the reaction path to radical formation. If nothing other than thermal motion occurs, there is a reasonable probability of return to starting materials (i.e., complete protection). A sudden input of energy (e.g., increased shear rate) will drive oxidation to the right. A common input is the delivery of a quantum of energy via incident light. This is particularly the case with ultraviolet (UV) light, which contains frequencies at which the C—Y bond becomes excited and more likely to fragment. In addition, it is likely that a number of sites will be struck either simultaneously, or too rapidly for the additive package to respond sequentially by migration. In addition to these factors is the likelihood that oxidation of the antioxidant will be promoted by UV radiation. Some antioxidants may even serve as sensitizers for UV-promoted degradation. This combination of factors has generated a number of additives that function as UV-light absorbers. Their general mechanism is to convert incoming light energy into heat.

Probably the most effective UV-light absorber is carbon black. Low levels of fine particle carbon black (e.g., ASTM Type N110 furnace black) can provide virtual opacity to incoming UV light at modest cost. The absorbed light generates electronic transitions in the graphitic-type black structure, which decay, with conversion to heat energy. Examples of commercial use include roofing, agricultural film, telephone drop wire, and so forth. Similar effects can be found with other fillers having high polarizability (dielectric constant). The most common is titanium dioxide. Sufficient titanium dioxide to provide visual opacity will greatly reduce the acceleration to oxidation associated with UV light. Most other high dielectric constant inorganic materials tend to be compounds of heavy or transition metals (oxides of lead, copper, iron) and, thus, of limited utility in many polymers.

The polymer may, on the other hand, have a built-in high polarizability (e.g., acrylics, polycarbonate, silicones, PVC). This will invariably result in a higher order of resistance to UV light than that of a hydrocarbon polymer. The mechanism of protection is entirely analogous to that provided by the above fillers: Incident UV light excites a polar group, either via pi → pi* or n → pi* electronic transitions. The excited state has a certain stability, permitting loss of energy by thermal vibration. The reduced-energy excited state then drops back to the ground state, losing a quantum of energy of lower frequency than the original incoming quantum, less able to contribute to degradation. In other words, the excited state functions similarly to the antioxidant–radical complex.

If the polymer needed for a given application does not have such built-in groups and cannot be filled with carbon black, or sufficient quantities of titanium dioxide, and must be used in the presence of UV light, it is necessary to provide a light-absorbing additive. These have been constructed with (at least) two considerations in mind: strong absorption of light in the UV region and stability of the excited state. The oldest and most widely used class consists of *ortho*-hydroxybenzophenones:

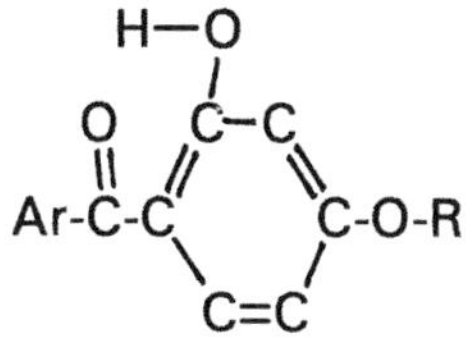

Absorption of a quantum of UV-light energy leads to a resonance stabilized quinoid excited state [15]:

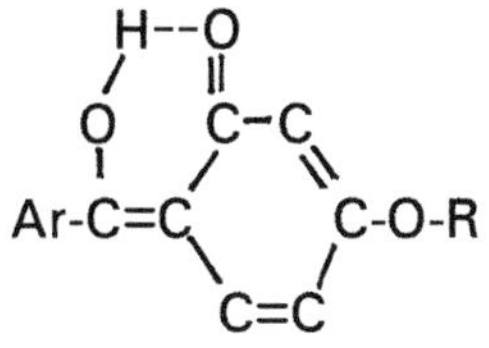

The diaryl ketone structure inhibits fragmentation of the excited state, enhancing its persistence. The above R— group can be hydrogen or methyl, providing polarity comparable to relatively polar polymers. Alternatively, it can be C_8 or C_{12}, for increased mobility in nonpolar systems. The second aromatic ring can be constructed with similar sidechains.

Ortho-hydroxybenzophenones are, unfortunately, readily oxidized. Including bulky groups that would provide steric hindrance would defeat their function by making it more difficult for the excited state to have the necessary planarity for resonance stabilization. Oxidative loss of the critical oxygen-bonded hydrogen atom leads to fragmentation, generating, in the presence of hydroperoxides, benzoic acid, substituted phenol, and the corresponding benzoate ester [16]:

$$Ar—CO—Ar—OR \rightarrow ArCOOH + HO—Ar'OR + Ar—COO—Ar'OR$$
$$|$$
$$O\cdot$$

(36)

A useful approach to combat this is to provide both aromatic rings with hydroxy groups ortho to the ketone linkage. Then, Reaction (36) tends to generate a substituted phenyl salicylate as the primary oxidation product:

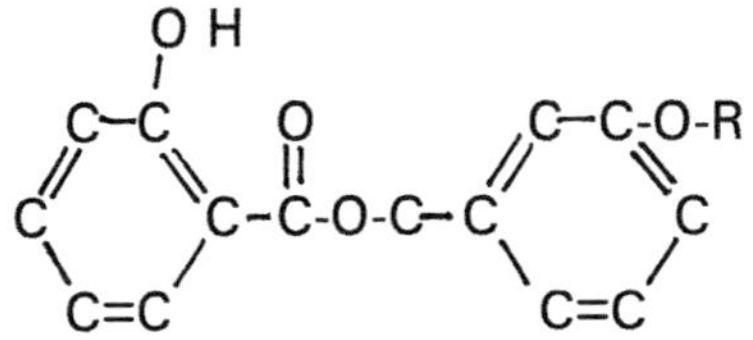

Phenyl salicylates have an analogous structure to *ortho*-hydroxybenzophenones and were actually the first organic UV-light absorbers used in practice (Eastman, 1945, cellulose acetate). Aryl salicylates are more readily oxidized and are less effective than hydroxybenzophenones; their use has not grown greatly. Nevertheless, the generation of a second UV-light absorber during oxidation of the first is a useful improvement.

At one time, hydroxybenzophenones were the most widely used organic UV-light absorbers; they remain among the most cost-effective. Over the past 30 years, they have lost market share to a similar, but generally more efficient class: *ortho*-hydroxylphenyl benzotriazoles:

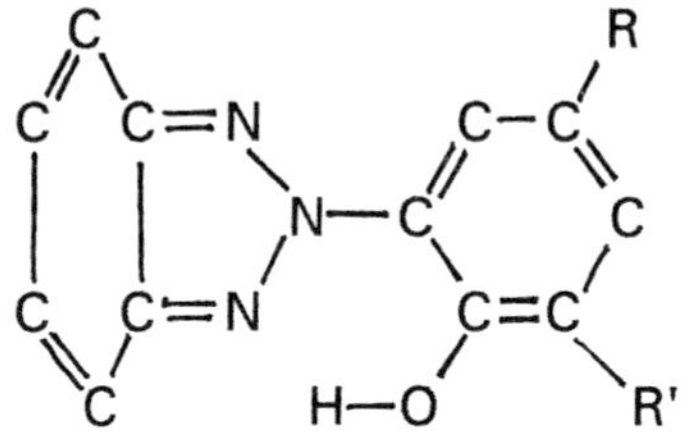

Absorption of a quantum of UV-light energy leads to an excited state with transfer of hydrogen, this time to nitrogen, creating a zwitterion, the negative charge resonance stabilized on the phenolic ring and the positive charge on the benzotriazole nucleus. In comparison to the analogous reaction with hydroxybenzophenones, the excited state appears to be more easily reached and longer lived, leading to inhibition of degradation at lower additive levels. This is fortunate, as benzotriazoles are much more expensive to synthesize.

The greater persistence of the excited state tends to overcome the fact that, in many systems, benzotriazoles are oxidized in the presence of UV light about as rapidly as hydroxybenzophenones [16]. The main reason, however, for the larger volume of the former now in commercial use is that their oxidation products are much lower in color than those of typical hydroxybenzophenones. Although the degradative chemistry of benzotriazoles does not appear to be well known, it is significant that commercial products invariably block the ortho and para positions of the phenolic ring with alkyl substituents (R and R′ above). In a number of cases, *t*-butyl and similar groups are used (without, in this system, affecting the primary function). Thus, development of quinoid chromophoric oxidation products is inhibited. Benzotriazoles have received FDA sanction for indirect food contact in polyolefins and PVC. They are, in addition, widely used in a broad variety of polymers.

Direct competition to benzotriazoles is offered by another class, cyanoacrylates:

$$\underset{Ar}{\overset{Ar}{\diagdown}}C=C\underset{CO_2R}{\overset{CN}{\diagup}} \longrightarrow \underset{Ar}{\overset{Ar}{\diagdown}}C \dotplus C^{-}\underset{CO_2R}{\overset{CN}{\diagup}}$$

In the excited state, the zwitterion produced has substantial resonance stabilization, permitting a low usage level of these expensive additives. A plus factor with cyanoacrylates is their good resistance to oxidation (particularly with R— as phenyl). A negative factor is the ease with which such structures can hydrolyze in the presence of hydroperoxides, yielding —COOH in place of —CN. This will decarboxylate, generating a product not only not useful but readily oxidized. Cyanoacrylates are generally limited to special applications.

Oxanilides are similar in behavior, but rather more resistant to hydrolysis:

$$\underset{}{R\text{-}Ar\text{-}NH\text{-}\overset{\overset{\displaystyle O}{\|}}{C}\text{-}\overset{\overset{\displaystyle O}{\|}}{C}\text{-}NH\text{-}Ar\text{-}OR'} \longrightarrow R\text{-}Ar\text{-}N^{+}H=\overset{\overset{\displaystyle HO}{|}}{C}\text{-}\overset{\overset{\displaystyle O^{-}}{|}}{C}=N\text{-}Ar\text{-}OR'$$

The excited state is resonance stabilized through hydrogen transfer between the two oxygens. This type of UV-light absorber is also efficient but, in common with cyanoacrylates, has been superseded by the development of hindered amines.

J. Excited-State Quenchers

A problem with UV-light absorbers of all types is that the protection afforded depends on the thickness of the article. In coatings or thin films, it is either not possible to interpose a sufficient amount of absorber to be effective or the levels used become prohibitive. There is, therefore, a need for antidegradants less dependent on path length. In some cases, this is met with the use of excited-state quenchers. This mode of action is simple in theory: An additive is used which will collide with or come into close contact with a molecule in the excited state from having absorbed a quantum of UV-light energy. The subsequent interaction will then cause the original excited species to drop back into the ground state (without fragmentation) and the second additive to enter an excited state. The premise is that the first excited state has a sufficient lifetime for this to occur. This is not likely in the case of C—Y (weak C—H or C—C) bonds. Their excited states, as well as those of hydroperoxides, have lifetimes typically that of one C—Y stretching vibration. In other words, the first time the bond stretches thermally, it dissociates. The excited state that persists is that of the UV-light absorber. The excited-state quencher may, therefore, be thought of as a synergist for the latter, catalyzing its decay to harmless products. This is much more efficient than depending on random collisions of the excited UV-light absorber to spend energy thermally.

Quencher synergists have been used mainly with hydroxybenzophenone UV-light absorbers. The most common species are dithiocarbamate salts of nickel:

$$(R)_2\text{-N-C} \underset{S}{\overset{S}{\diamond}} Ni \underset{S}{\overset{S}{\diamond}} \text{C-N-}(R)_2$$

In practice, this type of additive must be built around the presence of an easily excitable unpaired electron; that is, it requires a transition metal. The latter must be well coordinated so as not to cause decomposition of hydroperoxides into two radicals and catalyze thermal oxidation. The solution adopted in the above compound is to use a transition metal salt of a divalent sulfur compound; in other words, of a species which decomposes hydroperoxides in a beneficial manner, analogous to thiodipropionate esters. Although transition metal salts of acid phosphites, which should behave similarly, are not commercial, products such as the following are in use:

$$(R2N)_2\text{-P} \underset{S}{\overset{S}{\diamond}} Ni \underset{S}{\overset{S}{\diamond}} \text{P-}(R2N)_2$$

The crucial point is that the ionic hydroperoxide decomposer (P: or S:) must win the competition with the transition metal. Thus, the latter must be chosen so that two-electron transfers [e.g., Ni(II) to Ni(IV)] are preferred to one-electron processes, ruling out Cu, Fe, Co, Ti, and so forth. Nickel dithiocarbamates and related compounds are good enough divalent sulfur antioxidant synergists as to be used routinely for this purpose in synthetic elastomers, particularly with sulfur-based cross-linking systems and amine antioxidants.

Compounds in this category (containing metals with unpaired electrons) typically absorb in the visible region and are more or less strongly colored, the above examples exhibiting the familiar Ni(II) green. In addition, they also absorb in the UV range; that is, they are UV-light absorbers in themselves. This factor is made use of in combination UV absorber–quencher products such as

$$\text{A/O-CH}_2\text{-P} \underset{\displaystyle O \quad O}{\overset{\displaystyle R\text{-O} \quad O}{\diagdown \diagup \diagdown \diagup}} \text{Ni} \underset{\displaystyle O}{\overset{\displaystyle O \quad O\text{-R}}{\diagup \diagdown \diagup}} \text{P-CH}_2\text{-A/O}$$

As with the dithiocarbamates, R— groups are chosen to promote mobility in particular polymers. A/O, in the above, is intended to indicate a di-t-butylphenolic group. The above (Irgastab® 2002) can be considered a complete antioxidant/synergist/UV absorber package. Another example is the butyl amine complex with the nickel salt of a sulfur-bridged hindered phenol (Cyasorb® 1084). These products have been generally limited to use in polyolefins. A greater market share has been limited by the concurrent development of hindered amine stabilizers.

K. Hindered Amine Stabilizers

In the past 20 years, a number of antidegradants have emerged of the general structure:

$$(\text{CH}_3)_2\text{-C} \underset{\displaystyle C}{\overset{\displaystyle N}{\diagup \diagdown}} \text{C-(CH}_3)_2$$

In the above, R— may be hydrogen or an alkyl, alkoxy, or acyl group. R' (as usual) provides a mechanism for influencing mobility in the polymer and reducing volatility. It has been shown that additives of this type are neither UV-light absorbers nor excited-state quenchers [17]. Yet, they are effective in providing antidegradative protection to a variety of polymers under conditions of UV-light incidence, where combinations, instead, of hindered phenolic and synergists are very much less effective.

It is known that all of the hindered amine (N—R) additives react with hydroperoxide radicals:

$$\text{N—H} + \text{—COO·} \rightarrow \text{N—O·} + \text{—COH} \tag{37}$$

$$\text{N—R} + \text{—COO·} \rightarrow \text{N—O·} + \text{—COR} \tag{38}$$

$$\text{N—OR} + \text{COO·} \rightarrow \text{N—O·} + \text{—COOR} \tag{39}$$

Generation of the —N—O· nitroxyl radical is key to their success. In the above reactions, the hydroperoxide radical, a good hydrogen abstracter from the polymer, has been converted to a nitroxyl radical, a poor hydrogen abstracter from inactivated C—H. Thus, the improved quantum yield from hydroperoxide decomposition has been short-circuited. The nitroxyl radical, however, does react with polymer radicals:

$$N—O· + —C· \rightarrow N—O—C— \tag{40}$$

This generates a product that reacts with hydroperoxide radicals [Reaction (39)] to yield the nitroxyl radical. Thus, the additive (or its nitroxyl radical derivative) has catalyzed the recombination:

$$—C· + —COO· \rightarrow —COOC— \tag{41}$$

by complexing one (or both) of the reactants for a sufficient length of time. Hindered amines also react with peroxides:

$$N—H + —COOC— \rightarrow N—OH + —COC— \tag{42}$$
$$N—H + —COOH \rightarrow N—OH + —COH \tag{43}$$

The resultant hindered hydroxylamine (NOH) readily donates hydrogen to regenerate the nitroxyl radical. The latter is, therefore, extremely persistent. The overall conclusion is the replacement of chain-reaction degradation with chain-reaction interference with degradation.

Why, then, do hindered amines work in the presence of UV light when (existing) phenolic analogs do not? In beginning the discussion of polymer protection in the presence of UV light, three factors were mentioned:

1. Defeat of repetition of the antioxidant mechanism by the large number of sites initiated
2. Susceptibility of the antioxidant itself to degradative oxidation in the presence of UV light
3. Damage to the critical return-of-the-complex-to-starting-materials mechanism by sudden input of a quantum of energy.

Hindered amines counter all of these:

1. The chain-reaction protection mechanism provides the increased efficiency needed to counter increased initiation.
2. Oxidation, instead of destroying the additive, converts it to the nitroxyl radical.
3. Input of UV-light energy adds to the ease with which the nitroxyl radical is regenerated from products formed by reaction with radicals.

Hindered amines are often used in combination with benzotriazole UV-light absorbers. This commonly permits reduced levels of both components.

Despite the above, it should not be assumed that hindered amines are immune to degradative oxidation. A number of the first hindered amines (HALS) commercialized had the structure

$$
\begin{array}{c}
\text{NH} \\
(CH_3)_2\text{-C} \qquad \text{C-}(CH_3)_2 \\
\text{CH}_2 \quad \text{CH}_2 \\
\text{CH} \\
\text{O-CO-R}
\end{array}
$$

Eventually oxidation occurs at the 4-position (to nitrogen) with abstraction of that hydrogen by a radical, leading to formation of a ketone or an olefin at that site, with elimination of the ester "tail." Later, HALS additives have used linkages that are more difficult to eliminate under the reaction conditions, such as tertiary amines, in place of esters.

II. SELECTION CRITERIA

It is generally thought (by users) that additives should be nontoxic, colorless, odorless, easily handled, instantaneously incorporated, nonvolatile, totally stable, effective at vanishingly small levels, and free. Thus, all are, of course, imperfect. As regards antidegradants, two areas of function must be considered: protection of the composition during processing, including polymer storage prior to processing, and maintenance of expected service life. Some considerations affect both.

A. Considerations Affecting Service and Processing

The formulator shares responsibility in ensuring that the use of additives does not result in adverse exposure to workers or in contamination of the environment. For example, cadmium oxide, a hazardous material, is used as a heat stabilizer in certain elastomers. It should be added in the form of of a dust-free dispersion, such as a prill or paste, to prevent worker inhalation and should be used at levels at which extraction in service, or after, is consistent with pertinent regulations. The dispersion itself would be prepared under much more rigorous conditions of worker and environmental protection. With regard to ingredient toxicity, the formulator must, therefore, consider what types of additives are consistent with procedures and patterns of behavior at his or her site, and in future use. Often, this will rule out some ingredients; at other times, it will dictate changes to procedures.

In certain cases, usage will be limited by the extent of regulatory approval, such as with FDA sanction of ingredients for direct or indirect food contact. Such sanction does not relieve the formulator of further circumscribing the range of ingredients that will be used, if technical knowledge dictates.

Closely related to whether, or how much of, a given additive will be used is the extent to which it is cost-effective. Decisions in both areas are governed by the application. For example, because elastomers stabilized with cadmium oxide are valuable in oil well drilling operations and constitute a trivial fraction of the cost of such operations, it is reasonable to undertake the expense needed to use that ingredient without undue hazard. A second example, tris(nonylphenyl) phosphite (TNPP), is currently the only organic phosphite having FDA sanction in plasticized PVC for food contact applications in the United States. It is relatively inexpensive. If the price of TNPP were to escalate, other, more efficient organic phosphites would be submitted to obtain FDA sanction.

In most applications, the end user is not without alternatives. Synthetic polymers compete among themselves and with naturally occurring or derived sustances (wood, metals, paper, cotton, glass, etc.) The design of polymeric compositions must perpetually respond to the market situation.

B. Considerations Primarily Affecting Processing

Polymers are typically supplied to industry containing sufficient antidegradants to ensure stability during storage for some length of time, as described in the supplier's sales specification. This antidegradant content may be sufficient to permit fabrication of a given article, or it may not. It may enable certain types of fabrication processes, but not others, or a certain type of process with some added ingredients, but not with others. For example, a polyolefin may be delivered with sufficient antidegradant content to permit extrusion, but compositions containing substantial quantities of halogenated flame retardants may require further additives for stabilization during processing.

The major factor involved during processing is the mobility and compatibility of the additive (as discussed in Section I.E). Differences in performance between competitive materials trace more to this factor than any other. An additive that is highly mobile will tend to be efficient; that is, it will be able to repeat its useful function often during processing. On the other hand, high mobility may lead to exudation, particularly if processing involves loss of streamline flow or sudden changes in pressure or shear rate, or to loss through volatility, if processing goes beyond certain time–temperature parameters. There is competition between the remedies of adjusting processing to compensate for such effects and partial or total replacement with a less mobile analog.

An old first approximation is that, to be useful in processing, the antidegradant must have thermal stability greater than or at least as great, thermal stability as the polymer. Commercialization of highly heat-stable polymers has shown this to be an oversimplification. Stabilizers are used, for example, in fluoropolymers that are not nearly as heat stable as the polymer. What is needed is for the combination of polymer and additive, to be more stable, in a particular process, than the polymer itself.

An important consideration is the processing needed to incorporate the additive. If added at the polymerization stage, it may need to be soluble in the monomer or solvent or dispersable in an aqueous phase. If added in a finishing step, for example, because of reactivity with the polymerization catalyst system, a concentrate of the additive package in the polymer is typically added. Because polymer manufacture is an activity involving substantial capital investment, such decisions tend to be well thought out. The addition of additives during compounding can be an entirely different matter. Technical considerations (i.e., addition of antidegradants as early as possible in the mixing cycle) may not be given priority over factors such as when it is found convenient to do this or that, or which physical form is preferred. In such a situation, the technologist can (usually) only reflect that the manufacturing department that consistently subordinates good science to convenience and lack of expense will eventually have nothing to manufacture.

C. Considerations Primarily Affecting Service

Compatibility and mobility remain significant factors affecting the performance of antidegradants during the service life of the finished article. The antioxidant has been chosen to be an excellent competitor to the polymer in the attraction of radicals generated in the presence of oxygen. Similarly, the UV-light absorber has been chosen to be an excellent competitor for the quantum of light energy. Regeneration of the antioxidant is favored by close proximity of a secondary synergist. The function of the UV-light absorber is aided by the close proximity of a quencher or of a hindered amine. At ambient temperatures, the mobility of these species is much lower than during processing. It is, therefore, important that the needed association complexes be formed during mixing and fabrication. Hence, the ingredients in the additive package should be selected to maximize mutual compatibility, taking care that chemical reactions that could inactivate the ingredients or generate undesired side effects are avoided. For example, phenolic antioxidants form strongly yellow complexes with titanium if Ti—Cl groups are present. This can easily result with use of titanium dioxide in PVC or polyolefins having appreciable chloride residues. In this case, the effect can be countered by scavenging halide with organotin compounds. Subtle interactions between additives dictate that comparative performance be evaluated using actual compositions, containing all ingredients, rather than using models consisting only of the polymer and the specific additive.

Such construction of tests is also needed to ensure that the final composition will not lose sufficient additives through volatility, migration, or extraction to affect desired service life. These factors are also influenced by the extent to which various additives are associated with each other in the composition.

III. TESTING OF ANTIDEGRADANTS

Test methodology depends strongly on the objective. This may range from an improved understanding of the mechanism of degradation to simply demonstrating that a particular sample of a given composition is typical. The scientist interested in exploring the chemistry of degradation may study the reactions and kinetics of small-model compounds. A colleague needing to compare a series of new compounds as antioxidants may add various levels to standard polymer compositions and measure the uptake of oxygen at elevated temperatures. This test, to a certain extent, mirrors reality, in that, for a while, nothing happens. The induction period is considered a measure of oxidative stability. With polymers that lose small molecules, such as HCl from PVC, gas evolution as well as oxygen uptake may be studied. Often, such studies are combined with thermal analysis of the composition. These methods are popular with producers of antioxidants because much data can be collected rapidly. Unfortunately, they have a low correlation with practical tests run by users.

Protection during processing is most often assessed by measurement of changes to the relative viscosity of a polymeric composition using a procedure that mimics the process. For example, a sample may be extruded (or calendered, etc.) in the laboratory and its melt flow index (ASTM D1238) compared to the original. It may then be reprocessed several times as an acceleration of factory conditions and the same measurement obtained. Concurrently, discoloration may be observed and yellowness index (ASTM D1925) compared.

In many cases, it is convenient to explore process safety with a torque rheometer, typically adjusted to a higher temperature or shear rate than would be found in practice. In this way, changes in viscosity may be observed directly and compared to standards (i.e., other samples known to process well or poorly). At the same time, small samples may be removed periodically for analysis of discoloration. With simple compositions, chemical reactions can also be followed (e.g., gas evolution). In many cases, torque rheometer measurements can be correlated with factory processes, the major factor being the use of a shear rate consistent with the latter. Even instances where fineness of division is of major importance, such as with powder or plastisol molding or coating, rheometer testing (at low shear rates) can yield useful information.

Stability during service is most often investigated using circulating air ovens to provide temperature acceleration and various exposure devices to accelerate UV-light aging, either in itself or in combination with other events simulating an outdoor environment. It should be stessed that accelerated tests are entirely relative. Extrapolation to normal service temperatures is highly uncertain. At one time, it was thought that measurements of a property of interest could be made (in a convenient time frame) after exposure for various lengths of time at several elevated temperatures and extrapolated downward to service temperature by estimating a heat of activation and constructing an Arrhenius plot of the data. It is now known that activation energies are not constant or linear with temperature and that such methods are highly unreliable [18]. In fact, in consideration of previous discussions of mechanisms, it is likely that the addition of antidegradants ensures an activation energy that is complex in temperature dependence.

Oven or UV-light exposure can be followed by a number of measurements—discoloration often the simplest. Alternatively, effects on elongation or tensile, yield, impact, or flexural modulus can be determined. Which will be followed depends on the application.

Often, the most complex cases relate to combinations of processing and storage. For example, a film or sheet (and many other articles) may be produced by one means or another and stored for some time before being subject to postfabrication processing such as printing, metallization, or other decorating or converting. The fabrication step will most likely result in some conversion of synergist, scavenger, or perhaps antioxidant to reaction products. The extent to which these become less compatible can strongly influence subsequent processing. This is another area in which water, as a product of oxidation, can have undesired effects. It may, therefore, be necessary to include not only antidegradants for the polymer and protective agents for the antidegradants but also scavengers for their reaction products.

IV. APPLICATIONS

The following subsections constitute a survey of antidegradant usage common or, at least, commonly reported in the United States at the present time. Trends affecting this pattern are noted.

A. Polyolefins

Polypropylene compositions normally contain catalyst scavengers at a level of 0.05–0.1% of the polymer. These may consist of calcium stearate (or other carboxylate), hydrotalcite, or a blend of both. This level, however, may double or triple in flame retardant compositions containing halogenated additives, which may also contain 0.05–0.1% organotin maleate as an acid scavenger or as a phase transfer catalyst for hydrotalcite.

In addition, hindered phenols, most often of high molecular weight, are used at levels of 0.05–0.25%, depending on the application. This is supplemented by 0.1–0.25% of an organic phosphite synergist for best processing and color stability. In cases where stability of physical properties during service is of greater importance, the latter may be replaced by 0.1–0.3% of thiodipropionate ester synergists.

In cases where UV-light stability is important, the following systems are common: 0.3–0.5% hydroxybenzophenone, 0.2–0.4% benzotriazole, 0.1–0.3% hindered amine, and, most recently, 0.1% hindered amine plus 0.1% benzotriazole. The trend may be to systems such as the last (except for low-end products which will continue with hydroxybenzophenones).

Where the product is used as wire insulation, 0.1–0.2% of a specific metal detector is invariably added as well.

Polymers such as polyethylene, for which cross-linking predominates over chain scission, tend to require lower levels of antidegradants than analogs, for which the reverse is true (e.g., polypropylene). High-density polyethylene compositions typically contain 0.025–0.05% (based on the polymer) of an acid catalyst scavenger, again either calcium stearate, hydrotalcite, or a blend. Again, highly flame-retardant compositions use increased levels. Hindered phenolic antioxidants are commonly used at 0.02–0.05%, as are synergists. UV-light absorber systems reported include 0.25–0.5% hydroxybenzophenone, 0.2–0.4% benzotriazole, 0.05–0.15% hindered amine, and 0.05–0.1% hindered amine plus 0.05–0.1% benzotriazole. Again, the trend is probably to such combinations. The common wire insulation applications use about 0.1% metal deactivator, but 0.2–0.25% is used for foamed constructions.

Low-density and linear-low-density polyethylene are reported to use higher acid catalyst scavenger levels, 0.05–0.75% in the most recent processes and 0.1–0.15% in the more established solution processes. Antioxidant, metal deactivator, and synergist levels are similar to HDPE. UV-light protection systems include those mentioned for HDPE as well as 0.15–0.25% hydroxybenzophenone plus 0.15–0.25% nickel quencher, and various hindered amine–hydroxybenzophenone combinations (not reported as yet elsewhere).

The phenolic antioxidants most often reported used in polyolefins include the following:

Di-*t*-butyl-4-methylphenol	BHT
Octadecyl di-*t*-butylhydoxyhydrocinnamate	Irganox® 1076
Methylenebis(4-methyl-6-*t*-butylphenol)	Cyanox® 2246
Methylenebis(2,6-di-*t*-butylphenol)	Ethanox® 702
Thiobis(4-methyl-6-*t*-butylphenol)	Santonox®
Tetrakis[methylene (di-*t*-butylhydroxyphenyl) propionate]methane	Irganox® 1010
Tris(methylhydroxy-*t*-butylphenyl)butane	Topanol® CA

Divalent sulfur synergists mainly used are ditridecyl, dimyristyl, and distearyl thiodipropionate. Organophosphite synergists used in polyolefins include the following:

Tris(nonylphenyl) phosphite	TNPP, Naugard® P
Distearyl pentaerythritol diphosphite	Weston® 618
Tris(di-*t*-butylphenyl) phosphite	Irgafos® 168
Bis(di-*t*-butylphenyl) pentaerythritol diphosphite	Ultranox® 626

The most popular hydroxybenzophenones used in polyolefins are those with lauryl (e.g., Inhibitor DOPB®) or octyl (e.g., Cyasorb® 531) in the para position. These provide better mobility in nonpolar polyolefins than —OH or —O—methyl groups. With benzotriazoles, the same effect is provided by alkyl groups on the aromatic ring (e.g., Tinuvin® 327 and 328), although the original product with merely a single methyl group (Tinuvin® P) remains widely used. Hindered amines used in polyolefins include the dimeric products Tinuvin® 123, 765, and 770 and higher-molecular-weight products such as Tinuvin® 622, Chimassorb® 944, and Cyasorb® 3346. The most common metal deactivators include Irganox® MD1024, Eastman Inhibitor® OABH, and Naugard® XL-1. The first and last of these are antioxidant-protected metal chelating agents.

Stabilization in polybutene compositions forms a special case because the major use, water pipe, dictates low extractability. The additive package typically includes 0.1–0.2% of an acceptable phenolic antioxidant and synergist (e.g., FDA-sanctioned BHT and dilaurylthiodipropionate [DLTDP]) as well as a calcium carboxylate scavenger.

B. Styrenics

Crystal polystyrene (PS) generally contains 0.1–0.25% each of a hindered phenolic antioxidant and an organophosphite synergist, generally the same types as used in polyolefins. Impact-modified grades (HIPS) use about the same levels but also include 0.02–0.05% fine-particle zinc stearate as a lubricant and also to scavenge catalyst traces from the stereospecific polybutadiene impact modifier. With black or dark colored products, amine antioxidants, such as Naugard® 445 are often used with cost benefits. For UV-light resistance, about 0.5% of either a benzotriazole or a combination with a hindered amine is used.

With ABS, magnesium stearate is most often used instead of zinc, at 0.05–0.1%. Phenolic antioxidants are used at slightly higher levels, 0.1–0.35%, because of the polybutadiene content, with 0.2–0.5% of a phosphite or thiodipropionate. Of the latter, dilauryl thiodipropionate is the most widely used in all styrenics, with TNPP the most used phosphite. With dark products, amine antioxidants, such as Vanox® 1081 are common. UV-light protection is similar to impact styrene.

Styrenic thermoplastic elastomers (TPEs) tend to include 0.05–0.1% zinc stearate. Phenolic antioxidants are used at 0.2–0.4% in styrene–butadiene copolymers, where cross-linking predominates, and at up to 1% in styrene–isoprene, where chain scission prevails. These are accompanied by 0.2–0.5% of a synergist, either TNPP or DLTDP. Dark colored products, such as shoe components, use amine antioxidants (e.g., Wingstay® 100).

C. Condensation Polymers

The oldest stabilization system for polyamides is one of the most interesting: the combination of very low levels (10–50 ppm) of copper acetate and 0.1–0.15% potassium iodide. A serious degradative route for aliphatic polyamides is through hydroperoxide-assisted hydrolysis of the amide linkage. Amide hydrolysis normally requires strong acid. In the presence of hydroperoxides, hydrolysis is rapid at elevated temperatures, even at neutral pH. In this case, effective competition can be provided by adding traces of a hydroperoxide decomposer (e.g., copper plus halide) that normally would be avoided, thus directing oxidation away from the amide linkage, instead of to C—H bonds. If the hydrocarbon segments are short (e.g., Nylon 6 or 66), this can be effective. Thorough drying of the polymer prior to processing, as with all condensation polymers, is still a requirement.

Direction of oxidation to C—H results in rapid discoloration. If this can be tolerated, the copper acetate/potassium iodide system remains the most effective. It is most common with applications in which accelerated oven aging is carried out at 120–150°C. Dark colored applications can also make use of amine antioxidants, usually 0.5–1.0% of aromatic *para*-phenylene diamines such as AgeRite® White or Wingstay® 100. These are as effective as the above in testing at 100–120°C and are much less water extractable. Light colored products use hindered phenols, originally BHT, now often products such as Irganox® 1010 or 1076 are used the latter particularly with longer hydrocarbon segments). UV-light protection is provided by either 0.4–0.5% hydroxybenzophenone, usually a dihydroxybenzophenone such as Uvinul® 400, 0.2–0.3% benzotriazole such as Tinuvin® P, or combinations with a hindered amine.

Aromatic polyamides generally do not contain typical antidegradants because of the combination of very high processing temperatures and general resistance of the polymers to oxidation. Thermal stability is controlled mainly by modification of polymer structure.

With polyurethanes, antidegradants are added to the polyether or polyester ingredient, to protect it prior to reaction, as well as to the reaction product, if it remains thermoplastic. Probably BHT, at 0.2–0.5%, remains the most common additive. Dark colored products, particularly foam for concealed padding, such as in automotive interiors, most often use amines, such as octylated diphenylamine, at 0.05–0.2%. The use of combinations of BHT and amine antioxidants is apparently also common. Where discoloration is objectionable, a combination of a highly blocked phenolic antioxidant, such as the above for polyamides, is used at 0.1–0.3%, with a phosphite synergist (e.g., Irgafos® 168) at the same level. Polyurethanes requiring UV-light protection usually employ benzotriazoles, or combinations with hindered amines, at 0.5–1.0%. In thermoplastic or elasto-meric polyurethanes where the best resistance to thermal degradation is needed, traces of water should be scavenged by reaction with a polycarbodiimide, such as Stabaxol® P (a powder) or Ucarink® XL-29SE (a solution). These, used typi-cally at 0.05–0.25%, scavenge traces of all active hydrogen compounds, including acetic acid, and are thus useful in PUR/EVA blends.

Polyurethane foam is one of the most reactive substrates toward discolora-tion by oxides of nitrogen (NO_x). In general, all activated aromatics react to some extent with atmospheric NO_x to yield strongly colored quinoid-type reaction products. Phenols and aromatic amines react very rapidly. Levels of NO_x at which this reaction becomes a problem can result from outdoor exposure, where ozone levels are high, or indoor exposure to arc welding or continuous spark or corona discharge. Even highly blocked phenolic antioxidants can contribute to discolora-tion. The phenomenon, usually a brownish discoloration, is, for reasons not en-tirely clear, referred to in industry as "gas fading." This may stem from cases where discoloration replaced an original, desirable shade. In any event, the most common practical remedy has been to shift the phenolic–phosphite balance more strongly in favor of the phosphite synergist. The latter reacts directly with oxides of nitrogen, converting them to harmless, reduced products. For this application, the phosphite should be trialkyl (e.g., Weston® 618) rather than a phenolic deriva-tive.

Polyurethanes formed using organotin catalysts may be inherently stabilized with regard to HX scavenging in flame-retardant compositions containing haloge-nated additives. If not, it may be possible to adjust catalyst ratios to accomplish this end.

The stabilization of thermoplastic polyesters includes chelation of the esteri-fication catalyst, now usually antimony oxide. This is a case where use of a catalyst that is not readily removable during or after polymerization produces a product sufficiently superior (to a volatile catalyst, e.g., boron fluoride) as to justify dealing with its persistence. The problem is that an efficient esterification catalyst is also an efficient hydrolysis catalyst. Antimony chelation is typically accomplished with 0.02–0.1% of trimethyl or triphenyl phosphate, or triphenyl phosphite. The same systems are also effective with a manganese acetate esterifi-cation catalyst.

The above is usually supplemented with 0.1–0.3% each of a phenolic and phosphite synergist, of polarity similar to what would be used in polyamides.

Use of 0.5–1.0% hydroxybenzophenones for UV-light resistance has been reported but has probably been replaced in practice with hindered amines and blends.

Polycarbonates form a special case because of the need to preserve clarity. Hydrolysis generates turbidity; additives, therefore, should be, if not actually hydrophobic, at least themselves resistant to hydrolytic attack. A common additive package consists of 0.05–0.15% of a relatively hydrophobic antioxidant, such as Irganox® 1076, and a similar level of a hydrolysis-resistant phosphite, such as Sandostab® P-EPQ.

Polyacetals degrade through loss of small molecules: formic acid and formaldehyde. Formic acid is a strong enough acid to be readily scavenged by use of 0.1–0.3% calcium stearate or citrate. Similar levels of melamine or dicyandiamide are used to scavenge formaldehyde. Use of 0.2–0.3% of a phenolic antioxidant, such as Santowhite® Powder or BHT is common. There are occasional reports of the use of phosphite or thiodipropionate synergists. Polyacetals have poor UV-light resistance but can be protected substantially with 1–3% fine-particle carbon black. Combinations adding up to about 0.5% benzotriazole and hindered amine can provide limited outdoor exposure in nonblack articles.

D. Other Applications

Halogenated polymers form a special category of sufficient complexity to require a separate chapter (Heat Stabilizers, Chapter 7).

Cellulosic esters remain an important and widely used class of polymers. Incorporation of 0.2–0.5% of an antioxidant such as BHT, Ethanox® 702, or Topanol® CA is common. Both DLTDP and phosphites, primarily TNPP, have been used as synergists. For occasional outdoor applications, as with cellulose acetate butyrate, filled systems are preferable and have been supplemented with 0.25–0.5% hydroxybenzophenone UV-light absorbers. Neutral hindered amines, such as Tinuvin® 123 should also be effective.

Methacrylate polymers (e.g., PMMA) are commonly UV-light stabilized with 0.05–0.2% of a benzotriazole such as Tinuvin® P. Use of combinations with nonbasic hindered amines (e.g., Tinuvin® 123) might well prove advantageous. This product has found use in at least one other amine-sensitive area, thermoset melamine resins.

Hindered amines are also under study in a variety of engineering plastics such as polyphenylene ether and polysulfone and should be considered broadly for inclusion in additive packages for such polymers.

V. COMMERCIAL SOURCES

Source	Antidegradant trade names
Alcoa	SorbPlus (8)
American Cyanamid	Cyanox, Cyasorb (1, 5)
Argus	Mark (3–5, 7, 9, 10)
BASF	Uvinul (5)
Bayer	Vulcanox (1, 2, 5)
Ciba Geigy	Irganox, Irgastab, Irgafos, Tinuvin, Chimassorb (1, 4, 5, 7, 9)
Dover Chemical	Doverphos (4)
Eastman	Eastman Inhibitor (1, 5, 7)
Ethyl Corp.	Ethanox (1)
Ferro	UV-Chek (1, 5, 6)
GE Specialty Chemicals	Ultranox, Weston (4)
Goodyear	Wingstay, GoodRite (1, 2)
Hoechst Celanese	Hostanox, Hostavin (1, 5, 7)
Kyowa Chemical	Alcamlzer (8)
Monsanto	Santonox, Santowhite (1, 2)
Morton	Carstab (3, 9)
Olin	Wytox (1)
Sandoz	Sandostab, Sanduvor (1, 4, 5)
Synthetic Products	Synpro, Synpron (9, 10)
Uniroyal	Naugard, Polygard, Naugawhite (1, 2, 4, 7)
Vanderbilt	Vanox, AgeRite (1, 2)
Zeneca	Topanol, Nonox (1)

The number(s) in parenthesis indicates the type(s) of antidegradant

1.   Phenolic antioxidants
2.   Amine antioxidants
3.   Thiodipropionate synergists
4.   Organic phosphites
5.   UV-light absorbers
6.   Nickel quenchers
7.   Metal deactivators
8.   Hydrotalcites
9.   Organotin scavengers
10.   Metal stearates

As usage tends to change with development of new materials and extension to a wider range of applications, representatives of the above suppliers should be contacted for state-of-the-art recommendations.

REFERENCES

1. Chemical Marketing Reporter 26 (November 1991)
2. L Pauling. Nature of the Chemical Bond. Ithaca: Cornell University Press, 1948, pp 52–53.
3. KU Ingold. Chem Rev 61:563, 1961.
4. AJ Chalk, JF Smith. Trans Faraday Soc 53:1214, 1957.
5. A Pearson et al. J Am Chem Soc 85:354, 1963.
6. O Hahn et al. Angew Chem 69:669, 1957.
7. CD Cook et al. J Am Chem Soc 77:1783, 1955.
8. A Baghieri et al. Polym Degrad Stabil 5:145, 1983.
9. RF Grossman. J Vinyl & Additive Tech 3:5, 1997.
10. C Armstrong et al. Eur Polym J 15:241, 1979.
11. RC Mehrotra, R Bohra. Metal Carboxylates. New York: Academic Press, 1983, pp 52–55.
12. RF Grossman. J Vinyl Tech 14:11, 1992.
13. WT Reichle. Solid State Ionics 22:135, 1986.
14. S. Miyata. U.S. Patent 4,379,882 (1983).
15. JF Rabek, Photostabilization of Polymers. London: Elsevier, 1990.
16. JE Pickett, JE Moore. Polym Degrad Stabil 42:231, 1993.
17. NS Allen et al. Makromol Chem 179:1575, 1978.
18. RH Hansen et al. Org Coat Plast Prepr 34:97, 1974.

2
Colorants

Melvin M. Gerson†
Podell Industries, Inc., Clifton, New Jersey

John R. Graff†
Bayer Corporation, Haledon, New Jersey

Color in plastics is a desirable feature for marketability and identification. Although colors specified to fulfill these ends are based on esthetic principles, achieving the desired colors is based on physical, chemical, and engineering principles.

Colorants are defined as materials that will modify the light incident on a surface so that light of other wavelengths, at the same or lesser intensities, is reflected or transmitted [1]. Colorants are generally colored materials, although optical brighteners that reinforce color appearance may themselves be colorless. Colorants can be incorporated into the plastic mass or applied to the surface of the plastic as a coating or laminate.

There are basically two types of colorants: those that are soluble in the formulation (dyes) and those that are insoluble (pigments). Some pigments show slight solubility, although such manifestation is generally undesirable and considered a defect. Solubility leads to bleeding or migration, plateout, and crocking, and can disqualify an otherwise acceptable pigment.

I. DYES

Because solubility, by definition [2], refers to the dissociation of a chemical into individual molecules or ions, we can readily infer the properties of many dyes. The advantages and disadvantages of dyes use as follows:

---

† Deceased.

Advantages	Disadvantages
1. Excellent transparency	1. Migration or bleeding
2. Ease of incorporation into the formula	2. High degree of reactivity with other ingredients in the formula
3. Excellent tinting strength or coloring ability	3. High degree of reactivity with degradation products
	4. Poor lightfastness
	5. Lack of heat stability

The disadvantages of dyes outweigh their advantages, particularly in light-fastness and heat stability. The development of solvent soluble dyes suitable for use in plastics [3] is progressing slowly, and selected members of this class of colorants are finding special uses where they can be incorporated into formulas in very low concentrations and where lightfastness is not a requirement in end use.

II. PIGMENTS

The predominant method for coloring plastics is with *pigments.* For our purpose, we must define *pigments* as insoluble materials that have a particulate structure and that can selectively absorb and reflect the rays of incident light or confer opacity to the plastic, or both. This definition includes all the colored pigments that can be used in concentrations as low as 0.001 phr and the white pigments, such as titanium dioxide, which are sometimes used at concentrations of 20 phr or higher.

Plastics processing requires pigments with a special set of characteristics compared with other colorant-using industries. Pigments that in themselves will not withstand temperatures of 300–500° for varying lengths of time obviously cannot be used. Pigments containing any metallic cations normally found in stabilizers must be used carefully. In some cases, these pigments may enhance stability; in other cases, unexpected reactions occur and too much of a given cation may cause a loss of control over stability. It is hard to formulate rigid rules about suitability of use for any pigment. As pigment technology improves and as formulation change, many pigments previously considered unsatisfactory become usable.

Some pigments are unsuitable for use where electrical resistance is required because absorbed soluble salts that are difficult to wash out remain in the pigment matrix after drying; these are ionic in nature and electrically conductive. The electrical resistance of such pigments can be enhanced by preparing a pigment dispersion by the technique known as flushing. This process effects the transfer of the pigment precipitate from the water phase directly to an oil phase (plasticizer) without previous drying. The oil wets the surface of the pigment preferentially and becomes bound to it. The water separates into a distinct phase (carrying the water-soluble salts with it) and is carried off during the decantation phase of this operation [1].

The technology of pigments and the science of color is growing at a very rapid rate; therefore, descriptions of pigment use such as this chapter are only a review of what is known at present. The latest information on the subject is best obtained from the technical literature of pigment manufacturers.

III. PIGMENT DISPERSIONS

Pigments are furnished as dry powders. Before they can be used, they must be properly dispersed. Dispersion of pigment is a three-step process: (1) disintegration of the agglomerates that have formed during drying, packaging, and storing; (2) wetting of the surface and replacing the pigment–air interface with a pigment–vehicle interface; and (3) stabilization of the dispersion to prevent reflocculation [1,4,5].

The key to using a pigment most efficiently for the best color value and end-use properties lies in its proper dispersion. Money value derives from tinting strength in relation to cost. This property depends on the degree of dispersion—the greater the degree, the better in the effort to approach ultimate particle separation. Pigment dispersion requires shear for its accomplishment when "flushing" is not possible. This shear is defined as the sliding of two adjacent or connecting parts or layers on each other so that they move apart in a direction parallel to the plane of their contact (i.e., in opposite directions). Although such shear occurs with a high-intensity mixer, such as a Banbury® or similar type of continuous mixer, a two-roll mill, a calender, or an extruder, the time that stocks are exposed to shearing stress in such equipment is not enough to result in proper pigment dispersion. Specialized equipment such as three-roll ink mills, pebble mills, high-intensity high-speed mixers, and impingement mills are required for this latter purpose. Wetting agents and surfactants, acting by physico-chemical methods, can shorten and simplify this operation, but, basically, dispersion of pigments remains a mechanical operation [1,4–6]. The science of pigment dispersion has become a specialized technology, and as a result, the formulator has a series of predispersed pigments available in a variety of vehicles and forms. They may be prepared as pigment pastes (high-viscosity liquids) in plasticizers or as chips, powders, or pellets, which are handled like any other solid ingredient. The advantages of pigment dispersions are lack of dust, ease of handling, ability to be directly incorporated into the formula, and provision for standardization of shade and strength by the pigment dispersion manufacturer to a close tolerance. Testing and evaluating pigment dispersions are similar to testing and evaluating pigments themselves.

IV. PARTICLE SIZE, TINTING STRENGTH, AND TRANSPARENCY

The ultimate aim in using any pigment is to achieve the maximum in color value and in transparency or opacity. Pigments are manufactured to an optimum particle size for color value, tinting strength, and transparency–opacity characteristics. Agglomeration of these particles during manufacture tends to interfere with achieving this goal. Particle size is therefore a definite factor in the use of pigments and is controlled in the pigment manufacturing and dispersion operations. The desirable particle size varies from a low of about 20 nm with certain carbon blacks to μm for iron oxides, chrome oxides, and others. A better understanding of the significance of particle size is available by studying titanium dioxide as an opacifying pigment. The opacity property of a white pigment directly depends on its refractive index [4,6] and inversely depends on its particle size, up to a point. The refractive index of a pigment is an inherent property that depends on its crystal structure and chemical composition. The refractive index of titanium dioxide as the highest known among colorless (white) substances contributes to its excellent opacifying characteristics [7]. The other controlling factor is implied by the Mie theory [8,9]—that when the particle size of a pigment is reduced below one-half the wavelength of incident light, its ability to scatter light is reduced in proportion. This optimum particle size can vary with crystal structure, particle shape, and refractive index; it is of the order of 0.2 nm for titanium dioxide.

This property has become highly significant with the development of nano-particle (submicron) titanium dioxide. The particle size is so small that light is not scattered or absorbed to any great extent, thereby providing reasonable transparency to visible light. Nevertheless, because of its high dielectric constant, it remains an excellent ultraviolet (UV), light absorber and is the basis of the most effective sunblock preparations. Although much more expensive than ordinary grades of titanium dioxide, it is beginning to find use in UV-light-resistant clear plastics and coatings.

V. PROPERTIES OF PIGMENTS

The capacity for color, transparency or opacity, and tinting strength of a pigment is generally inherent in the pigment and independent of the formulation (excluding

solubility considerations). Every other property affecting pigment suitability—heat stability, crocking, and so on—is in some way affected by the choice
of the other ingredients or may affect this choice. Consequently no one should
evaluate a pigment (or pigment dispersion) without using the pigment in a valid
formulation and comparing results from a similar formulation uncolored or pigmented with a similar pigment as a reference standard.

   The most important properties that can be affected or changed by the presence of a given pigment are discussed in the following subsections. Also described
are the test techniques generally used for evaluating colorants on each of these
properties.

A. Heat Stability

The colorant is incorporated into a compound (with or without white pigment)
and subjected to heat for varying periods. It is compared with a similar compound,
without the colorant, that has been treated in the same way. Pigments that fade
or change color indicate a lack of inherent heat stability. Pigments that accelerate
the browning or darkening of the compound indicate an undesirable reactivity.

B. Light Stability

The light stability of a pigment may vary with its use in masstone or in tint tone.
This property is evaluated in both forms and compared with similar unpigmented
stocks. Exposure in accelerated light-aging or weathering devices is generally
the method (see Section V.K). Interpreting the results is similar to judging heat
stability. Loss of color indicates lack of inherent lightfastness. Premature browning or brittleness of the stock indicates reactivity with the formulation.

C. Sulfide Staining

The pigmented compound is exposed to an atmosphere rich in hydrogen sulfide
fumes and high in humidity. The change in color (usually significant darkening)
indicates sensitivity to this chemical, which is usually present in industrial fumes.
Because certain stabilizers, fire retardants, and other additives will cause this
darkening, an unpigmented compound of the same formulation should be tested
simultaneously for comparison.

D. Migration, Bleeding, and Blooming

Prepare a sheet or film of the formulation and place it in contact with a similar
white or uncolored sheet under a weight of 1 lb/in.^{2} This test is usually conducted
for 5 h at 180°F. Other times and temperatures may be used if they better represent

end-use requirements. The transfer of color to the uncolored sheet indicates migration or bleeding. Pigments that migrate will usually bloom (appear on the surface
as a fine powder) after a period of storage at room temperature. Pigments that
are slightly soluble in liquid ingredients of the formulation show this defect.
Pigments that sublime during heating may also bloom regardless of the presence
of plasticizers.

E. Crocking (Dry or Wet)

Crocking refers to ruboff, and the procedures used are those specified for textiles
in accordance with AATCC [10]. A piece of white, unsized fabric is rubbed
across the plastic surface under constant pressure. After the specified number of
rubs, the fabric is examined visually for color pickup.

F. Ease of Incorporation

Evaluating ease of incorporation must be consistent with the methods of processing to be used. An excellent test is to band a white compound on a two-roll mill.
Add the pigment or dispersion to be tested and mill for 3 min with constant
stripping and rebanding. Strip the sheet, allow to cool to room temperature, and
reband for 3 min. The sheet is examined for color streaks and other evidence of
unincorporated pigment. With polymers that do not mill well, the resin and color
dispersion should be blended and passed through a mixing extruder under conditions comparable to plant operation.

G. Electrical Characteristics

Because many pigments may be slightly ionic or polar or contain soluble salts
as a result of the manufacturing operation, their electrical characteristics should
be evaluated in accordance with National Electrical Manufacturers Association
(NEMA) procedures [11].

H. Plateout

Mill a colored sheet on a two-roll mill for at least 5 min. Follow this immediately
with a pure white compound on the same mill and determine the degree of color
imparted to the white sheet by pigment left on the rolls. When blooming, bleeding,
or migrating pigments exhibit this phenomenon, they should be eliminated from
the formula. Plateout of pigments can be eliminated by varying the amount of
lubricant, changing the lubricant, or incorporating a small amount of a finely
divided absorptive extender pigment such as silica or aluminum hydrate.

I. Resistance to Acid, Alkali, and Soap

Two or three drops of the test solution are placed on the surface of the pigmented compound for periods up to 3h. The compound is washed clean and the exposed spot is examined for change in color.

J. Oil Absorption and Rheology

Flow characteristics of pigmented plastics are particularly important in plastisol or extrusion technology. They may be evaluated by practical processing tests or by tests of their consistency over a period with a Brookfield viscometer or torque rheometer. The rheology of pigmented compounds is generally influenced by the oil absorption of the pigment. This property of a pigment is defined as the amount of vehicle required to surround each particle of the pigment completely and fill its interstices [12]. It is measured by determining the amount of vehicle required to change a pigment–vehicle mixture from a noncontinuous appearance to a homogeneous, wet-appearing mass. This will vary with the vehicle and its wetting ability, the pigment particle size and shape, the surface treatment applied during pigment manufacture, and the surface charge on the pigment [4].

K. Outdoor Durability

Outdoor durability includes lightfastness and also refers to chalking (formation of a fine powder on the surface of the plastic), the effect of weathering on tensile strength and embrittlement, and so on. These phenomena must be considered as the interrelation between the pigment particles and the polymer Matrix, and the degradation products on exposure [13]. Although some evaluation can be made using accelerated weathering devices, actual continuous exposure for 12 or more months in Arizona or Florida is believed to represent the severest conditions of humidity, sunlight, ultraviolet light, and weather that plastics are likely to be subjected to in actual service [14]. The names and addresses of commercial weathering stations in Florida and Arizona are listed in Ref. 15.

VI. PIGMENT CHARACTERISTICS

Color, tinting strength, tint tone, chemical resistance, lightfastness, heat stability, and so on are characteristics of the individual pigment, its particle size, its pigmentary shape, and any impurities that may be present as a result of the manufacturing process. Any general listing of pigments and their properties, as in the *Color Index* [3], are at best an average of the properties of those pigments within any given family or class that are commercially available. Individual products, even from the same manufacturer, may vary in particle size and shape, or in impurities present because of raw materials used or of surface treatments to the pigments, su as additives to control oil absorption, decrease flocculation tendencies, inhibit chalking (in the case of titanium dioxide), or improve transparency. These variations can result in marked differences in performance among the same pigments when they are evaluated in any of the performance tests described earlier. This is nowhere more apparent than in the electrical characteristics, rheology, weathering, and heat- and light-stability tests discussed. Most surfactants are innocuous but account for the differences in ease of dispersion and rheology that can be found between pigments of the same type from different manufacturers.

A common modification of pigments is accomplished by resination. Solubilized resin is added to the precipitation tank before neutralization and precipitation of the pigment. When precipitation of the pigment occurs, the resin is also precipitated with and on the pigment.

Resination of pigments at less than 5% may be considered a surface treatment. Resination from 5% to 30% by weight of some pigments is used to modify the optical or working characteristics to a significant degree. Resination improves transparency and makes the color more brilliant. Unfortunately, resinated pigments are not as heat stable as the base unresinated pigment and cannot be processed at equivalently high temperatures.

The designation "toner" refers to pigments that are almost 100% colorant. "Lakes" refer to pigments precipitated in the presence of an inert, extender pigment of very low refractive index, such as aluminum hydrate or barium sulfate. The most common forms of these extended pigments are either 60% color pigment and 40% extender, or 40% color pigment and 60% extender. They have the triple advantage of ease of dispersion, improved rheology, and less sensitivity to weighing errors. They are commonly used with the more expensive pigments such as cadmiums, vat oranges, vat yellows, and vat blue. The Food Drug & Cosmetic Colorant (FD&C) pigments are all prepared by precipitating the approved FD&C dyestuff on such an extender (usually aluminum hydrate). The characteristics of the extenders used are such that they have little effect on the color or transparency of the pigment.

VII. PIGMENTS, COLOR, AND STRENGTH

The color of a pigment should be evaluated only when the pigment is used in a representative formulation that is characterized by concentration and thickness of film or sheet consistent with the end-use requirements [1]. One of the significant tests must apply to cost of the formulation. Pigments used vary in price from a low of about 40¢ per pound for some black pigments to $40 per pound for some esoteric organic pigments derived from vat dyestuffs.

When a colorant is used in a formula and is observed in reflected light (light is incident on the surface and reflected to the eye), the *masstone* of the colorant is evaluated. When the same formulation is observed in transmitted light (i.e., the light is observed through the pigmented sheet or film), the *transparency* and *undertone* of the colorant is evaluated. When the colorant is evaluated in a combination with a white pigment (generally at a ratio of 5 parts of colorant to 95 parts of white pigment), the *tint tone* of the colorant is evaluated. It must be understood that the undertone of a pigment (when observed in transparency) is not necessarily the same as the tint tone of the pigment when observed in combination with white [1].

Definitions of *color* and methods for numerical evaluation of the color properties of a pigment are discussed in Section XVI.A.

VIII. CLASSIFICATION OF PIGMENTS

Two general methods useful in classifying pigments involve chemical composition and color. Any general study of pigments should begin with their chemical composition, as this generally defines their performance properties. Individual studies can then be made, based on the color requirements of the particular problem.

IX. INORGANIC PIGMENTS: OXIDES (HYDRATED OXIDES)

Many of the most durable pigments are metallic oxides. Because this class of compounds is generally unreactive, the properties that are most often manifested are excellent heat stability, light stability, chemical inertness, lack of bleeding and migration, desirable electrical characteristics, and very low absorption. These properties are improved by calcining, and those oxides that have been so treated are especially suitable.

A. Titanium Dioxide

The oxides of titanium exist commercially in two crystal forms: anatase and rutile. Both crystals are tetragonal and in an octahedral pattern. In anatase, the octahedra are packed so that 4 of the 12 edges of each octahedron are shared

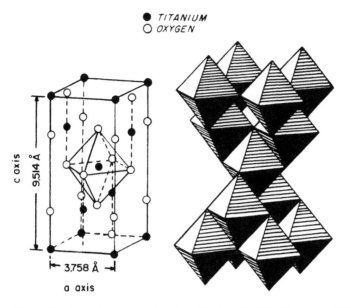

Figure 1   Crystal structure of anatase titanium dioxide. (Courtesy of Titanium Pigments Division, NL Industries, Inc.)

with adjacent octahedra; in rutile, 2 such edges are shared [7]. See Figs. 1 and 2.

Rutile pigments have the highest refractive indices of white pigments and are much more resistant to chalking than anatase. Anatase pigments have generally a slightly bluer shade (and therefore appear whiter); they are slightly lower in refractive index and are slightly easier to disperse.

Surface treatments on titanium dioxide pigments may consist of zinc oxide, aluminum oxide, silicon dioxide, or other oxides deposited in monomolecular layers by vapor-phase calcination. In many instances, these surface treatments can affect heat and light stability.

Titanium dioxides, because of their excellent tinting strength and very high opacity, are the whitest pigments known. The older white pigments such as zinc sulfide, lead carbonate, and lithopone are rarely used.

B.   Antimony Oxide

Antimony oxide is used because of the fire retardancy it imparts. Because it will cause a whitening effect and loss of transparency, it must also be considered a

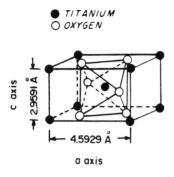

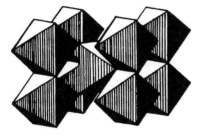

Figure 2   Crystal structure of rutile titanium dioxide. (Courtesy of Titanium Pigments Division, NL Industries, Inc.)

white pigment. Where transparency is required, antimony pentoxide rather than the more common and lower-cost trioxide is used. The pentoxide is supplied as submicron particles. Recently, nanoparticle antimony trioxide has been introduced as a fire retardant.

C.   Oxides of Lead, Zinc, and Silicon

Lead oxides, zinc oxides, and silicon oxides are all of little value because of their refractive indices (i.e., poor opacity), low tinting strength, and reactivity in vinyl formulations. Silicon oxides are so low in refractive index that they are classed as inert, extender, or filler pigments.

D.   Iron Oxides and Hydrates

The compounds in the class containing oxides and hydrates of iron vary in color from light yellow, through red, to a dark brown and black. The iron oxides that are mined often contain many impurities (such as magnesium, manganese, zinc, and aluminum in ionic form) that may be reactive in use. When these metallic ions are calcined, they are converted to their oxide form and are quite unreactive at temperatures below 400°F. Hydrated iron oxides, which are yellow, are also stable to temperatures below 400°F. Red iron oxide pigments mostly are manufactured synthetically and consist of at least 88% Fe_2O_3. Synthetic manufacture results in a high degree of purity, low reactivity, and a wide variety of shades and undertones.

Brown iron oxide pigments, burnt siennas and burnt umbers, are calcined natural pigments that are combinations of Fe_2O_3 with significant amounts of manganese dioxides, silicon oxides, or aluminum oxides. Synthetic brown oxides are also prepared by calcining iron oxide, so that a blend of FeO and Fe_2O_3 forms in the proper proportion and assumes a desired crystal form. Black iron oxide is prepared in a similar fashion. Overheating or too strenuous oxidation in any of these calcination processes will cause a reversion of the desired pigment to the red Fe_2O_3.

Iron oxides are desirable because of their excellent durability, inertness, and low cost. Their chief disadvantages lie in their high specific gravity, relative coarseness or large particle size (when compared with organic pigments), low tinting strength, and lack of brilliance (see Section XVI.A). The yellow iron oxides are more properly characterized as tan pigments. The red iron oxides form brick red, brown, or brown maroon colors, in contrast with the brighter cadmium salts and organic red pigments.

E. Chromium Oxides and Hydrates

The only other oxide pigments that are commercially important are the green oxides and hydrates of chromium. Chromium oxide, Cr_2O_3, is probably the least reactive and most lightfast green pigment known. It is low in brightness, opacity, and tinting strength, and very difficult to disperse. Hydrated chromium oxide (chrome hydrate) is $Cr_2O_3 \cdot 2H_2O$. This pigment has all the stability characteristics of chromium oxide (except that it will not withstand ceramic processing temperatures), is much easier to disperse, is more transparent, has higher tinting strength, and is much more brilliant in color. Although hydrated chromium oxide is more expensive, it is a much more practical pigment than chromium oxide. The utility of chromium-based pigments has been strongly reduced by the perception of toxicity associated with ''heavy'' metals. Thus, a valuable, lightfast green pigment has been largely lost to the plastics industry.

F. Miscellaneous Colored Oxides

Almost all other colored oxides have been evaluated for use as pigments. Many of them, such as the cobalt oxides, have found use in ceramic processes in which their ability to resist temperatures to 1500°F are of value. They are, however, deficient in brilliance and tinting strength when compared with other pigments available for use in plastics.

X. ELEMENTAL PIGMENTS

A. Carbon Blacks

The generic term *carbon black* refers to a large series of colorants based on the formation of complex substances that are chemically largely elemental carbon during the controlled, incomplete combustion of natural gas or oil. Bone black (animal black or drop black) results from the controlled combustion of animal bones, with the organic matter forming elemental carbon on a calcium carbonate–calcium phosphate substrate [16–18].

Three main forms of carbon black pigments are available [19], depending on the feedstocks and method of manufacture, characterized as follows:

Channel blacks[a]	Furnace blacks	Thermal blacks
Small particle size	Larger particle size	Largest particle size
High jetness	Low jetness	Lowest jetness
Acidic surface	Alkaline surface	—
High oil absorption	Low oil absorption	Low oil absorption
Brown undertone	Blue undertone (recommended for tinting)	Bluest undertone

[a] No longer manufactured in the United States, but available through imports.

The older forms of carbon black, often referred to as pigment blacks, lamp blacks, acetylene blacks, and so on, have largely been supplanted by these newer, more economical forms of black pigments.

Because of the variety of conditions subject to control during the manufacture of carbon blacks, these pigments can vary in particle size, adsorbed gases and moisture, porosity, structure (linking of carbon atoms into chains), and surface area. This results in a wide variation in masstone and tint-tone color, electrical conductivity, oil absorption, ease of dispersion, and rheological properties. Channel carbon blacks are recommended for UV stabilization for outdoor exposure [1,13,19] because of their ability to react with free radicals and also their ability to absorb ultraviolet radiation and render it harmless.

Carbon black pigments are so varied in their properties and are the subject of so much investigation [20] that they are best described by their individual characteristics (Table 1) and by an explanation of their choice for some specific uses. Jetness of black pigments can be estimated by a nigrometer reading (the lower the reading, the higher the jetness). As jetness increases in masstone, its tinting strength decreases, and oil absorption, ease of dispersion, rheological problems, and cost increase.

An example of the compromises that must be made can be found in considering pigments for the two separate applications: phonograph records and electrical insulation. Phonograph records use carbon black as a filler for its low pound–volume cost, and the very small particle size of channel black helps to minimize scratches and other surface noise. The regular color channel black (Table 1) traditionally has been used because of its low cost (compared with the smaller sizes), relatively high jetness, and high electrical conductivity, shown by its ability to resist the formation of electrostatic charges. However, owing to problems in connection with air pollution, channel black pigments have been phased out of production and have been replaced with furnace blacks of approximately equal surface characteristics and particle size.

The carbon black best designed for use in electrical insulation (highest electrical resistance) is the long flow furnace black (Table 1). Two related characteristics contribute primarily to the excellent electrical properties of the latter pigment, its ease of dispersion and its low structure. Ease of dispersion relates to the pigment's ability to be wetted by a vehicle (resin or plasticizer) and to have its envelope of air completely removed. Because organic vehicles are much less conductive than air and have a much lower tendency to ionize under high voltage, this prohibits formation of paths for electrical charges. Long flow furnace black is also described as a "low-structure" black. The individual carbon particles are discrete and not connected (Figs. 3 and 4). This results in easier separation of the pigment particles during dispersion and prevents formation of long pigment chains, which can transmit electrical current. On the contrary, the most conductive carbon black pigments has very high oil absorption and the lowest amount of adsorbed materials on its surface.

These grades of channel black are also being replaced, principally by furnace blacks having similar characteristics and properties.

B. Bone Blacks

Bone blacks are still used in masstone to achieve high jetness with a blue undertone. The presence of calcium carbonate and calcium phosphate gives them a color effect that other black pigments cannot duplicate. The presence of other salts and impurities makes it difficult to stabilize formulations that contain bone

Table 1 Typical Characteristics of Some Black Pigments

Pigments	Nigro-meter	Surface area (m²/g)	Particle diameter (μm)	Oil absorption (lb oil/100 lb pigment)	Tinting strength	Volatile content (%)	Fixed carbon (%)	pH	Characteristics
High color channel	64	850	12	375	165	13.0	87.0	3.0	Jettest black in common use
Medium color channel (A)	71	380	16	160	175	5.0	95.0	5.0	Economical jet black
Medium color channel (B)	74	320	17	155	185	5.0	95.0	5.0	Economical jet black for UV stability; low electrical conductance
Regular channel	81	130	22	125	182	5.0	95.0	5.0	General-purpose black; Phonograph records
Long flow channel	84	295	28	88	170	12.0	88.0	3.5	Easy to disperse; very low oil absorption; highest electrical resistance
Conductive furnace	94	190	29	250	140	2.0	98.0	7.5	Most electrically conductive black; for floors, for antistatic effects
Oil furnace (A)	84	86	25	85	227	1.0	99.0	7.5	Excellent low-cost channel black substitute; high tint strength and low oil absorption
Oil furnace (B)	90	42	42	72	187	1.0	99.0	8.0	High tint strength, blue tone, low oil absorption; used for tinting
Gas furnace	99	23	80	70	100	1.0	99.0	9.5	Very blue tone; very low oil absorption; most popular black for tinting
Thermal	110	6	470	33	35	0.5	99.5	9.5	Lowest oil absorption; bluest tone

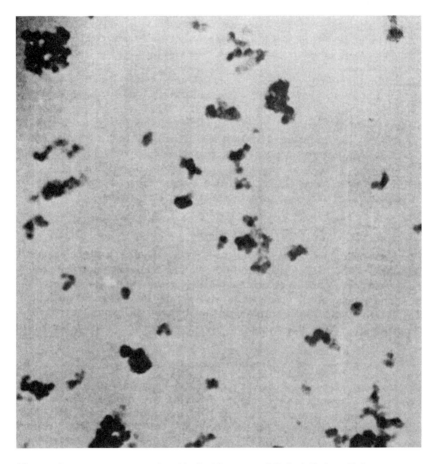

Figure 3 Low-structure carbon black. (Courtesy of United Carbon Co.)

black and to control for color during processing. Bone black should therefore never be used for tinting.

C. Metallic Pigments

The other types of pigments based on elemental chemicals are the metallic flake pigments. Although many metallic elements are hammered into flakes and used in the paint, ceramic, and printing ink industries [16,21], the plastics industry limits itself to aluminum, colored aluminum, copper, and bronze flakes (Table 2). All the metallic pigments are prepared by hammering the metal into thin

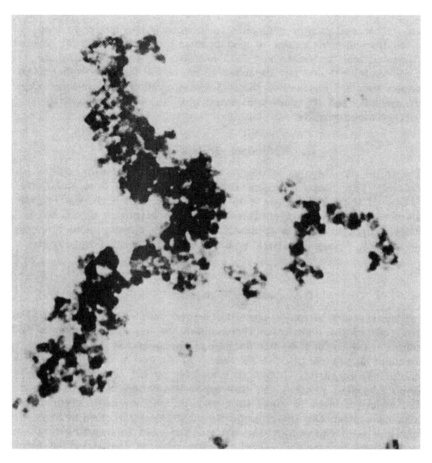

Figure 4 High-structure carbon black. (Courtesy of United Carbon Co.)

Table 2 Elemental Metal Pigments

Commercial	Metal composition
Aluminum bronze	Aluminum
Copper bronze	Copper
Pale gold bronze	Cu, 92%; Zn, 6%; Al, 2%
Rich pale gold bronze	Cu, 90%; Zn, 9.25%, Al, 0.75%
Rich gold bronze	Cu, 77%; Zn, 22%; Al, 1%
Green gold bronze	Cu, 68.75%; Zn, 31%; Al, 0.25%

sheets in the presence of a lubricant such as stearic acid. The most desirable effect is achieved when the flakes orient themselves in a plane parallel to the surface of the film, and the reflectance is similar to the specular, or mirrorlike, reflectance from the metal itself. Using coarser flakes (approximately 200 mesh), leafing grade powders (the leafing characteristic is enhanced by adding a maximum of lubricant), and slow processing at slightly higher than normal roll temperatures (on a calender) or die temperature (in an extruder) will enhance this metallic effect.

The copper alloys were first used for gold effects. Because the metals of these alloys are reactive, the stability of these compounds is limited. The metals tend to react during processing and with degradation products when exposed to light and humidity. Aluminum flakes are much less reactive and normally show only a slight duiling on long-term exposure to light and humidity. One can render both types of flakes less reactive by coating them with a very thin coating of a thermosetting resin. Although this enhances the stability of the copper alloys, it does not permit them to equal the aluminum pigments in durability. Durability of the aluminum pigments is also enhanced by this treatment.

A whole gamut of metallic colors and tones can be obtained by using transparent pigments in combination with aluminum flakes. Gold, copper, brass, and bronze colors of any degree of lightfastness and heat stability can be prepared, depending on whether one uses low-cost transparent yellows with or without an azo red. More lightfast formulations will result with the higher-cost, more durable organic pigments. The effect created is a yellow, gold, or bronze color distributed throughout a film. Although the aluminum flakes can be incorporated to give a metallic effect colored to the degree desired, the optical effect is not quite the same as that obtained from bronze powders. The increase in durability and processability may be worth the compromise, however. Another type of metallic effect is obtained by coating very thin (less than 0.001 in.) aluminum foil on both sides with a thermoplastic or thermosetting resin that has been pigmented with transparent, durable, nonbleeding pigments. The coated foil is then cut into very small flakes. They are available in two grades: "glitter," from 0.005 to 0.010 in. in diameter, and "flitter," from 0.015 to 0.100 in. in diameter. Flitters and glitters can be incorporated into clear, colored, transparent, pastel, or white film to give unusual color effects. They are seldom used as the sole colorant of a film, but are used to give a speckled effect, especially in floor coverings.

XI. METALLIC SALTS

These brightly colored inorganic pigments, metallic salts, are used because of their excellent heat stability, lack of tendency to migration and crocking, and good to excellent lightfastness. Their chief deficiencies are their very high specific gravity (high pound–volume cost), relatively large particle size, and low tinting strength compared with organic pigments. A technique for improving the chemical resistance and the lightfastness of these pigments has been devised that consists of coating each pigment particle with a coating of fumed silica. This permits the use of some pigments which have had borderline lightfastness and resistance to soap, alkali, and acid.

A. Iron Blues and Chrome Greens

The iron-containing salts such as Prussian blues and chrome greens (coprecipitates of ferric ferrocyanide and lead chromate) contain the reactive Fe cation and significantly affect heat and light stability.

B. Chrome Yellow and Orange Pigments

The chrome yellow and orange pigments compose a series based on lead chromate. Medium chrome yellow is lead chromate, with the lighter shades being mixtures of lead chromate and lead sulfate. Chrome oranges are basic lead chromate and are sensitive to acids. The chrome yellows vary in shade from light greenish yellow to a reddish yellow. They are attacked by acids and alkali and will turn black in the presence of hydrogen sulfide.

C. Molybdate Oranges

Molybdate oranges are complex mixtures of lead chromate, lead sulfate, and lead molybdate. They have excellent tinting strength and heat and light stability. One of their chief uses is in combination with high-cost organic red and maroon pigments for formulations that are bright in color, have good stability except to acids and alkalis, have high opacity, and are relatively low in cost. They will turn black in the presence of hydrogen sulfide.

D. Cadmium Pigments

Cadmium pigments vary in shade from the bright yellow of cadmium sulfide, through the reds of the cadmium sulfoselenides, to the deep maroon of cadmium selenides. Because they do not darken in the presence of H_2S, they are used for outdoor exposure (except in yellow tints) and where the maximum in heat stability is required. They are available as CP (chemically pure) and lithopone extended grades. In this case, the lithopone refers to barium sulfate, which is inert. Although simple chemical formulas can be written for their composition, x-ray diffraction studies indicate that the term *coprecipitation,* frequently used to describe their preparation, implies a complex crystal or solid solution that provides for unexpected chemical stability [22].

Inorganic pigments that contain cadmium, lead, mercury, or chromium now find only minor use in plastics in North America and Europe because of regulatory pressure limiting their manufacture and disposal. The pigments themselves are generally not implicated because of very low solubility, but manufacture involves intermediates that are clearly hazardous. In some parts of the world, there is continued use of these pigments. As with lead-based stabilizers, it is entirely possible to engineer facilities to provide safe manufacture and use if required and if cost-effective.

E. Mercury–Cadmium Pigments

A newer series of pigments based on coprecipitation of red mercury sulfide with cadmium sulfide and selenides is analogous in properties and similar in shade to the cadmium orange, red, and maroon series. Their chief advantage is a lower cost than the cadmium series [22]. Their stability equals the cadmium pigments except where exposure to moisture or to acids is a problem. These pigments are also available in the CP grades and lithopone grades.

F. Ultramarine Blue

Ultramarine blue, an inorganic pigment made from calcining sulfur, clay, and a reducing agent at high temperatures, has never been identified chemically. It has a bright, reddish blue color. Ultramarine blues are chemically stable except to acids. This pigment is not recommended where lightfastness is a main requirement.

G. Mineral Violet

Manganese violet, permanent violet, and mineral violet are synonyms for manganese ammonium phosphate. The chemical stability of this substance is good, except to alkali and moisture. Its lightfastness is excellent in masstone and tint tone for interior use, but not good enough for exterior use. It has largely been supplanted by the newer organic violet pigments, which have better money value (color strength per unit cost).

H. Miscellaneous Chemical Salts

Strontium chromate, zinc chromate, cobalt salts, colored silicates, and other colored chemical salts are used as pigments in other industries. They are generally low in money value and brightness and are chiefly recommended for the ceramic industries. The exceptions are the various zinc chromate salts that are used in solution and plastisol coatings to provide corrosion resistance for steel, aluminum, and magnesium metal substrates.

The characteristics of some inorganic pigments are given in Table 3.

XII. ORGANIC PIGMENTS

Organic pigments compose a group of colorants that embraces a wide variety of chemicals having an equally wide variety of costs and properties. They are brilliant in color and have good tinting strength. Vesce [23] has categorized these pigments by chemical class. New classes must be continually added to this list because of new developments [3] (Table 4).

Organic pigments are classified according to chemical structure. Similarity of structure generally implies a similarity of properties. Table 4 presents the most common classification of these pigments.

A. Basic Colorants

Basic colorants consist of basic dyestuffs precipitated with heteropoly acids. Even the "permanent" types are so poor in resistance to heat, light, and solvents that they are not recommended.

B. Insoluble Azo Pigments

Colorants that are made by coupling various aromatic compounds to form monoazo compounds characterized by the —N=N—linkage make up the insoluble azo pigments. They have no groups capable of forming salts with metals. As a class, they bleed, migrate, and crock in most formulations because of their solubility. They have poor heat stability and may sublime out of the compound during processing.

C. Bis-Azo Pigments

Bis-azo pigments are formed from o-dichlorobenzidine, which can be diazotized and coupled at both ends of the molecule with other intermediates. They do not have as great a tendency to bleed as the monoazo class, and their heat stability and lightfastness is better.

Table 3 Inorganic Pigments Suitable for Vinyl Compounding

	White					Green				Yellow			
	Titanium dioxide, rutile	Titanium dioxide, anatase	Zinc oxide	Antimony oxide	Zinc sulfide	Chromium oxide	Hydrated chrome oxide	Yellow iron oxide, natural	Yellow iron oxide, synthetic	Natural siennas, ochres	Chrome yellows	Cadmium sulfides	Nickel titanate
Chemical class	OX	OX	OX	OX	MS	OX	OX	OX	OX	OX	MS	MS	MS
Brightness	H	H	VH	H	VH	VL	M	VL	VL	VL	H	H	H
Opacity	VH	H	L	VL	M	L	T	VH	VH	VH	H	H	VH
Lightfastness													
Masstone	VH	VH	VH	VH	VH	VH	VH	VH	VH	VH	MH	H	VH
Tint tone	VH	VH	VH	VH	VH	VH	VH	VH	VH	VH	M	L	VH
Heat stability	VH	VH	VH	VH	VH	VH	VH	L–M	H	L	H	H	VH
Resistance to reactivity with compound variables	H–VH	VH	VL	VH	VL	VH	VH	VL	H	VL	H	H	VH
Migration resistance	VH	VH	VH	VH	VH	VH	VH	VH	VH	VH	VH	VH	VH
Acid resistance	VH	VH	VL	H	L	VH	VH	L–M	VH	L–M	L	L	VH
Alkali resistance	VH	VH	M	H	M	VH	VH	VH	VH	VH	L	H	VH
Resistance to sulfide stain	VH	VH	VH	VH	VH	VH	VH	VH	VH	VH	VL	VH	VH
Weathering resistance	VH	M	M	H	L	VH	VH	L	VH	VL	M	VH	VH
Tint strength	VH	H	L	VL	L	VL	VL	VL	L	VL	L	L	VL
Economics of use	VH	H	L	VL	L	L	L	H	H	H	H	L	VL

Note: H, high; L, low; V, very; M, median, moderate; T, transparent; R, reactive in vinyls. Chemical class key: OX, oxide; MS, metallic salt; E, elemental.

	Oranges			Reds, maroons, violet				Blue		Brown		Black		
Molybdenum orange	Cadmium sulfoselenide	Mercury cadmiums	Red iron oxide, synthetic	Cadmium sulfoselenides and selenides	Mercury sulfoselenides and selenides	Mineral violet	Ultramarine blue	Cobalt blue	Brown iron oxide	Burnt sienna	Carbon black, channel, furnace, lamp	Bone black	Black iron oxide
MS	MS	MS	OX	MS	MS	MS	MS	MS	OX	OX	E	E	OX
VH	VH	H	VL	H	H	MH	VH	M	VL	L	VL	VL	VL
VH	H	H	VH	H	H	M	M	L	H	T	VH	L	VL
H	VH	VH	VH	VH	H	H	H	VH	VH	VH	VH	R	VH
H	M	M	VH	M	L–M	M	R	VH	VH	VH	VH	R	VH
VH	VH	VH	VH	VH	VH	VH	VH	VH	H	H	VH	R	H
H	VH	VH	VH	VH	VH	VH	VL	VH	H	H	VH	VL	H
VH	VH	VH	VH	VH	VH	VH	VH	VH	VH	VH	VH	VH	VH
L	L	L	VH	L	L	H	VL	VH	H	VH	VH	VH	H
L	H	H	VH	H	H	L	VH	VH	VH	VH	VH	VH	VH
VL	VH	VH	VH	VH	VH	VH	VH	VH	VH	VH	VH	VH	VH
M	H	L–M	VH	VH	MH	L	VL	VH	VH	VH	VH	L	H
L	L	L	L	L	L	VL	VL	VL	L	L	VH	L	VL
H	L	M	H	L	M	VL	L	VL	VH	VH	VH	L	H

Table 4 Classification of Organic Pigments

Class	Typical structure	Typical common names	General properties
Basic	Precipitated basic dyes with tannic acid, phosphotungstic or phosphomolybdic acid	Auramines, Methyl Violet, Rhodamines, Victoria Blue, Methylene Blue	Not recommended: shows poor heat stability and lightfastness.
Insoluble azo	TOLUIDINE RED	Clorinated paras, Fire Red, Para Red, Permanent Orange Toluidines, Arylamide Reds, Naphthols, Diansidine Orange, Nitraniline Oranges	Not recommended: shows bleeding and migration, crocking, poor heat stability, and plateout.
	HANSA G	Hansa	Not recommended: shows bleeding, migration, crocking, plateout, sublimation, poor lightfastness [24]
Bis-azo	DIARYLIDE YELLOW	Diarylide yellows	Most members of this class bleed, migrate, crock, and plate out. The AAOT type is used for flooring. The AAOA type is used for transparent yellows. All have poor lightfastness in tint tone and fair lightfastness in masstone.

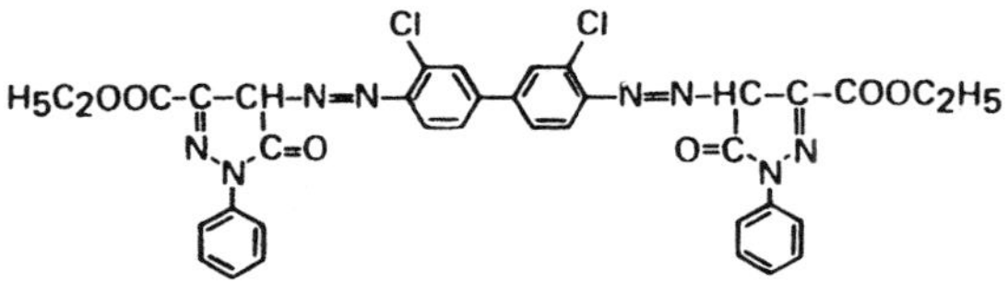

Soluble (precipitable) azo (pigments are formed by precipitation with calcium, barium, or manganese ion; all are nonbleeding and nonmigrating)

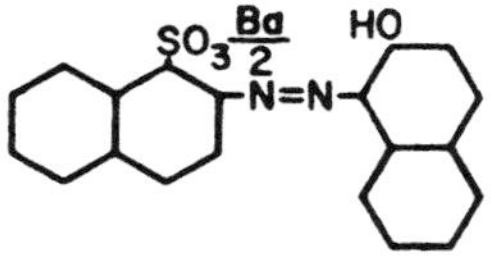

H₅C₂OOC-C-C-CH-N=N

PYRAZALONE RED

$H_5C_2OOC-C-C-CH-N=N$ ... N=N-HC-C-COOC$_2$H$_5$

Diarylide Orange, Pyrazalone Reds

Poor heat stability for long-term processing. Slight migration at room temperature; noticeable migration at high temperatures. Poor lightfastness in tint tone; fair to good lightfastness in masstone.

BARIUM LITHOL

Lithols, Persian Orange, Red Lake C

Sensitive to acids and alkalis. Poor heat stability and lightfastness in masstone. Poor lightfastness in tint tone.

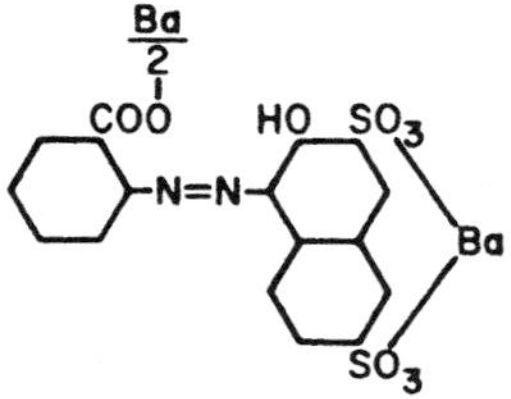

PIGMENT SCARLET

Pigment Scarlet

Sensitive to acids and alkalis. Good short-term heat stability. Poor long-term stability. Fair to good masstone lightfastness. Poor tint tone lightfastness. Permanent Red 2B is subject to plateout.

(*continued*)

Table 4 Continued

Class	Typical structure	Typical common names	General properties
	PERMANENT RED 2B	Permanent Red 2B, BON Reds and Maroons, Lithol Rubine	Excellent except for possible migration and bleeding.
Miscellaneous	**NICKEL AZO YELLOW**	Metallized chelated azos	Excellent, except for poor lightfastness. Many are excellent; some are good to light and exterior exposure. Available in yellow to deep red and maroon shades.
		Bis-azo, condensation [25].	
Condensation acid class	**DISAZO**	Alkali Blue, Acid Violet, Peacock Blue, Eosine, Quinoline Yellow	Not recommended: shows poor heat stability and lightfastness.

Vat pigments

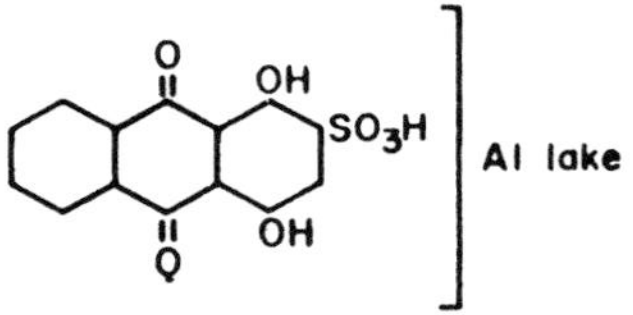

ALIZARINE MAROON

INDANTHRENE BLUE

THIOINDIGO

Alizarine, Madder Lake

Good masstone lightfastness; poor tint-tone lightfastness.

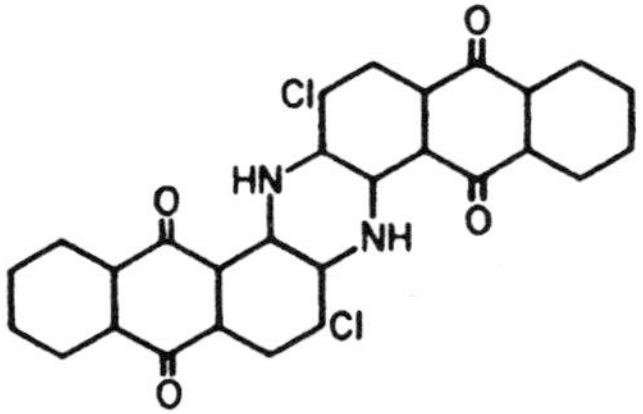

Anthrapyrimidine, Pyranthrones, Perylenes, Anthrimides, Indanthrone, Flavanthrone, Isoviolanthrone

Nonbleeding and nonmigrating, have excellent heat stability and lightfastness in masstone and tint tone. Stable to all types of chemical agents except reducing agents. Rated excellent in brilliance and transparency.

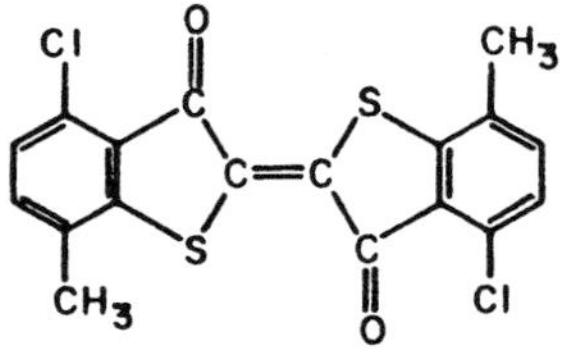

Indigos, Thioindigo Reds, Thioindigo Maroons

Generally excellent lightfastness. Some numbers show slight bleeding. Used primarily for tinting.

(continued)

Table 4 Continued

Class	Typical structure	Typical common names	General properties
Phthalocyanines [26,27]		Phthalocyanine blues, vary in shade from very red undertone to very green undertone	Beta crystal form and alpha crystal forms with Cl or long-chain alkyl substituents on the benzene rings are stable to heat and solvents. Unmodified alpha crystal form is unstable. Heat- and solvent-stable forms meet all the criteria for excellent performance.
		Phthalocyanine greens, vary in shade from blue undertone to yellow undertone with the number of chlorine atoms present. When bromine is used, the shade is a very bright yellow green.	
	PHTHALO GREEN		Excellent performance.

	Structure	Description
Miscellaneous class		
Dioxazines [28]	 **CARBAZOLE VIOLET**	Stable in all requirements.
Quinacridones	 **QUINACRIDONE VIOLET**	Stable in all requirements. Available in orange, red, maroon, violet shades.
Isoindolinones	**ISOINDOLINONE**	Excellent stability.

D. Soluble (Precipitable) Azos

If azo pigments are formed from intermediates containing free phenolic, carboxylic, or sulfonic acid groups, they will be soluble in most solvents, including water. Adding calcium, barium, or manganese ions will precipitate these compounds to form nonbleeding pigments with good heat stability, fair to good lightfastness in masstone, and poor lightfastness in tint tone. Their resistance to alkali and to acid is poor because of the presence of the metal ion. This group of pigments in the orange to red and maroon shades are among the most popular pigments because of their relatively low cost and excellent tinting strength.

E. Miscellaneous Azos

The two main members of this class are the chelated azos and the condensed azos. Azo pigments, chelated with a metal atom, have excellent characteristics. Their increasing tendency to bleed and migrate, however, with increasing plasticizer concentration limits their use. Condensed azo pigments are formed by condensing two molecules of pigment from the insoluble azo group into a single molecule. This increase in size and molecular weight results in a series of brilliant red, yellow, and orange pigments with excellent properties.

F. Condensation Acid Pigments

Dyestuffs that can be precipitated onto a mordanting agent suitable for pigment use, such as aluminum hydrate, form this class of pigments. They are too water soluble and poor in heat stability for widespread use.

G. Anthraquinone and Vat Pigments

Vat dyestuffs are colorless in their reduced or hydroquinone form

HYDROQUINOID
(UNCOLORED)

and colored in their oxidized or quinone form

QUINOID
(COLORED)

More than 400 such compounds are listed in the *Color Index,* and they have all been investigated for use as pigments. Those selected are prepared in their quinoid form, surface treated for proper texture, and washed free of impurities. They have been chosen for their brilliance of color and for their ability to meet criteria. They are the most expensive of the pigments available and are used because of their excellent lightfastness.

H. Phthalocyanine Pigments

The phthalocyanines are a series of blue and green pigments that have excellent stability to light, heat, and chemical reagents. They are insoluble in water and in the usual ingredients of formulations. The tendency of the early phthalocyanines to crystallize under heat with a change in hue and in tinting strength has been overcome by adding substituents to the benzene ring and by careful processing in manufacture. Very red-shade blue pigments are available from the unsubstituted phthalocyanines, to blue-shade greens when 14 to 16 chlorine atoms are added to the possible 18 positions on the rings. Yellow-shade greens are made when bromine is substituted for part of the chlorine in the latter case. The phthalocyanines, as a class of pigments, appear to have most, if not all, of the properties considered desirable for use.

I. Miscellaneous Organic Pigments

The technique of tailoring an organic molecule to provide the color and stability characteristics required for processing has progressed to a fine science. The newer pigments such as the isoindolinones and quinacridones fall into this class. The specific capabilities and uses are best described by the individual manufacturers, who continually modify and improve their products. Specific compositions are listed in the *Color Index* [3] and in the references shown for each group. Detailed recommendations for use are constantly being revised because of these improvements. Basic lists for evaluation of pigments appear in the technical literature on a regular basis [29–31] (see Table 5).

Table 5   Organic Pigments

	Common pigment name	Chemical class	Brightness	Transparency	Lightfastness—masstone	Lightfastness—tint tone	Heat stability	Migration resistance	Acid resistance	Alkali resistance	Weathering resistance	Tint strength	Economics of use
Green	Phthalocyanine green, all types	PCN	VH	VH	VH	VH	VH	VH	VH	VH	VH	VH	VH
Yellows	Flavanthrone	AN	VH	VH	VH	VH	VH	VH	VH	VH	VH	VH	L
	Diarylide yellow OT	DIS	VH	M	M	VL	VH	L	VH	VH	VL	VH	VH
	Diarylide yellow AAOA	DIS	VH	VH	M	VL	VH	L	VH	VH	VL	VH	VH
	Pigment yellow 83	DIS	VH	VH	M	M	H	VH	VH	VH	L	VH	VH
	Isoindolinones	MV	VH	VH	VH	VH	VH	M	VH	VH	H	VH	ML
	Nickel-azo	MA	H	M	VH	VH	VH	M	L	M	M–H	VH	M
	Anthrapyrimidine	AN	VH	VH	VH	VH	VH	H	VH	VH	H	VH	L
	Bis-azo, condensation	MA	VH	VH	M–H	M–H	H	M	VH	VH	M–H	VH	M
Oranges	Diarylide orange	DIS	VH	VH	L	VL	VL	VL	H	H	VL	H	H
	Nitraniline oranges	AZ	VH	VH	VL	VL	VL	L	VH	VH	VL	VH	H
	Dianisidine orange	DIS	VH	H	L	L	H	L	VH	VH	L	VH	H
	Vat orange GR	AN	VH	VH	L–M	H	H	VH	VH	VH	H	VH	VL
	Vat orange RK	AN	VH	VH	L–M	L–M	L–M	M	VH	VH	M	VH	VL
	Isoindolinones	MV	VH	VH	VH	VH	H	VH	VH	VH	VH	VH	L

Colorant	Class										
Lithols	SA	VH	L	VL	VL	VL	VL	L	VL	VH	VH
Permanent red 2B	SA	VH	M–H	M	VL	M–H	VH	L	VL	VH	VH
Lithol rubine	SA	VH	VH	L	VL	L	VH	L	VL	VH	VH
Pigment scarlet	SA	VH	VH	L	VL	H	VH	L	VL	M	VH
Red lake C	SA	VH	VH	VL	VL	VL	VH	VL	VL	VH	VH
Bon reds and maroons	SA	VH	VH	M	VL	M	VH	M	VL	VH	M–H
Pyrazalone	DIS	H	VH	H	VL	M	L–M	VH	VH	VH	H
Quinacridones	MV	VH	H	VH	VH	VH	VH	VH	VL	H	L
Alizarine maroon	AN	VH	VH	H	VL	H	VH	M	M	M	M
Perylenes	AN	VH	VH	H	H	VH	VH	VH	VL	VH	L
Isoindolinone	MV	VH	VH	H	H	VH	VH	VH	VH	VH	L
Bis-azo, condensation	MA	VH	VH	H	H	VH	VH	VH	H	VH	L
Thioindigos	AN	HH	VH	H	H	M–H	M–H	VH	M	VH	L
Phthalocyanine	PCN	VH	VH	VH	VH	VH	VH	VH	VH	VH	VH
Indanthrone	AN	VH	VH	VH	VH	H	H	VH	VH	VH	M
Quinacridone	MV	VH	VH	VH	VH	VH	VH	VH	VH	VH	L
Carbazole	MV	VH	VH	VH	H	VH	VH	VH	VH	VH	M
Thioindigos	AN	VH	VH	H	M	M–H	M–H	VH	M	VH	L

Red and Maroons (Lithols through Thioindigos)
Blues and Violets (Phthalocyanine through Thioindigos)

Note: Key to chemical class: PCN, phthalocyanine; AN, anthraquinone; BIS, bis-azo; MA, miscellaneous azo; MV, miscellaneous vat; AZ, azo; SA, soluble azo. Ratings: V, very; H, high; M, median. Pigments not readily identified by common names are listed by *Color Index* number where possible [3].

XIII. PEARLESCENT AND OTHER INTERFERENCE PIGMENTS

A series of pigments is available in which the pigments exist as thin, transparent flakes and yield color effects due to reflection from both surfaces of the particle and the consequent reinforcement or interference of these light rays [1,32]. The effect is a clean, silvery multicolored appearance, which can be modified with small amounts of durable colors. Natural pearlescence is too fragile and costly for use, although it is occasionally used in plastisols. The same visual effect is obtained through pigments made from lead carbonate, lead phosphate, lead arsenate, and bismuth oxychloride. Of these, only the last is now considered nonhazardous.

A new series of interference pigments made from titanium-coated mica has also been introduced. These pigments are easier to stabilize for heat and light than some of the older forms.

All these pigment particles are subject to fracture, with consequent loss of "pearlescence" when subjected to too much shear. For the best appearance, they should be processed with a minimum of intensive shear and in such a way that the particles orient in a plane parallel to the plastic surface.

XIV. FLUORESCENT AND OTHER OPTICALLY ACTIVE PIGMENTS

A. Fluorescent Pigments

Fluorescent pigments [1,33] are based on fluorescent dyes that have been incorporated into a thermosetting resin matrix, cured, and ground to a suitable fine powder. When the dye and resin are carefully chosen, pigments with excellent heat stability and resistance to bleeding and migration result. The original fluorescent pigments were deficient in lightfastness. Newer pigments are reported to withstand 12 months or more of exterior exposure, however. Their fluorescence derives from their ability to absorb ultraviolet or visible light and reemit it at other visual wavelengths. This has the effect of a brighter color being reflected from the surface than would be expected from light incident on it. Fluorescent pigments are only of value when they are observed under a strong source of daylight or incandescent illumination reinforced with ultraviolet light. Although fluorescent lightbulbs can be used, optimum color value is best manifested under natural daylight. Individual fluorescent pigments can be blended for variation in color effects. When they are combined with standard color pigments, UV absorption of these latter pigments severely inhibits the efficiency of fluorescence.

B. Luminescent and Phosphorescent Pigments

The series of luminescent and phosphorescent pigments is based on the variations possible in ZnS, CdS, and ZnO crystals, which can store UV or visible light and reemit visible light when all sources of light are removed. These pigments are highly fragile and are liable to fracture in processing operations. At best, the reemitted light is very weak in intensity and is of relatively short duration. These pigments are used in acrylic systems or in vinyl–acrylic top coats as novelty coatings for various commercial products [1].

C. Emission Pigments

Pigments that emit light without needing any external light source are available from the vast source of radioactive materials now available. Their price is prohibitive and their use regulated by the Atomic Energy Commission.

D. Optical Brighteners

The unusual group of colorless, transparent pigments composed of the optical brighteners can absorb ultraviolet light and reemit it in the visible wavelengths without affecting transparency of the film. They usually reemit light in the blue end of the spectrum and effectively make transparent and white appear brighter and whiter (more blue white). Although the brighteners are relatively expensive, amounts of the order of 0.001–0.01 phr are usually sufficient. Unfortunately, their lightfastness is limited and short periods of exterior exposure destroy their utility [34].

XV. SPECIAL PIGMENT PROBLEMS

A. Heat Sealing and Electrical Resistance

When heat sealing is to be accomplished by directly applying heat and pressure, the pigment will have no effect on the process so long as it does not migrate or bloom. When dielectric heat sealing is used, pigments of minimum conductance are required. Pigments that will pass the NEMA requirements [11] are the most desirable.

B. Toxicity of Pigments

There are basically four categories of use conditions of interest to formulators:

1. Products designated as toys or other objects that children might put in their mouths

2. Products to come in contact with food, such as food packaging, bread wrappers, and candy box separators
3. Products to be used in packaging cosmetic products, like soaps or shampoos
4. Products for which the ultimate in nontoxicity is required, such as bottles for salad oil, beer, and other liquid food products and for medical prosthetic devices

Each category requires a different set of pigments, and the cost of coloring increases significantly with increasing rigidity of specification. If the manufacturer can prove that there is no possibility for the pigment to be ingested—either because of an impermeable barrier between the pigments and the material to be eaten or because of the insolubility and inertness of the pigment—then any pigment can be used [21]. Because this is difficult to prove, formulators follow the individual requirements of the four groups out lined as set down by the U.S. Food and Drug Administration or other regulatory agencies.

1. *Toys.* Although individual state laws have varying requirements related to toxicity, the U.S. Department of Commerce Commercial Standard for Artist Materials to be used in Schools (C8-136-61) and the ANSI Standard [35] provide general guidelines for choosing pigments. In general, they place stringent limits on the amount of lead, mercury, bismuth, soluble barium, cadmium, antimony, and arsenic that can be used. This specification effectively proscribes the use of chrome yellows, molybdate oranges, cadmium pigments, bismuth, and synthetic pearlescence.

2. *Food Packaging.* In 1958, an amended form of prior legislation known as the Food Additives Amendment was passed by Congress [36]. This statute covers materials that are likely to become part of food as a result of food processing or packaging. The Food and Drug Administration regularly publishes lists of pigments (as well as other materials) that may be used for this application. Suitably prepared titanium dioxide, iron oxides, and ultramarine blue can be used in this application, as well as in foods, without regard to certification [37]. Other pigments can be used provided that they will not become part of the food. A series of test solvents and test procedures has been prepared that simulate exposure to various types of foods [38].

3. *Colors for Cosmetics.* Suitable lists of colorants that can be used in products that come in contact with the skin have been prepared by the FDA under the Food, Drug, and Cosmetic Act of 1938, as amended in 1960 [38].* Provisions for certification of purity have been made. These D&C colors are prepared as

---

* Copies of Ref. 38 and its amendments, and copies of the food additive amendments of the periodic changes in their regulations, can be obtained by writing to the Food and Drug Administration, Washington, DC.

lakes on aluminum hydrate for incorporation. Their use as colorants in cosmetic packages exempts the formulator from complying with "the proof of nontoxicity" when noncertified colors are used. Unfortunately, these colorants are seriously deficient in lightfastness and heat stability and require special processing conditions for satisfactory use.

4. *Food Colors.* A list of FD&C colorants suitable for use in coloring food has also been prepared by the FDA. This list is smaller than the D& list, and the laked colors on this list are also deficient in heat and light stability. The FD&C certified colorants can be used without restriction in any application where toxicity is a problem.

C. Outdoor Durability

The use of plastics for outdoor applications is constantly increasing. Colorants can be important in increasing this use. Carbon black pigments will enhance the durability of any plastic because of their absorption of ultraviolet light. These short wavelengths of light in the sunlight radiation spectrum are the ones that most seriously affect vinyls.

Pastel colors and white should be formulated with the nonchalking grades of titanium dioxide to provide for maximum life and prevention of the disfiguring white powder, or "chalk," that can appear with other grades. Because many pigments have poorer lightfastness in tint tone than in masstone, only pigments that have been thoroughly evaluated with titanium dioxide should be used.

Outdoor durability refers to exposure to all of the deteriorating factors that may affect a plastic composition. Light is only one such factor; with humidity or moisture, acid attack from industrial fumes and alkali and soap from cleansing agents representing other factors that affect weatherability. Pigments that show suitable resistance to all of these factors should therefore be chosen. If a compromise must be made, those pigments that darken or change hue to a darker value are preferable to those that lighten or fade.

XVI. COLOR AND CHEMICAL COMPOSITION

Color is the term that we give to the human eye's reaction when observing certain wavelengths of light. Colorants have the ability to accept incident light, absorb portions of this light, and reflect or transmit other portions. For example, when a colorant absorbs all of the light except blue light and when it reflects or transmits this blue light, we have a blue colorant. When a colorant absorbs all (or most) of the light equally, we are dealing with a black colorant. White colorants reflect equally all wavelengths of the incident radiation.

The color that we see depends on both the absorption characteristics of the colorant and the medium that surrounds it. Chemical groups that cause a stress on the nucleus of the molecule and consequent response of electrons to the energy of the incident light will be colored [1,4,28,39,40]. Interference between incident and reflected light can occur when pigment particles have two dimensions significantly different from the third, as with pearlescent or other flake-type pigments. Dichroism occurs when pigment crystals have significantly different refractive indices along different axes [1,4] and when these pigments can orient themselves in specific planes. Interference pigments and dichroic pigments are difficult to evaluate for color because of the difficulty in duplicating particle orientation.

A. Color and Color Measurement

Color results from the effect on the eye of light modified by colorants [41]. It depends on the following:

1. A light source
2. A modifier for the light source
3. An observer to interpret the modified light

Visible light is a physical phenomenon defined as that portion of the electromagnetic spectrum between 4000 and 7000 Å (Fig. 5). The physicist can characterize the ability of a pigment to modify light by observing its spectrophotometric curve (Fig. 6). The spectrophotometer measures the absorption of incident light on a colorant at each wavelength of visible light.

Spectrophotometric curves are not a measure of color because they do not provide a value for the type of light under which the colorant is viewed and because they do not provide a value for the human eye–brain capability to see color. The eye sees three attributes of color that Munsell has defined [10] (Fig. 7).

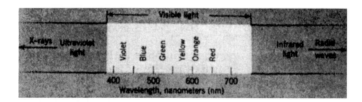

Figure 5 Visible light spectrum.

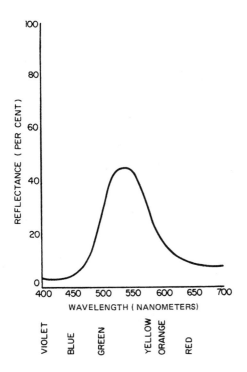

Figure 6 Spectrophotometric curve of a green pigment, by Davidson and Hemmendinger. (Courtesy of *Color Engineering.*)

1. Hue: the quality of color we describe as red, yellow, green and so forth.
2. Value: the quality of color we describe as light or dark when compared with white (Munsell value of 10), through grays, to black (Munsell value of 0).
3. Chroma: the quality of color that distinguishes it from a gray of the same value. The terms *saturation* and *purity* are sometimes used to describe this quality.

In defining color, one must consider the characteristics of the light under which the colored object is viewed. A red colorant (Fig. 8) illuminated by a blue light (curve C, Fig. 9) will appear quite dark and different from its appearance under a red light (curve A, Fig. 9). All these variables have been combined by the International Commission on Illumination into the CIE system (Commission International de l'Eclairage) of colorimetry [41,42]. This system develops its three-dimensional characteristics from the three lights (Fig. 10), which when

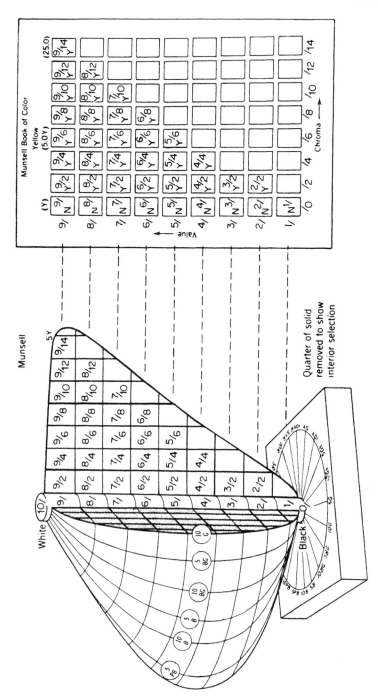

Figure 7 Munsell color system. A system of specifying objective colors on scales of hue, value, and chroma, exemplified by a collection of chips forming an atlas of charts that show scales for which 2 of the 3 variables are constant, the hue scales containing 5 principal and five intermediate hues, the value scale containing 10 steps from black to white, and the chroma scales showing up to 16 steps from the equivalent gray. All three scales are intended to represent equal visual (not physical) intervals for a normal observer and daylight viewing with gray to white surroundings. (From ISCC *Newsletter*, no. 156 (Nov.–Dec. 1961.) D. Nickerson.)

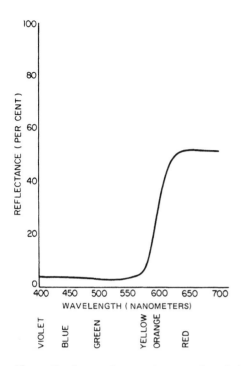

Figure 8   Spectrophotometric curve of a red pigment, by Davidson and Hemmendinger. (Courtesy of *Color Engineering.*)

properly combined will produce white light; the definition of *white light* is ''a bright light containing all the wavelengths of visible light at equal intensity.''

The CIE measurements can be made with three-filter colorimeters that measure the light reflected or transmitted by a colored object under one of the standard sources of light (Fig. 9). These measurements are designated the tristimulus values, X, Y, and Z. They are converted into tristimulus coordinates or trichromatic coefficients, x, y, and z, by the equations

$$X = \frac{X}{X + Y + Z}$$

$$y = \frac{Y}{X + Y + Z}$$

$$z = \frac{Z}{X + Y + Z}$$

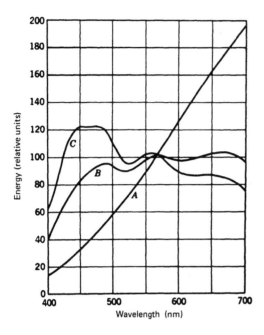

Figure 9 Illuminants A, B, and C. (From Ref. 43; reprinted by permission of John Wiley & Sons, Inc., New York.)

Tristimulus values are defined as the product of the relative energy of the light source used, the reflectance or transmittance of the object, and the tristimulus value of the equal-energy spectrum colors (Fig. 10). The tristimulus coordinates are the percentages of each component for the particular color observed [47]. These data can then be plotted on the CIE chromaticity diagram (Fig. 11).

Hue can be estimated from "dominant wavelength." Chroma is estimate from the value for "purity." The third dimension of value can be estimate directly from the Y reading determined on the colorimeter. The value obtained for Y has been determined to represent closely the brightness of a color as viewed by the human eye [41,42].

The significance of the CIE system is that it permits the establishment of numerical definitions for color and for tolerance variations from color standards.

Computer programs (analog and digital) based on the measurements obtained from a spectrophotometric curve, the definitions of the standard light sources, and the definition of the CIE equal-energy colors have been devised [41,45–47].

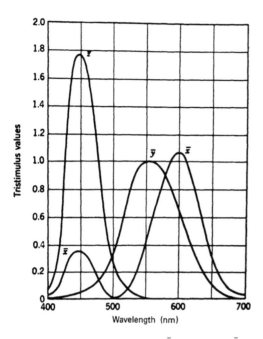

Figure 10 Characteristics of x̄ (red light), ȳ (green light), and z̄ (blue light) for the spectrum colora. (From Ref. 43; reprinted by permission of John Wiley & Sons, Inc., New York.)

B. Metamerism

Metamerism is a defect in a color match that is defined as the change in color relation between two colored objects when the conditions of viewing are changed. The phenomenon was first noted when colors were matched in artificial light (curve A, Fig. 9) and then observed to be different under daylight (curve C, Fig. 9). Spectrophotometric curves of two gray color chips (Fig. 12) illustrate two colors that are close matches under tungsten light but mismatches under fluorescent light and daylight. The "simple gray" of this set of colors becomes redder in contrast to the "complex gray" as the incident light becomes bluer. When this phenomenon occurs during routine color-matching procedures, it invariably indicates that there is a difference in the type of colorant used in the two samples. This procedure has often been used for quality control analysis and pigment identification. Metamerism can be a serious problem to a formulator attempting to match a colored object (paint, paper, textile, ceramic) that has been made with

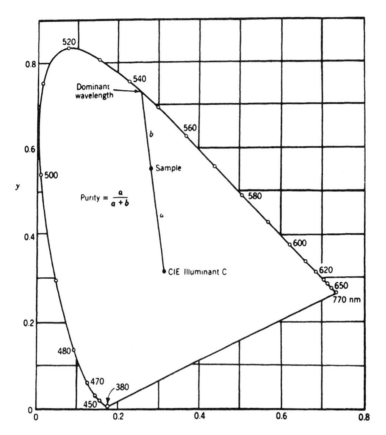

Figure 11 A typical chromaticity diagram. (From Ref. 44; Reprinted by permission of Interscience Division, John Wiley & Sons, Inc., New York.)

colorants unsuitable for use in plastics. Ad equate solutions to these problems can be arrived at by analyzing the spectrophotometric curves of colorants that are suitable and combining them to eliminate most of the metamerism [48].

Subsequent studies indicate that metamerism may also occur with differences in angle of light incidence or light reflection, with differences in observers, or with any change in conditions of color observation [41,49].

C. Visual Color Matching

The ultimate end of the coloration of compounds and the extensive color characterization procedures in the preparation of "suitable colors" for the

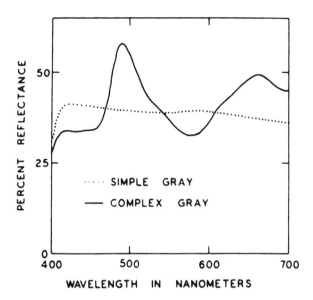

Figure 12 Spectrophotometric curves of a metameric pair of colors that will be similar in color in daylight but significantly different in blue (fluorescent tube) light and red (tungsten) light. (By W. C. Granville; courtesy of *Color Engineering*.)

purpose desired and the reproduction of these same colors. The original duplication of the desired color and its continued production is an engineering problem. Our engineering limits for suitable duplication of color are those that the human eye can distinguish. Although the human eye sees color essentially in terms of

1. Hue: dominant wavelength
2. Value: brightness, lightfastness
3. Chroma: saturation, purity

other factors in the character of the light that is transmitted to the eye are important. Color can be affected by (1) a matte versus a press-polished surface, (2) the depth of embossed pattern, and (3) whether the colorants are at the surface (high concentration of pigment in a thin film) or submerged within the film, sheet, or slab (low concentration of pigment in a thin film or slab) [1,32,41,50]. *Constant care* must be taken, therefore, to ensure that colors are evaluated with all these other variables omitted. Color samples for comparison to the standard meet the following:

1. Be as close to each other in surface characteristics as possible
2. Be viewed with lights of the same characteristics as those under which they will be used
3. Be prepared on clean equipment and free of contamination from other colored objects
4. Be evaluated in terms of their end-use requirements.

This concept has been introduced as the Look–Think method of color evaluation and is valid regardless of whether the evaluation is made by eye or by instrument [48].

XVII. FACTORS AFFECTING COLORANTS

Plastics are used in a variety of applications and fabricated into finished materials on a broad range of equipment. Both the application and the fabrication technique can influence the choice of pigments. Not all pigments of the same chemical type perform in the same way. Variations in manufacture, surface treatments, and additives may, for example, make a particular Pigment Blue 15 more desirable than another. A consistent, repeatable testing procedure is, therefore, useful in making proper pigment selections.

Colorants may be classified in three categories: (1) pigments, (2) dyes, and (3) special effect (i.e., metal flakes, mica, pearlescent). A pigment is an insoluble coloring material, whereas a dye is soluble in the polymer matrix. Although solubility of a colorant offers ease of incorporation (no shear forces needed to provide dispersion), dyes often are avoided because they tend to migrate. Migration is the movement of a colorant from one material into another material with which it is in contact. A classic example of migration is seen when a piece of stained wood is painted. In a short time, the dye in the stain will move into the paint, causing discoloration. Some pigments may also be partially soluble and exhibit some tendency to migrate.

Pigments are classified as either inorganic or organic. The most commonly employed inorganic pigments are titanium dioxide, iron oxide reds and yellows, lead chromate yellows, and molybdate oranges. Also used, but less frequently, are cadmium reds, yellows, and oranges, selected mixed metal oxide pigments such as nickel titanate (greenish yellow) and chrome titanate (reddish yellow), and ultramarine blue.

Organic pigments are available in a wide range of properties, prices, colors, and chemical classes. With the exception of pigments having poor heat stability or tendencies to migrate, bloom, or plate out, most may be used in plastics. End-use properties such as lightfastness, weathering properties, electrical conductivity, or chemical resistance may be additional limiting factors. The pigments most commonly used are selected monoazo types, bis-azos, phthalocyanine blues and greens, quinacridone reds and violets, diarylide yellows, selected vat pigments, isoindolinone and isoindoline yellows, and quinophthalone yellows.

A. Pigment Incorporation

The process used may influence the choice of pigment incorporation technique. As with most polymers, there are three basic approaches to incorporating color:

1. Dry color
2. Color concentrates
3. Precolored compound

Although processors occasionally purchase dry color, it is most often put into a color concentrate which has a carrier. Today, many processors rely on a combination of in-house concentrate preparation and outside purchases. Plasticizer concentrates have in part been displaced by homopolymer chip concentrates. Dry concentrates reduce waste, do not settle or separate, and are cleaner to handle then pastes. Pastes, on the other hand, can be readily mixed together, and even before a colored compound is processed, the color of the paste may give a reasonable indication of what to expect in the final product. Pastes are the preferred material for coloring plastisols and organosols.

Dry color may be used in compounding when intensive mixers are used, but housekeeping and health considerations must be taken into account. The use of color concentrates eliminates the housekeeping problems associated with dry color and assures the processor that the color has been developed to optimize its properties.

Precolored compound finds decreasing use in industry. The rigid PVC industry, which up until the late 1970s used precolored compound extensively, now relies on in-house compounding or color concentrate.

B. Selecting Pigments

With plastics that are processed at fairly low temperatures and easily compounded on a two-roll mill, it is relatively simple to evaluate pigment performance. A standardized program for pigment evaluation is valuable. Coupled with colorimetric data, such an evaluation will provide the user with colorant choices based on fastness properties, color properties, and economics.

The following factors may be considered in the course of evaluating pigments:

1. Heat stability
2. Lightfastness/weatherfastness
3. Migration resistance
4. Tinting strength
5. Money value
6. Chemical resistance
7. Blooming
8. Dispersibility/dispersion behavior
9. Behavior at low concentration
10. Plateout

Some aspects of pigment behavior are not constant and will vary depending on heat history, dispersion technique, concentration, and other additives. If feasible, an evaluation program will be designed to examine these variables.

C. Heat Stability

When processing can be held to 400°F (205°C), there are relatively few pigments which cannot be used because of poor heat stability. The monoazo pigments, in general, should be considered problematic with respect to heat stability. Included here are such products as toluidine red, pyrazolone red and orange, dinitraniline and dianisidine oranges, Red Lake C, and Permanent Red 2B. Although some of these pigments have adequate heat stability for low-temperature processing, they are marginal to unsatisfactory above 200°C. Residence time and concentration may be critical. In addition to questionable heat stability, most of the azo pigments exhibit other deficiencies: migration, blooming, plateout, and poor lightfastness. Among the products listed above, only the pyrazolones and the 2B reds find fairly widespread application but only in noncritical applications where migration, plateout, and limited lightfastness can be tolerated. Heat-stability testing of pigments should be carried out at the highest temperature and the longest residence time that the compound may reasonably be expected to encounter. By also including intermediate points, a profile of the pigment's behavior can be derived. Several concentrations should also be tested. Many pigments have satisfactory heat stability at concentrations of 0.1% and higher, but are deficient below 0.1%.

Heat history is cumulative. Therefore, reworking of scrap or off-standard material may cause a color failure. In reworking scrap, particular attention should be paid to pigment concentration. It may be inadvertently reduced below a critical minimum. This, coupled with additional heat history, may give rise to migration or reduced lightfastness with normally satisfactory pigments.

Depending on the equipment available, heat-stability testing may be conducted in a variety of ways:

1. Prolonged two-roll milling with samples being taken at appropriate intervals
2. Static oven aging at varying times and/or temperatures
3. Compression molding at varying times and temperatures

The first method, although accurate, does not usually employ the highest temperature to which the compound may be subjected. The last two require care to ensure accuracy and reproducibility because ovens and presses are subject to rapid temperature changes or uneven heat distribution. A control in the form of an unpigmented sample should always be run.

Poor heat stability may appear as a darkening of the color (usually in masstones), a loss of color strength (tints or very low concentration masstones), or discoloration (usually yellowing or browning). If there is an apparent increase in color strength or a change in hue to a different but not undesirable color, it is likely that the pigment has dissolved and has, in effect, become a dye. A migration test on this heat-stressed material will confirm this.

D. Migration Resistance

Colorant migration may occur when the colorant is soluble in a polymer system. As long as the colored material does not come into contact with another material in which the colorant is also soluble, migration may not be a problem. Migrating colors should not be used where printing is part of the process (e.g., in wall covering). A migrating color in a substrate can move into the ink, causing discoloration. Conversely, a migrating pigment in the ink would move into the substrate. Laminated or bonded constructions, particularly of different colors, should not employ migrating colors. It is common practice to color adhesives, either for identification or to aid in applying a uniform coat. If such adhesives are used with plasticized polymers, migrating colors should not be used in the adhesive.

Depending on the degree of solubility and the severity of the conditions, migration may occur in a matter of minutes or it may take days. The most commonly used migration test is the ''sandwich'' test. A piece of colored plastic is placed against a piece of white. Surfaces of both samples must be smooth to ensure contact. Placed between heavy metal plates, the sample is oven-aged at 80°C for 24 h. On removal from the oven, the colored material is peeled away from the white. A stain on the white is evidence of migration. The intensity of the stain may be assessed by using a gray scale or other appropriate means. Because migration continues until equilibrium is reached, assessment of migration should be done promptly. Otherwise, the stain which is found only on the surface of the white sample will appear to lessen in intensity over time.

It is recommended that several pigment concentrations, especially a very low one, be checked for migration. With some pigments, migration is a problem only at very low concentrations. In such a case, it is thought that, at this low concentration, the pigment is fully dissolved. At a higher concentration, it is only partially dissolved and is continously dissolving and recrystallizing. In the former state, migration may occur readily, where as in the latter case migration is virtually nonexistant.

In the preceding section on heat stability, it was mentioned that prolonged heat history may cause a pigment to go into solution. This will usually show up as increased color intensity or transparency. Sometimes, a shade shift may occur. A migration test on such material is in order.

E. Lightfastness/Weatherability

The ability of a colorant to hold its shade on exposure to light or weather is frequently critical. As with so many properties, lightfastness is often dependent on the pigment concentration and sometimes on heat history. A general rule of thumb is that as the concentration of an organic pigment goes down, especially in tints, lightfastness goes down also. With inorganic pigments, the reverse is likely to occur. Masstones more often will darken, although tints are more stable.

Countless studies have been conducted which attempt to correlate nature weathering with accelerated or artificial weathering [51,52]. Because weather and climate are variable, correlations cannot be precise. The most commonly used tests for lightfastness employ either a carbon arc, a fluorescent sun lamp, or a xenon arc as a light source [53]. The carbon arc lamp emits some UV-light energy at wavelengths not present at the Earth's surface. Because of this changes due to exposure will occur two to three times faster under carbon arc illumination than under sunlight or xenon illumination, but it can be argued that the low wavelengths of the carbon arc provide an unrealistic environment. Many polymers are particularly sensitive to the UV wavelengths emitted by the carbon arc lamp. The carbon arc light source has a long history of use in the United States, but it has been largely replaced by the xenon arc light source. Xenon units offer precise temperature and humidity control and maintain uniform light energy output over the life of the lamp.

Although lightfastness testing is useful for materials used indoors, it is imperative to test actual weathering characteristics if a product is to be used outdoors. Some pigments, both organic and inorganic, will exhibit excellent lightfastness, but a combination of light and moisture will cause rapid failure. Cadmium pigments, for example, exhibit very good lightfastness but poor weatherability. A commonly employed cycle in an accelerated weathering test is 102 min of light exposure at 145°F (63°C) followed by 18 min of darkness with water spray. Other cycles are available.

Even "accelerated" testing requires time. For products requiring good light or weatherfastness, a minimum exposure would be 500–1000 h in a carbon arc unit and twice that in a xenon unit. In isolated cases, as for building materials, exposure periods of 5000–7000 h may be appropriate.

Regardless of the exposure interval, an assessment of appearance change is required. Several methods are employed:

1. Description, for instance, slight fade, moderate darkening.
2. Color measurement data versus an unexposed control.
3. Gray Scale rating from 1 through 5, where 5 means no change and 1 means severe change. Directionality of change may be indicated by an appropriate letter, such as d (drker), r (redder), or y (yellower). These visual methods are reliable only when carried out under controlled conditions with a standard source of illumination, such as with a Macbeth® cabinet.
4. Spectrophotometrically using a colorimeter (see Section XVII.J).

Chemical resistance of pigments can be important both in process and in end use. Some colorants (metal complex types, for example) may react with other additives. This usually results in an unmistakable color change as a "new" colorant is formed. When stabilizers and pigments contain barium, calcium, tin, nickel, magnesium, zinc, lead, or sulfur, possibilities exist for reactions to occur. Careful testing is suggested if changes in additives are contemplated.

Chemical resistance in end use may also be important. Sulfide staining of lead-based pigments is well known. Alkali resistance is necessary for materials subject to washing. Other end-use environments may require specialized testing.

F. Blooming

Blooming is the phenomenon of a pigment or other additive coming to the surface. Depending on the pigment, blooming may occur rapidly (i.e., immediately after processing, or slowly over a period of weeks or months). Whatever the case, when the compound is wiped or rubbed, the color comes off. Blooming pigments include nephthol and toluidine reds, dianisidine and dinitranili oranges, Hansa yellows, and some diarylide yellows. Occasionally, blooming may occur only under certain conditions involving other additives, heat history, or pigment concentration.

Testing for blooming should be done after an extended storage period. Blooming may also appear after a relatively short exposure in an accelerated lightfastness test. If samples develop a blotchy or mottled appearance, test for blooming.

G. Plateout

Plateout refers to the tendency of certain pigments to form deposits on metal parts (calendar and mill rolls, embossing cylinders, extruder screws). In a continuous process, this deposition can build up to a point where finished product quality is affected. Embossing cylinders can become so caked that definition is lost. The need for subsequent cleanup and the possibility of contaminating a following run are problems created by the plateout phenomenon.

Additives such as TiO_2, fillers, lubricants, and stabilizers can reduce or eliminate plateout or, if not chosen carefully, may contribute to intensification. To test for plateout, incorporate the pigment into a standard, unfilled compound on a two-roll mill. Following the milling, run a standard compound containing a known quantity of TiO_2 as a "cleaner" batch. The developed color intensity of the cleaner batch is an indication of plateout. If comparative tests are being run, exercise care that the pigment or pigment dispersion does not cling to mill end plates. It may subsequently wind up in the "cleaner batch" and distort the test results.

As a rule, pigments which exhibit severe plateout tendencies should be avoided. However, slight plateout may often be tolerated. Many classical azo pigments, Permanent Red 2B, Hansa yellow, some diarylide yellows, toluidine red, and dianisidine orange will plate out.

H. Low-Concentration Behavior

Under the sections on heat stability, lightfastness, and migration, the low-concentration behavior of pigments was mentioned. Too often, pigment evaluations are carried out on the basis of a single masstone and tint. Many pigments, the organics especially, may exhibit anomalous behavior at very low concentrations. Heat stability and lightfastness frequently decrease as pigment concentration goes down. Less common, and therefore often overlooked, is the tendency of a pigment to go into solution or behave as a dye at a very low concentration. Low concentration can be defined as less than 0.1%, although with some pigments, the anomalous behavior at low concentrations may not appear until the pigment loading is much lower. There are a number of instances where pigments can be employed inadvertently at levels that may give rise to problems:

1. Reworking scrap of unknown composition.
2. Reducing the total pigment loading of a formulation because there appears to be no justification for using so much expensive pigment. The reason may have been to keep one of the components above a critical minimum level, but three color formulators later, the original reason may have been forgotten!

3. Increasing a part thickness, which may allow a desired color to be achieved with less pigment.
4. Eliminating or reducing filler or flame retardant in a formulation, which may make it possible to reduce pigment loading with no color change.
5. Simply creating a new color match without regard to loading of individual pigments.
6. Using a contaminated pigment or pigment concentrate. As a result of careless housekeeping or poor cleaning of dispersion equipment, pigments which were not intended may wind up in a finished product.

It is recommended that evaluation of a pigment include a tint containing 0.01% pigment plus 1.00 or 2.00% titanium dioxide. Although testing such a concentration will not pinpoint the critical minimum concentration, simply knowing in advance what performance to expect can alert the formulator to excercise care.

I. Pigment Dispersions

Pure pigments are available as either a dry powder or presscake. The latter form, which normally contains about 70–75% water, is commonly used in preparing some types of printing inks or in making high-quality polyolefin color concentrates. The water in the presscake is displaced by an oil or carrier resin. Because the pigment particles are not dried and agglomerated, dispersions of very high quality can be achieved. However, for most applications, this quality is not necessary and the cost of the process is not justified. Therefore, dry pigment is usually used in making dispersions.

Dispersion of a pigment involves incorporating the pigment into a plasticizer or resinous carrier and then grinding or shearing the mixture in such a way that pigment aggregates or agglomerates are broken down and the primary pigment particles are encapsulated or wetted by the carrier.

In preparing dispersions, two methods are common:

1. Dispersing into suitable plasticizer(s) on a three-roll mill or paint mill
2. Dispersing into a concentrate on a two-roll mill or intensive kneader such as a Banbury mixer

J. Color Measurement and Computer Color Matching

Computer color matching and auxiliary programs have taken much of the guesswork and drudgery out of the color-matching process. They have not completely replaced—visual color matching. In the imperfect world of color theory and perception, it is suggested that a computer color-matching system operated by a skilled visual color matcher will give the best results. A computer programmer/

operator with no knowledge of pigments may arrive at seemingly satisfactory answers, but technical and practical considerations related to the process will be overlooked. For example, a color match containing three different yellow pigments might offer low cost and adequate fastness properties, but production color control would be extremely difficult, if not impossible, to achieve.

The computer color-matching library is the key to success in establishing a reliable system. The best hardware and software available cannot compensate for careless sample preparation. It is tempting when building a computer color-matching library to prepare samples from material at hand. The following precautions will save time and frustration:

1. Select a resin which is free of gels. Although not a problem in visual color matching, gels affect the optics of a system and may provide erratic data.
2. If a filler is to be used, treat it as a pigment. The color of some fillers is variable. Reserve enough material to eliminate this variable.
3. Do not take pigments from production stock. Ask your pigment suppliers for standards. Record batch numbers of all pigments being entered in the computer.
4. Use a dispersion technique which represents your production method; it should be repeatable.
5. Learn how to make black, white, and gray before working with chromatic colors. Time spent in this exercise will pay dividends later.
6. When colors are added, start with the commonly used inorganics such as chrome yellow, molybdate orange, and iron oxide. Because they are opaque and relatively free of dispersion problems, success with them is easier than with a shear-sensitive organic pigment. For environmentally less objectionable choices, see Section XVII.L.
7. Housekeeping and sample preparation procedures must be critically reviewed. Sample contamination is often the cause when satisfactory calibration data cannot be obtained. Sources of contamination range from the obvious (dirty mills or mixer) to the less obvious (a dirty knife or spatula, a dirty filter in an exhaust hood, or careless handling of a finished sample). One should assume that everything is a potential source of contamination. Train operators to keep raw materials covered, equipment clean, and samples protected.

K. Colorimeter Measurements

Colorimeter measurements began as a comparison of the development of yellow coloration in white plastics by means of spectrophotometric absorption and reported as a Yellowness Index, per ASTM D 1925. This still finds some use in tracking degradation from heat or UV-light exposure but has been superseded by more sophisticated methods. There are now a range of colorimeters on the market that can compare the color of a sample to white or other standards and report the total color shift, ΔE (also in some articles as DE), and the shifts of the components, Δa, Δb, and ΔL. Very few observers can discern, for example, ΔE changes as low as 1 unit visually, whereas instruments are 10–100 times more sensitive. Colorant suppliers, compounders of color concentrates, and users, such as manufacturers of vinyl siding, now rely almost completely on such measurements. The technique is now also used to track discoloration arising from polymer degradation.

L. Recent Pigments, Current Usage

The following is a summary of current supplier recommendations for organic and inorganic pigments.

Organic Pigments

Red
Permanent Red 2B	PS, PE, PP, flexible PVC, cellulosics
Pigment Scarlet	PS, PE, PP, flexible PVC, cellulosics
Red Lake C	PS, PE, PP, ABS,
Perylene Red	PE, PP, ABS, acrylics, acetals, PVC, cellulosics, thermoset polyesters, polyurethanes
Quinacridone Red	PE, PP
Bis-azo condensates	PS, PE, PP, acrylics, acetals, flexible PVC, polyesters epoxies, cellulosics, phenolics

Orange
Bis-azo	Same as bis-azo red
Diarylide	Polyurethanes
Pyrazolone	PS, PE, PP, polyurethanes
Isoindolinone	PS, PE, PP, acetals, cellulosics, polyesters, epoxies
Anthranthrone	PS, PE, PP, PVC, cellulosics, polyesters, urethanes

Yellow
   Bis-azo   Same as red and orange
   Diarylide   PS, PE, PP, acetals, cellulosics, polyesters,
                    epoxies
   Flavanthrone   PVC, polyesters, epoxies, urethanes
   Isoindolinone   Same as corresponding orange
   Hansa   Phenolics

Green
   Phthalocyanine All except fluoroplastics and some polyamides

Blue
   Phthalocyanine Same as corresponding green
   Indanthrone   PS, PE, PP, PVC, cellulosics

Violet
   Quinacridone   PE, PP, acrylics, flexible PVC
   Dioxazine   PS, PE, PP, PVC, acrylics, acetals, urethanes
Inorganic
   Pigments

Red
   Ultramarine   PS, PE, PP, ABS, PVC, cellulosics, epoxies,
                    urethanes, polyesters, phenolics, but not bright
   Cadmium   All, but perceived as hazardous

Orange
   Molybdate   PS, PE, PP, PVC, cellulosics, polyesters,
                    urethanes, phenolics, but perceived as
                    hazardous
   Cadmium   All plastics, but thought hazardous

Yellow
   Chrome   All except acrylics, acetals, polyamides and
                    fluoroplastics, but perceived as hazardous
   Iron oxide   PS, PE, PVC, polyamides, acetals, cellulosics,
                    polyesters, but relatively dull.
   Cadmium   All plastics, but thought hazardous
   Metal   All plastics, but relatively dull
     complexes

Green
    Chrome   PS, PE, polyesters, but thought hazardous
    Hydrated   Same plus PP, PVC, urethanes, polyesters, but
      chrome   thought hazardous
    Chromium   Almost all, but thought hazardous and dull
      oxide
    Metal   Almost all, but relatively dull
      complexes

Blue
    Ultramarine   All except fluoroplastics, but weak
    Metal   Almost all, but relatively dull
      complexes

Violet
    Ultramarine   Same as with Ultramarine Blue
    Metal   Same as with blue and green
      complexes

As can be readily seen from the above, regulation has inhibited the use of the most effective inorganic pigments that provide the greatest thermal stability and resistance to aging and weathering. The most popular compromise has been to use a combination of a nonhazardous (but relatively dull) inorganic pigment, such as an iron oxide or the metal complex pigments used in ceramics, plus a suitable bright organic to get a color match. On aging or weathering, the color will slowly fade from a bright to a dull shade, but this is generally preferable to fading to off-white. The technique merely expands a practice used for many years in garden hose covers and similar items. The area of heavy metal replacement in pigment selection is fast moving and should be followed in the current literature.

REFERENCES

1. P Papillo. Mod Plast 44(12):131, 1967.
2. BS Hopkins and JC Badler. Essentials of Chemistry. Boston: Heath, 1946.
3. American Association of Textile Colorists and Chemists, Color Index. Durham, NC. American Association of Textile Colorist and Chemists, 1971.
4. T Patterson. Pigments. An Introduction to their Physical Chemistry. New York: Eisevier, 1957.
5. TC Patton. Paint Flow and Pigment Dispersion. New York: Interscience, 1964.
6. JJ Mattiello. Protective and Decorative Coatings. New York: Wiley, 1946, Vol. 4.
7. Titanium Pigment Corporation, Handbook, Titanium Pigment Co. 1965. New York: Ronald Press.
8. G Mie. Ann Phys 25:377, 1908.

9. CJ Taylor. J Oil Colour Chem Assoc 49:1063, December 1986.

10. American Association of Textile Colorists and Chemists. Standard Test Method 8–61.

11. National Electrical Manufacturers Association. Standards for Electrical Resistivity.

12. American Society for Testing and Materials. ASTM, D281-31. Philadelphia: 1986.

13. JB De Coste and RH Hansen. SPE J 18:431 April 1952.

14. HC Felsher, WJ Hanau. Pigmented plastics for outdoor exposure. Society of Plastics Engineers Retec, Coloring of Plastics III. 1956.

15. Florida East Coast Exposure: Subtropical Testing Service, Inc., 8920 S.W. 120 Street, Miami, FL. Florida West Coast Exposure: Suncoast Testing Service, Inc., P.O. Box 2347, Sarasota, FL. Arizona Exposure: Desert Sunshine Exposure Tests, 7740 N. 15 Avenue Phoenix, AZ.

16. JJ Mattialo. Protective and Decorative Coatings, New York: Wiley, 1946, Vol. 2.

17. I Drogin. Color Eng 5:4, July–August 1967.

18. I Drogin. Color Eng 2:3, March 1964.

19. SG Rober Jr. Carbon black pigments in plastics. Society of Plastics Engineers Retec, Coloring of Plastics III, 1966.

20. LJ Venuto, WM Hess. Am Ink Maker, October–November 1967.

21. DE Brody. Paint Varnish Prod, July 1966.

22. WG Huckle, GF Swigert, SE Wiberly. Ind Eng Chem Prod Res Dev 5:362–366, December 1966.

23. VC Vesce. Off Digest 31:419, December 1959.

24. HC Yao, P Resnick. Degradation reactions of azo pigments. American Chemical Society, Division of Organic Coatings and Plastics Chemistry, New York Meeting. 1963, Vol 23, 23. 486.

25. H Gaertner. J Oil Colour Chem Assoc 46:13, January 1963.

26. RF Hill. J Oil Colour Chem Assoc 48:603, July 1965.

27. FM Smith, JD Easton. J Oil Colour Chem Assoc 49:614, August 1966.

28. A Pugin. Off Digest 37:782, July 1965.

29. TB Reeve. "Pigment colors for vinyl coatings," Color Eng 19, July–August 1966.

30. JE Simpson. Mod Plast 40(12):90 1962.

31. JE Simpson. "Coloring Plastics," series of articles in Mod Plast 44, 1965.

32. LM Greenstein, HA Miller. SPE Tech Paper 13:1021, 1967.

33. Z Kazenas. Paint Ind Mag February 1960.

34. H Zussman. Mod Plast Encyc 43(10-a):490, 495, 1966.

35. American National Standards Association, Inc., 1430 Broadway. New York, NY.

36. U.S. Congress Public Law 85-929.

37. MJ Dunn. SPE Tech Papers 14, 1968.

38. Part 121—Food Additives—Title 21—Food and Drugs. Federal Food and Cosmetic Act. Section 121.2514, Paragraph 34, Subpart F—Food Additives, 13.3–13.8.

39. MJ Dunn. Color and constitution. Coloring of Plastics II, 1965.

40. LS Pratt. Chemistry and Physics of Organic Pigments. New York: Wiley. 1947.

41. FW Billmeyer Jr., M Saltzman. Principles of Color Technology. New York: Wiley–Interscience, 1966.

42. DB Judd, G Wyezecki. Color in Business. Science, and Industry. New York: Wiley, 1963.

43. RW Burnham, et al. Color: A Guide to Basic Facts and Concepts. New York: Wiley–Interscience, 1963.
44. FW Billmeyer Jr., M Saltzman. Principles of Color Technology. New York: Wiley–Interscience, 1966.
45. CG Leete, JR Lythe. Color Eng 27, January–February 1966.
46. E Allen. J Paint Technol 39:368, June 1967.
47. NK Blackwood, FW Billmeyer Jr. Color Eng 24, March–April.
48. RK Winey, WV Longley. SPE Tech Paper 13:1110, 1967.
49. Color Eng, May–June 1967, special metamerism issue.
50. RS Hunter. SPE J 23(2):51, 1967.
51. A Stoloff. Interim Report of the Vinyl Siding Division Weathering Committee. Society of the Plastics Industry, 1980.
52. A Stoloff. Weathering Committee Report, Vinyl Siding Institute, Society of the Plastics Industry, Washington, 1980.
53. ASTM D1499, D1501, G53, G23. American Society for Testing and Materials. Philadelphia, 1986.

3
Coupling Agents

Richard F. Grossman
Halstab Division, The Hammond Group, Hammond, Indiana

I. INTRODUCTION

Many of the properties provided by fillers relate to the character of the particle surface, specifically to surface area, structure, and the nature of reactive groups. To optimize the balance of properties of a given composition, it is often useful to make changes to the filler surface. Several types of additives are used in this regard. If the primary purpose is to lower the extent of interaction between the polymer and the filler (i.e., to promote flow during processing), the additive is an internal lubricant. If, on the other hand, the primary effect desired is to increase the extent of interaction, either to speed incorporation or to modify properties, the additive may be a dispersion aid or a coupling agent. Dispersion aids function by lowering the energy barrier to close approach of the additive to the polymer. Invariably, this type of additive contains a section having dipolar or van der Waals attraction to the polymer and another with strong specific interaction with the filler. Coupling agents, although often of similar structure in terms of attraction to the filler and the polymer, chemically bond to one or the other, sometimes both.

II. MECHANISM OF COUPLING TO THE FILLER

Coupling agents have the general formula R′—O—M—R—X. M (implying metal) is usually silicon, titanium, or zirconium. Thus, M is not divalent, as shown for simplicity, but usually tetravalent: $(R′—O—)_a—M—(R—X)_{4-a}$. X is a functional group which interacts strongly with, or bonds chemically to, a particular polymer. R denotes an organic group that links the functional group, X, to the metal, M. The group R has, in well-chosen situations, thermal, oxidative, and solvolytic stabilities comparable to those of the base polymer. R′—O— forms a good leaving group; that is, it may be displaced readily from the metal by —OH groups present on the filler surface. The coupling agent may react with —OH directly:

$$\text{(filler)—OH} + \text{R′—O—M—R—X} \rightarrow \text{(filler)—O—M—R—X}$$
$$+ \text{R′—OH}$$

or may react with traces of moisture at the filler surface, and then condense with the filler:

$$H_2O + \text{R′—O—M—R—X} \rightarrow \text{HO—M—R—X} + \text{R′—OH}$$
$$\text{(filler)—OH} + \text{HO—M—R—X} \rightarrow \text{(filler)—O—M—R—X} + H_2O$$

The hydrolyzed (or partially hydrolyzed) coupling agent may also react with itself:

$$2\text{HO—M—R—X} \rightarrow \text{X—R—M—O—M—R—X} + H_2O$$

This not only consumes the coupling agent to no useful end, but, considering that coupling agents are actually multivalent,

$$(\text{R′—O—})_x\text{—M—}(\text{—R—X})_y$$

condensation may lead to polymeric products. These can appear as gel structure or lumps. Therefore, exposure to moisture must be controlled. It is also important that traces of water present be at the filler surface rather than throughout the compound, so that the filler coupling reaction can compete with self-condensation. Finally, filler coupling must be rapid compared to self-condensation. This factor often governs which fillers will respond to a given coupling agent. In many systems, the two-step hydrolysis sequence dominates. In these cases, the addition of a desiccant, such as calcium oxide, may remove enough water to inhibit the coupling reaction. In such cases, it is necessary to react the filler with the coupling agent prior to interaction with the polymer. Thus, there are two useful routes:

1. Addition of the coupling agent to the filler prior to mixing
2. Addition of the coupling agent during mixing

The first alternative is most commonly carried out by filler suppliers. This is particularly attractive in cases where the pure coupling agent may present a hazard because of flammability or reactivity. An enclosed sigma blade mixer is normally used. Provision for maintaining constant temperature, automatic injection of liquids, and, in some cases, an inert atmosphere are common. Such pretreatment should be considered by the compounder if a particular filler is not available with the desired coupling agent, if the filler treatment must occur before the coupling agent contacts the polymer, and if the coupling agent involves hazard in the pure state.

The second alternative offers the greatest economy but may not be as precise as pretreatment. It is often, however, entirely adequate. In cases where the coupling agent is used at low levels, the addition in the form of a dispersion is desirable. Encapsulation with inert waxes is common. The point of addition, whether of dispersion or pure coupling agent, is normally with the filler. Addition at late stages in the mix should be avoided because of the low flash points of many coupling agents. In most cases, coupling to the filler results in an increase in reinforcement at a given filler level, with improvements in modulus, tensile, and flexural strengths. The gains can be very substantial, at times up to 50% greater than without coupling. The greatest improvements are to be found with fillers that, from shape and size considerations, have strong reinforcement potential, but which, unreacted, have limited attraction to the base polymer. Filler–coupling agent bonds tend to be more labile than typical polymer cross-links. Bond breaking and reformation at the filler surface under stress often occur. As a result, it is sometimes found that both ultimate elongation and impact resistance are improved. The ability of the filler–coupling agent bond to rearrange depends on M, in the structure R′—M—R—X, not only forming strong —M—O— bonds but also in having unoccupied orbitals (i.e., having strength as a Lewis acid, so as to be able to coordinate with one oxygen while leaving another). As a result, the metals used are chosen from the center of the periodic table, where Lewis-acid activity is prominent. If M is titanium or zirconium, unoccupied d orbitals are readily available, favoring bond reformation. Some technologists consider titanates and zirconates to be primarily dispersion aids because of the ease with which bonds to the filler may be reformed or hydrolyzed; others rank them as coupling agents because of the chemical, rather than physical nature of the bonds formed. When M is silicon, unoccupied orbitals are higher in energy; the bonds formed during coupling are more permanent and less subject to hydrolysis.

III. INTERACTION WITH THE POLYMER

Coupling agents were originally designed for use with thermosetting polymers. The —R—X group was chosen to react with the cross-linking agent. Thus, —X would be a vinyl or acrylyl group for use with peroxide-initiated, free-radical cross-linking of polyolefins or unsaturated polyesters. Similarly, —X would be chosen as an amine or epoxy group for use with thermosetting urethanes or epoxies. Mercaptide end groups are used in conjunction with sulfur vulcanization of elastomers. More recently, coupling agents have been developed that function with thermoplastic polymers. Several modes of interaction with the polymer are in use:

1. Coordination to the polymer by dipole interaction. This is significant only with poly(vinyl chloride) (PVC) and other polar thermoplastics. The extent of dispersion force interaction of coupling agents with polyolefins and other nonpolar polymers is likely to be no greater than that of untreated filler. This is not to say that filler treatment with coupling agents in nonpolar thermoplastics may not lead to useful effects (e.g., from function as an internal lubricant).
2. Grafting of the coupling agent to the polymer. Coupling agent–activator systems have been designed that are capable of grafting to nonpolar polymers without formation of polymer cross-links.
3. Self-condensation to form low polymers that trap chains of the original polymer, yielding a thermoplastic interpenetrating network (IPN).

Both of the latter mechanisms require close control of the concentration and mode of addition of the coupling agent. Usually, all of the above involve the addition of the coupling agent to the polymer before coupling to filler. In the case of PVC, this may be done by direct absorption into the resin or inclusion of the coupling agent with the plasticizer (in the case of flexible compounds). As a general rule, coupling agents should not be added simultaneously with liquid stabilizers, organophosphites, or epoxidized oils without previously determining that unwanted reactions with such components do not occur. In the case of pelletized resins, coupling agents are conveniently added to the mixer by liquid injection or by the addition of concentrates. If economically justified, some may also be added by pellet absorption, as is sometimes done with organic peroxides. Another instance in which the coupling agent is added to the polymer rather than to the filler is when coupling is desired, not to an internal filler but to an external surface. Here, the function of the coupling agent is that of an adhesion promoter. In most cases, adhesion to the external surface is a step that occurs some interval after mixing. In such a situation, the formulator must choose ingredients carefully to prevent dissipation of the coupling agent through internal coupling. Sometimes, this potential problem may be circumvented by treating the external surface directly with the coupling agent, as is often done with glass fiber. This is analogous to pretreatment of filler. Often such pretreatment is complemented by further addition of the coupling agent to the compound; there are coupling agents designed to react with, for example, glass fiber previously treated with another coupling agent.

IV. APPLICATIONS IN VINYL

An important application for coupling agents in vinyl is their use as adhesion promoters. This applies mainly to vinyl plastisols, with some usage in coatings.

Current practice centers on silane coupling agents of the type $(R'\!-\!O\!-\!)_3\!-\!Si\!-\!R\!-\!X$, where the $R'\!-\!O\!-$ group is methoxy or ethoxy and $-X$ is either an amino or a mercapto functional group. When the plastisol or coating is heated to fusion, the $Si\!-\!R\!-\!X$ functionality can displace labile chlorine much like a stabilizer:

$-Si\!-\!R\!-\!SH$
$+ -CH\!=\!CH\!-\!CHCH_2\!-\!Cl \rightarrow -CH\!=\!CH\!-\!CH\!- CH_2S\!-\!R\!-\!Sl\!-$
$+ HCl$

The reaction proceeds the same way with $-Si\!-\!R\!-\!NH_2$. As with stabilizers, mercaptide linkages provide better heat stability and color than additives containing amine groups. The mercaptosilanes have, on the other hand, odors that some find uningratiating or intolerable. This is probably the factor that has led to most compounds using aminosilanes. Interaction with mercapto- or aminosilanes provides improved adhesion to polar surfaces bearing $-OH$ groups through formation of $-O\!-\!Si$ bonds. In addition, silane treatment may improve adhesion to polar surfaces not capable of forming bonds by simply scavenging surface moisture that impedes wetting out. Use levels of 0.2–1.0 phr have been reported. Silane coupling agents containing more than a single amine group must be used with care in vinyl compounds, as cross-linking may result. In certain circumstances, this may be desirable, such as in building plastisol viscosity. Little work has been reported until recently regarding the use of coupling agents with fillers in vinyl. Probably, this stems from the prevalent use of calcium carbonate as the major filler. Calcium carbonate provides essentially no surface $-OH$ groups for reaction with silanes. Further, many grades used in vinyl are surface treated with calcium stearate, reducing (intentionally) their surface activity. Recently, this lack of reactivity has been overcome by use of multifunctional coupling agents that self-condense, trapping filler surface irregularities within their web. These coupling agents are primarily polymeric silanes, used at 0.2–1.0 phr. The effects are as follows:

1. Loss of properties caused by dilution with filler is often reduced.
2. The level of filler that can be tolerated is often increased.
3. The ease of processing at a given level of filler is often improved.

Polymeric silanes are introduced into the compound by direct addition to the polymer–filler blend. The coupling agent should be added to the blend and absorbed prior to addition of stabilizer. Otherwise, the latter may essentially monopolize the polymer–filler interface. Similarly, lubricant addition is also delayed. Certain reactive lubricants, such as stearic acid, may have to be replaced with others free of active hydrogen. Coupling agents for vinyl are marketed by the following:

Supplier	Type	Grades
Mannchem	Mercaptozirconate	MOD S
	Aminozirconate	MOD A
Dow Corning	Mercaptosilane	Z-6062
	Aminosilane	Z-6020, 6026
	Polyaminosilane	Z-6050
Huls America	Mercaptosilane	CM8500
	Aminosilane	CA0750
	Polyaminosilane	CT2910
	Polymeric silane	CPS076
Kenrich	Neoalkoxytitanate	LICA 38, KR TPP
	Neoalkoxyzirconate	KZ TPP
PCR Corp.	Mercaptosilane	Prosil 196
	Aminosilane	Prosil 221
Witco	Mercaptosilane	A-189, A-1891
	Aminosilane	A-1100, A-1110
	Polyaminosilane	A-1120, A-1130

V. APPLICATIONS IN POLYOLEFINS

Vinyl and acrylyl silanes have been used for many years in thermoset formulations based on crosslinkable polyethylene (PE), ethylene vinyl acetate (EVA), chlorinated polyethylene (CPE), and blends with other polymers. In these cases, the method of bonding to the polymer resembles peroxide-initiated crosslinking, with polymer radicals reacting with, for example, the coupling agent's vinyl group, instead of terminating with a second polymer radical:

$$R\cdot + CH_2{=}CH{-}Si{-}(OR')_3 \rightarrow R{-}CH_2{-}C\cdot H{-}Si{-}(OR')_3$$

The radical formed is relatively stable and tends not to form a vinyl silane homopolymer under typical process conditions. This type of coupling agent is also of use in thermoplastic polyolefins, particularly in the case of reinforcement with glass fiber. Pretreatment of glass fiber with vinyl or acrylyl silanes leads to improved adhesion, especially in wet environments, to polyolefins, including thermoplastic polyolefin (TPO) elastomers. Typical usage is in the range of 0.2–1.0 phr; products of this type are supplied by the following:

Company	Type	Grades
Mannchem	Acrylyl zirconate	MOD M, M-1
Dow Corning	Acrylyl silane	Z-6030
	Vinyl silane	Z-6082
Huls America	Acrylyl silane	CM-8550
	Vinyl silane	CV-5000, 5010
PCR Corp.	Acrylyl silane	Prosil 248
Witco	Acrylyl silane	A-174, 175
	Vinyl silane	A-172

There is some use of reactive silanes and analogous materials that contain long-chain alkyl groups as the —R—X segment. These interact with polyolefins and other nonpolar polymers through dispersion or van der Waals forces and may assist wetting and dispersion. Although these are proper filler treatments functioning as internal lubricants, it seems, nonetheless, reasonable to include them here:

Company	Type	Grades
Mannchem	Zirconate	MOD F
Huls America	Silane	CO-9745

Reported use levels are in the range of 0.5–2.0 per 100 of filler. There has been recent work with multifunctional and polymeric coupling agents in polyolefins, paralleling that discussed in the foregoing as regards vinyl polymers. With both of these approaches, the coupling agent is designed to condense with itself at the polymer–filler interface, trapping or entangling the polymer with the filler surface. As with vinyl, this may lead to improved properties associated with better filler–polymer compatibility. Products sold in this area include the following:

Company	Type	Grades
Huls America	Polymeric silane	CPS-078.5
Kenrich Petrochemical	Neoalkoxy titanate	LICA 12

The use levels and precautions are the same as those mentioned previously for vinyl.

VI. GRAFTING COUPLING AGENTS TO POLYOLEFINS

It has been known for some time that vinyl and acrylyl silanes may be grafted to polyolefins either via electron beam irradiation or through reactions initiated by thermal decomposition of organic peroxides. In an ingenious extension of this, workers at Midland Silicones, Ltd., later Dow Corning, were able to graft vinyltrimethoxysilane;

$$CH_2{=}CH{-}Si{-}(OCH_3)_3$$

to low density polyethylene (LDPE) using traces of dicumyl peroxide, and then to crosslink the silane-grafted polymer with water, with dibutyltin dilaurate as a soluble Lewis acid catalyst. The combination of water and the acid catalyst hydrolyzes and condenses the $-Si-OCH_3$ linkage to form $-Si-O-Si-$ crosslinks. As might be anticipated from the properties of silicone polymers, these crosslinks are highly heat resistant. In typical practice, about 2 phr vinyltrimethoxysilane (VTMOS) and 0.05–0.3 phr dicumyl or similar peroxide are used to prepare graft polymer in an inert atmosphere, generally using a compounding extruder or mixer. The crosslinkable grafted polymer is packaged under inert gas to prevent premature crosslinking and may be stable for many months. A concentrate is also prepared, containing about 1 phr dibutyltin dilaurate plus whatever antioxidant, pigment, or other additives may be needed in the final compound. The two products are pellet blended prior to use, often in a 95/5 ratio of grafted polymer to concentrate. The final product, generally insulated wire, is then crosslinked by passage through, or standing in, water. Comparable technology was also developed by Union Carbide. Experiments have shown that the technique is useful with EVA, CPE, high density polyethylene (HDPE), and ethylene propylene (EP) elastomers.

Copolymerization with unsaturated monomers, such as with EPDM, hinders the graft reaction, presumably by competition for radicals generated. By analogy, it was anticipated that incorporation of filler would hinder the cross-linking reaction, with filler coupling competing for hydrolyzed $-Si-O$ linkages. Originally, this proved true, particularly with fillers having a surface where $-OH$ is prevalent (e.g., certain carbon blacks, aluminum trihydrate). Surprisingly, it was found that increased concentration of VTMOS overcame much of this, at least in LDPE and EVA. Both filled and unfilled compounds for wire coverings are available commercially. Even though vinyl silane is used in great excess compared to organic peroxide, there is invariably some crosslinking of polymer, unless the peroxide is specially selected to have low crosslinking efficiency. As a result, polyolefins are chosen having a higher melt index than normally desired, knowing that this will be reduced by the grafting procedure. An interesting variation of this process, applicable to ethylene acrylate copolymers such as EEA, was introduced by Union Carbide. They found that a number of silanes could be grafted to EEA and similar polymers by ionic, rather than free radical, reactions. Using tetra-isopropyl titanate as the catalyst, many silanes bearing —SH, —NH$_2$, activated vinyl, or ester groups as the —Si—R—X segment, can be grafted using exchange and condensation reactions. This procedure, although requiring an active copolymer such as EEA, avoids any peroxide crosslinking of the base polymer. Materials used in these processes are supplied by the following:

Company	Type	Grades
Dow Corning	Vinyltrimethoxysilane	Q9-6300
Huls America	Vinyltrimethoxysilane	CV4917
Witco	Vinyltrimethoxysilane	A-171
	Complete grafting package	Silcat R

The above work is significant not only to the technologist desiring a cross-linked product, who, after all, may draw upon alternative technology, but also to those who can employ the grafting principle instead to filler coupling. Indeed, this has followed as a spin-off benefit. Systems have appeared recently that may be used to graft coupling agents to a variety of polyolefins using activators designed to enhance filler coupling in preference to polymer cross-linking. There are two types now in use:

1. Azidosilane coupling agent, a self-activating type, initially developed for filled polypropylene (PP) by Hercules, but found generally useful in polyolefins. Usually 1 part per 100 filler is used of the 50% active coupling agent that is supplied.
2. Proprietary two-part coupling agents developed by Union Carbide. Overall, 1–2 parts per 100 parts of filler are used, but the ratio of silane to activator depends on the choice of polyolefin and filler, the lower activator levels being used for PP. Polypropylene presents the opposite hazard compared to PE; free radical activation for grafting easily induces chain scission with loss of molecular weight and viscosity. When properly used, azidosilane coupling agents do not have this effect

to any significant extent. Similarly, when properly used, the proprietary two-part systems appear highly effective. It is suggested that new programs be conducted cooperatively with the respective suppliers:

Company	Type	Grades
Hercules	Azidosilane	S3076S
Witco	Silane grafting agent	PC-1A, 2A
	Activator for above	PC-1B
	Silane for use with ATH	FR-1A
	Activator for above	FR-1B

The above treatments improve tensile strength, flexural strength, and Izod impact strength of glass- and mica-filled polyolefins. In the case of alumina trihydrate (ATH), ultimate elongation is also improved. As in many other areas, polymer modification competes with compounding for the solution of specific problems. The technologist should be aware that several suppliers have developed grafted polyolefins in which the grafted group functions as a filler coupling agent or an adhesion promoter. The agents bulk grafted include acrylic acid and maleic anhydride. Examples of commercial products are as follows:

Company	Type	Grades
British Petroleum	PP, PE (acrylic acid)	Polybond
Himont	PP (maleic anhydride)	PC-072

VII. APPLICATIONS IN STYRENIC POLYMERS

A number of different Si—R—X groups provide high compatibility with styrenics. Although not strongly polar [except for styrene acrylonitrile (SAN)] as compared to vinyl, the polarizable nature of the aromatic ring permits effective coordination to many groups of comparable solubility parameter. The following is a guideline to available types and applications:

Company	Type	Use	Grades
Mannchem	Acrylylzirconate	ABS	MOD M, M-1
Dow Corning	Acrylylsilane	ABS	Z-6030
	Epoxysilane	PS, ABS	Z-6040
	Ureidosilane	SAN, ABS	Z-6032
Huls America	Aromatic silane	PS	CP0320
	Epoxysilane	PS, ABS	CG6720
	Ureidosilane	SAN, ABS	CS1590
Kenrich	Neozirconate	PS	NZ 12/L
	Aminotitanate	ABS	L 97/A
PCR Corp.	Acrylylsilane	ABS	Prosil 248
Witco	Acrylylsilane	ABS	A-174
	Epoxysilane	PS, ABS	A-186, 187
	Ureidosilane	SAN, ABS	A-1160

VIII. CHOICE OF FILLERS

A cursory knowledge of mineralogy is all that is needed to match coupling agents to particular fillers. Metals whose oxides are sufficiently compatible to generate naturally occurring minerals are usually a good match with regard to the filler–coupling agent bond. The obvious examples are where the metals are identical: Organic titanates are a clear choice for use with titanium dioxide. Organic zirconates are similarly a clear choice for the zirconate fillers occasionally used in thermoplastics. Silane coupling agents were first used with silicas, silicates, and aluminosilicates (clays). Many compatible combinations were also found. Aluminum plus silicon is obvious from the existence of numerous aluminosilicates. Silanes couple strongly to hydrated aluminum oxide. Zirconium and titanium also form mixed oxides with silicon and aluminum; organic zirconates and titanates are effective with silicas, silicates, and aluminosilicates. Organic zirconates may be used with titanium dioxide and organic titanates with zirconate fillers. The above may suggest unlimited flexibility regarding the type of coupling agent to try with a given filler. This is by no means always true. With calcium carbonate, as described earlier in connection with vinyl, special-purpose coupling agents must be used. The same choices would also be useful with magnesium carbonate. Relatively active metal oxides are well served by coupling agents bearing carboxy, polyamine, and quaternary amine functional groups:

Company	Type	Grades
Mannchem	Carboxyzirconate	MOD C
Dow Corning	Thiouroniumsilane	Z-5456
	Polyaminosilane	Z-6050
Huls America	Quaternaryaminosilane	T2909.7
	Polyaminosilane	CA0699
Kenrich	Carboxytitanate	LICA 01
Witco	Polyaminosilane	A-1130

BIBLIOGRAPHY

In the area of coupling agents, the most useful references are bulletins issued by suppliers. Those that the thermoplastics compounder should have available currently include the following:

Company	Publication
Mannchem 26 Avenue C Woonsocket, RI 02895	*Performance Chemicals* (zirconates)
Dow Corning Midland, MI 48640	*Silane Coupling Agents* *Selection Guide to Organosilane Chemicals* *Siopas Technical Manual*
Huls America 2570 Pearl Buck Rd. Bristol, PA 19007	*Silanes, Coupling Agents* *Silanes, Property Profile*
Kenrich P.O. Box 32 Bayonne, NJ 07002	*Ken-React Reference Manual* *Increased Productivity Using Titanate and* *Zirconate Coupling Agents*
PCR Corp. P.O. Box 1466 Gainesville, FL 32602	*Prosil Coupling Agents*
Witco Corp. Old Ridgebury Rd. Danbury, CT 06817	*Organofunctional Silanes*

4
Fiber Reinforcement

Robert A. Schweitzer
Owens Corning, Granville, Ohio

Albert W. Winterman
BASF, Wyandotte, Michigan

Richard F. Grossman
Halstab Division, The Hammond Group, Hammond, Indiana

I. INTRODUCTION

Thermoplastic strength and dimensional stability, and in many cases other properties, are significantly improved by adding any of a variety of fibers and specialty fillers. Glass fiber-reinforced thermoplastics, for example, have found application in nearly every area of industry because of their unique combination of properties. The composites offer not only high strength and dimensional stability but also light weight, corrosion resistance, excellent dialectric properties, and design flexibility.

Although this chapter concentrates on glass fiber reinforcements, which find by far the most widespread use in reinforced thermoplastics, we briefly discuss carbon, asbestos, and organic fibers, and spherical glass fillers, because of their use in circumstances where their particular characteristics offer significant advantages.

II. GLASS-FIBER REINFORCEMENT

A. Development

Although fillers such as wood flour, mica, and glass were introduced to strengthen phenolics in the 1930s, fibers were first added as strengtheners in 1951. Rexford

Brandt, founder of Fiberfil, Inc. (DSM), found that specific material properties required for a military land-mine case could be met by compounding glass fibers into polystyrene [1]. The resulting product had nearly twice the tensile strength and four times the rigidity of unreinforced polystyrene. This was the first commercially available glass-fiber-reinforced thermoplastic. It contained 30% reinforcement (by weight), using fibers $\frac{3}{8}$ 1 in. long [2].

During the same period, designers were beginning to use Nylon $\frac{6}{6}$ until then a purely decorative plastic, as a molding material in engineering applications. The material's good toughness, high wear resistance, and low coefficient of friction made such functional use possible. However, the compound still lacked the rigidity, strength, and dimensional stability needed for some uses. Again, inclusion of glass fibers to reinforce the matrix improved these properties and extended the range of use for this material.

From that start, the technology for producing glass-fiber-reinforced thermoplastics and the number of uses for these composites have expanded rapidly. These materials are now widely used in the automotive industry for high-volume parts requiring fine finishes—body components, headlamp and taillight housings, instrument panels—as well as other components such as engine cover housings, insulated tanks, and fender liners. Another major use is the complex molded parts that have replaced die castings and sheet metal assemblies in many appliances. The automotive and appliance industries find reinforced thermoplastics attractive because of the design flexibility they provide, the very high production rates that can be attained, and the ability to form intricate shapes. Reinforced thermoplastics are also finding increasing use in the construction industry as structural forms and for such things as pipe work, drainage systems, and ducting. A broad spectrum of other markets—agricultural, aviation, chemical processing, electronics, marine, and recreational, to name a few—are using reinforced thermoplastics in a variety of products.

B. Benefits

Glass fibers effectively reinforce thermoplastics, enhancing their performance characteristics and improving their cost/performance ratio. Such reinforcement often gives lower-cost thermoplastic resins better performance than is obtained with more expensive resins. Fiber-glass-reinforcement is attractive because it provides the following properties:

> *High strength and stiffness.* With glass-fiber reinforcement, thermoplastics can be designed to meet a wide range of tensile, flexural, and impact

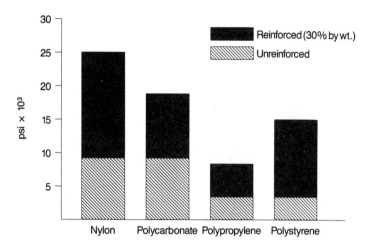

Figure 1 Tensile strength.

strength requirements. Figure 1 shows the effect of glass-fiber reinforcement on tensile strength in four thermoplastics. Figures 2 and 3 show effects on flexural modulus (stiffness) and Izod impact strength for the same compounds.

Light weight. Fiber-glass-reinforced thermoplastics have better strength-to-weight ratios than unreinforced plastic and most metals.

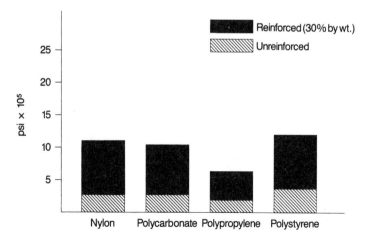

Figure 2 Flexural modulus.

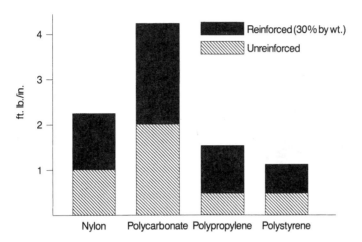

Figure 3   Izod Impact, unnotched.

Design flexibility. The compounds can be formed into any shape the designer desires—simple or complex, large or small—accounting for the broad range of applications.

Dimensional stability. Parts retain their shape even under severe mechanical and environmental stress.

Higher heat-deflection temperature. Crystalline resins, in particular, show large elevations in heat-deflection temperature (HDT) with reinforcement. Nylon 6/6 for example, has an increase in HDT from 170 to 495°F (76°C to 260°C) with 33% by weight reinforcement. Tables 1 and 2 show

Table 1   DTUL Response for Crystalline Polymers

Polymer	DTUL at 264 psi, 20% glass (°F)	Increase over base polymer (°F)
Acetal copolymer	325	95
Polypropylene	250	110
Linear (HD) polyethylene	260	140
PCO-72 (modified polypropylene)	300	160
Thermoplastic polyester	405	250
Nylon 6	425[a]	305
Nylon 6/6	490[a]	330

[a] At 30% glass by weight.

Table 2   DTUL Response for Amorphous Polymers

Polymer	DTUL at 264 psi 20% glass (°F)	Increase over base polymer (°F)
Acrylonitrile-butadiene-styrene	215	25
Styrene-acrylonitrile	215	20
Polystyrene	220	20
Noryl	290	25
Polycarbonate	290	20
Polysulfone	365[a]	20

[a] At 30% glass by weight.

reinforcement effects on distortion temperature under load (DTUL) for crystalline and amorphous polymers, respectively.

High dielectric strength. Glass-fiber-reinforced thermoplastics have outstanding electrical properties, which are useful for insulating current-carrying components.

Corrosion resistance. Rust and corrosion are not problems, and with the right resin system, resistance to almost any chemical environment is possible.

Parts consolidation. With the many improved properties, these compounds can be molded into a wide variety of structures that replace functional assemblies made from many parts and fasteners. These applications provide significant savings in materials and in fabrication, forming, and assembly operations.

Less finishing. Where reinforced composites replace more complex metal parts, color can be molded in, eliminating painting in many cases, and providing a long-lasting quality appearance.

Moderate tooling cost. Tooling usually represents a very small part of the product unit cost for glass-fiber-reinforced thermoplastic parts.

C.   Limitations

Glass-reinforced thermoplastics also have certain disadvantages and limitations [2].

Higher processing temperatures and injection pressures. Normally, processing temperatures are from 20°F to 50°F (11°C to 27°C) higher than those of nonreinforced compounds, and injection pressures must be 10–40% higher.

Machine wear. The abrasiveness of the fibers and the corrosiveness of the sizing materials used significantly increases wear on injection molding and extrusion machines.

Lower impact strength. The ambient notched impact strength of tough thermoplastics is generally reduced when glass reinforcement is added. Impact modifiers can often correct this.

Anisotropic properties. Glass-fiber compounds exhibit anisotropic properties (i.e., strength and impact properties are much higher parallel to the direction of flow than they are perpendicular to the flow) because of the orientation of the glass fibers during processing. For example, tensile strengths of nylon 6/6 and polypropylene with 40% chopped-fiber-reinforcement were 29% and 26% higher, respectively, in the direction of flow [3].

Loss of transparency. With glass fibers in the matrix, polymers are not transparent, but retain translucency.

Finish. The high gloss finishes of some thermoplastics cannot be obtained when glass reinforcement is added.

Cost. Glass-reinforced thermoplastics tend to be more expensive than nonreinforced materials.

Higher specific gravity. Glass fibers increase the specific gravity of thermoplastics.

III. PRINCIPLES OF GLASS-FIBER REINFORCEMENT

When high-strength, high-modulus glass fibers are added to a low-modulus resin and are well dispersed, the plastic flow of the matrix under stress transfers the load to the fibers. This results in a high-strength, high-modulus composite. Besides high strength and high modulus, the fibers provide a large contact area, assuring good adhesion with the polymer matrix. Other innate characteristics of glass fibers that make them attractive for thermoplastic reinforcement are their elasticity, good thermal properties, dimensional stability, excellent chemical and moisture resistance, excellent electrical properties, and high performance at relatively low cost.

A. Fiber Fabrication

Glass fibers are fabricated from silica and other ingredients, primarily oxides, which are melted in a furnace at temperatures exceeding 2300°F (1260°C). The molten glass flows through spinerettes and is stretched to a diameter of either 10

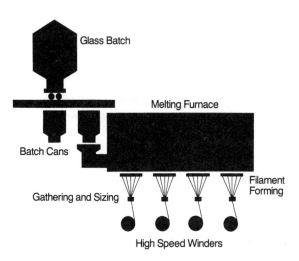

Glass Batch

Melting Furnace

Batch Cans

Gathering and Sizing

Filament
Forming

High Speed Winders

Figure 4   Direct melt continuous filament process.

or 13 μm for most applications (see Fig. 4). The diameter is determined by the drawing speed, which will be several thousand meters per minute (4).

Immediately below the bushings, an organic coating, or sizing (see Section III.H) is applied to the filaments. It contains agents to protect the fibers during processing and to promote binding to the polymer system. After the sizing is added, filaments are gathered into strands, generally containing 200–2000 filaments each. During further processing, a large number of strands may be wound together into a continuous "roving." The strands may be cut to short lengths (chopped fiber) or hammer milled to form very short fibers (milled fiber). These products and their significance in reinforcing thermoplastics are discussed later.

The characteristics and properties of reinforced thermoplastics are affected by several aspects of the glass fibers. These aspects include the type of glass, the fiber form, fiber length, the amount of fiber used, the fiber diameter, the arrangement of fibers in the matrix, and the fiber sizing. Obviously, the resin used is also a key factor, determining chemical, mechanical, and electrical properties, as well as color and surface finish. Thermoplastics are available to meet a variety of property requirements that can be enhanced with the addition of fiber reinforcement.

B.   Relationship of Fiber to Production Processes

Although this chapter is concerned only with the technology of fiber reinforcements themselves, it is recognized that their introduction into the matrix is a first step in the production process. In discussing reinforcement technology, we need to place it in the context of the complete production process. This relationship is seen in four basic principles for using reinforced thermoplastics that are true for most applications:

Mechanical strength of the finished part depends heavily on the combined effect of the amount, type, and arrangement of glass-fiber reinforcement in the part.

Chemical, electrical, and thermal performances of the part depend on the choice and formulation of resin and other additives to the plastic.

Material selection, plus design and production requirements (part size, shape, and production rates desired), determine which process technology is best suited to the application.

Cost/performance ratio, or total value received, results from sound design, adequate tooling, careful production planning, and judicious selection of materials and processes.

With that background, let us consider briefly the important fiber characteristics.

C. Type of Glass

The major portion of reinforcing fibers are made from E glass, a lime–alumina–borosilicate glass designed for production of continuous fibers. It provides high-temperature resistance, which is important in the stability of glass-reinforced thermoplastics, and superior electrical characteristics, as well as good strength and chemical resistance. The tensile strength of S glass, which is used in high-performance structural applications, particularly in the aerospace and defense markets, is about 33% higher than that of E glass. For special purposes, fiber can also be made from C glass, which is highly resistant to acid corrosion. Composition and basic properties of these glasses are shown in Table 3.

D. Fiber Form and Length

Manufacturers compounding their own reinforced thermoplastics generally use either chopped or milled fibers. Whereas continuous roving may also be fed into an extruder, and there is no significant difference in mechanical performance of the resulting composite, chopped fibers are generally more adaptable to conventional extrusion equipment.

Chopped fibers are available in various lengths from $\frac{1}{8}$ to 2 in. The shorter fibers ($\frac{1}{8}$ to $\frac{1}{4}$ in.) are best suited for injection molding. Longer fibers are blended with thermosetting resins for compression and transfer molding.

Table 3 Composition (wt%) of Glasses Used in Fiber Manufacture and Their Basic Properties in Fiber Form

Constituent	E Glass	C Glass	A Glass	S Glass
SiO_2	55.2	65	72.0	65.0
Al_2O_3	14.8	4	2.5	25.0
B_2O_3	7.3	5	0.5	
MgO	3.3	3	0.9	10.0
CaO	18.7	14	9.0	
Na_2O	0.3	8.5	12.5	
K_2O	0.2		1.5	
Fe_2O_3	0.3	0.5	0.5	
F_2	0.3			
Tensile strength of single fiber at 25°C (kg/mm^2)	370	310[a]	310	430
Tensile strength of strand (kg/mm^2)	175–275	160–235[a]	160–235[a]	210–320[a]
Young's modulus of fiber at 25°C (kg/mm^2)	7700	7400[a]	7400	8800
Density (g/cm^3)	2.53	2.46[a]	2.46	2.45
Refractive index, nD	1.550		1.542	
Coefficient of linear thermal expansion per °C $\times$ 10^6	5	8[a]	9	5[a]
Dielectric constant at 25°C and 10^{10} Hz	6.11			
Loss tangent at 25°C and 10^{10} Hz	0.006			
Volume resistivity (Ω cm)	10^{15}	10^{10}		

[a] Estimated or extrapolated value.

The fiber will degrade during processing whatever its starting length, but compounders need to preserve as much length as possible for the potential improvement in composite properties. One source suggests that the optimum length of fiber reinforcements is about 1–2 mm (0.04–0.08 in.) [5]. Put another way, the fibers need to have an aspect ratio (length to diameter) of 50–100 to be useful reinforcers. Fiber efficiency for strength is about 50% at an aspect ratio of 50 and about 80% at 100 [2]. Fiber efficiency for uniaxially oriented short fibers is defined as percentage effectiveness in providing reinforcing action relative to uniaxially oriented continuous fibers at the same volume percent loading.

An electronic image analysis study of fiber lengths [6] has shown that the initial chopped fiber length (in this case $\frac{1}{8}$, $\frac{3}{16}$, or $\frac{1}{4}$ in.) has little effect on fiber length in the final composite. The results also show considerably less degradation of fiber length in polypropylene than in a polyester [polybutylene terephthalate (PBT)]. The final fiber lengths in extrusion compounded PBT molded bars were 0.009, 0.010, and 0.011 in., respectively, for the $\frac{1}{8}$, $\frac{3}{16}$, and $\frac{1}{4}$ in. starting lengths. The major portion of fiber-length reduction occurred in the extruder. For propylene, which was also direct molded, final fiber lengths were 0.026, 0.024, and 0.028 in., respectively.

Milled fibers are $\frac{1}{32}$–$\frac{1}{8}$ in. long and provide similar flexural modulus and dimensional stability, but much lower impact strengths, when compared to chopped fibers. Milled fibers are used to reinforce thermoplastic parts with low- and moderate-strength requirements and to reinforce caulks and adhesives. They are often specified when dimensional stability under the head load is a critical requirement. Stability can be obtained without sacrificing the good processibility and surface finish of milled fiber components.

Continuous fibers in the form of strand roving and various glass-fiber packages can also be used for reinforcement in special cases. The special packages include the following:

Woven roving is a heavy, drapable fabric made from continuous-strand roving that imparts high strength to large molded parts and costs less than conventional woven fabrics.
Woven fabrics are made from glass-fiber yarns and are available with a variety of properties and in different widths and lengths.
Reinforcing mats are made from either chopped or continuous strands laid in a swirl pattern. These mats are used for medium-strength parts with a uniform cross section.
Surfacing mats are used in conjunction with reinforcing mats and fabrics to provide a good surface finish.
Combination mats consist of one ply of woven roving, chemically bonded to a chopped-strand mat to form a strong, drapable reinforcement that gives tailored fiber orientation.

These products are fabricated by glass-fiber producers and introduced to the resin during the molding process by the manufacturer.

E. Amount of Glass

The amount of glass used to reinforce a thermoplastic generally determines the extent to which composite properties are modified from those of the resin toward those of the fiber. The average loading of glass fibers used in thermoplastic compounds is about 25% by weight. Although strength generally improves as the amount of glass in the composite increases, at about 30% loading, the curve begins to flatten. With higher glass content, the resin's flow rate is slowed and material processing becomes much more difficult. Thus, except in special circumstances where the need for greater strength overrides cost penalties in processing, maximum loading is usually about 30%. When strength requirements are low, less glass may be used to reduce material costs. On the other hand, flexural modulus increases linearly with the volume fraction loading of glass. Where this is the key property, high loadings might be considered.

F. Arrangement of Fibers

The way in which fibers are positioned or oriented in a thermoplastic component will determine the direction and level of strength. There are three types of orientation:

Unidirectional. All fibers are running in the same direction. This arrangement provides the greatest strength in the direction of the fibers. With this orientation, up to 80% loading by weight is possible.

Bidirectional. Some fibers are positioned at an angle to the rest, as with woven fabric. This provides different strength levels in each direction of fiber orientation. Up to 75% loading by weight can be obtained.

Multidirectional. Fibers are running in all directions with essentially equal strength in all directions. Used with chopped and milled fibers and mats, from 10% to 50% loading by weight can be obtained. Manufacturers compounding their own composites will obtain this orientation.

The relationship between the amount of glass reinforcement and the arrangement of the fibers is easily seen. The more directionally oriented the fibers, the greater the reinforcement loadings possible, and, therefore, the greater the strength obtainable in the finished part—in the direction of the fibers. The more random the arrangement, the lower the reinforcement loading and, therefore, the lower the resulting strength.

Figure 5 illustrates how part strength is affected by glass-reinforcement percentage by weight, as well as by fiber orientation. Also shown are reinforcement products that yield directional arrangements indicated, together with widely used processing methods with which they are compatible.

G. Fiber Diameter

Two fiber diameters are used for most reinforcement: ''G'' (10 μm or 0.00035 in.) and ''K'' (13 μm or 0.00053 in.). In some resins, the finer diameter fiber will provide superior mechanical properties. For example, nylon 6/6 shows a 10 to 15% increase in tensile and flexural strength and a 25% increase in impact strength with ''G'' fibers. Flexural strength in polypropylene increases with small diameter fibers.[2]

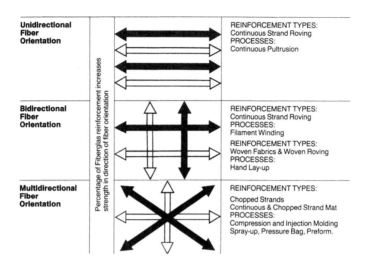

Figure 5 Arrangement of fibers.

H. Sizing

Three types of sizing material are added to glass fibers used for reinforcement, usually amounting to 0.3–1.0% by weight. These sizing materials are identified as follows:

> *Coupling agents.* These organosilanes promote maximum adhesion between the glass surface and the matrix resin (not chemical bonding in most cases). Good adhesion is necessary to ensure the effective transfer of stress to the fibers, the mechanism that gives the composites their high strengths.
>
> *Film formers.* These polymeric materials protect the glass against scratching during production and handling.
>
> *Processing aids.* These lubricants allow the processor to handle, convey, blend, meter, and disperse the glass in the matrix (ideally, all filaments are dispersed from bundles of 800–2000 filaments).

The size may also contain an antistatic agent. Manufacturers' descriptions of glass fibers indicate the specific sizing used. Table 4 shows the recommended reinforcing fibers (Owens–Corning Fiberglas descriptions) for various thermoplastics.

Table 4 Recommended Reinforcements

Polymers	Extrusion compounding[a]	Direct molding[a]
Acrylonitrile–butadiene–styrene	414, 408	885
Acetal	473, 408	
Acrylics	408	885
Acrylonitrile	414, 408	
Epoxies	405	
Fluoroplastics	739 Milled fibers	
Phenylene oxide	497	
Polyamides (nylons)	464	
Polycarbonate		
Low glass content	415	
High glass content	473	
Polybutylene terephthalate	408	
Polyethylene terephthalate	492	
Polyethylene and its copolymers	415	
Polyphenylene sulfide	497	
Polypropylene	457	885
Polystyrene and its copolymers	414, 408	P386E
Thermoplastic polyurethane	473	
Polyvinyl chloride and its copolymers	497	

[a] Numbers are designations of Fiberglas brand reinforcements.

IV. COMPOUNDING

There are two basic ways of producing a reinforced thermoplastic: extrusion compounding and direct molding. For either method, the equipment used should provide the following [2]:

    Steady-state running conditions
    Reproducible compounding conditions
    Easy cleaning
    Versatility
    Ability to adequately disperse fibers and other additives
    Capability to expose each particle to short and equal stresses
    Precise temperature control

In *extrusion compounding,* the glass fiber is dispersed in the molten polymer during extrusion. The type of extruder used, the fiber form, and the point on the extruder at which fiber is introduced may differ.

Both single-screw and twin-screw extruders are used for thermoplastic compounding [1]. Because of the hardness and abrasiveness of the glass fibers, conventional machines must be modified for this compounding process. In a single-screw machine, the screw can be optimized for specific polymers, or a general-purpose screw suitable for a number of polymers may be used. Reinforced thermoplastic compounding requires a screw with deep feed sections and comparatively high temperatures. The equipment is vented to remove volatiles arising from the polymer melt or from the sizing.

The machines used for thermoplastic compounding must have high-horsepower drives and relatively small screw diameters, as extrusion of the glass-fiber composite requires much more power per pound of output than do unfilled or lightly filled materials.

Single-screw extruders have a relative low capital cost, are versatile (hence suitable for short runs), and are easy to maintain and less complex to clean. However, wear is a problem, and they offer limited feed variants. These machines are used for a wide variety of thermoplastics.

The twin-screw extruder, with two meshing screws effecting positive pump action, can compound and extrude bulky feed stock more efficiently. With their ability to process glass reinforcements, these machines have found widespread commercial use in recent years. Unlike single-screw machines, which rely on high head pressures to produce a homogeneous melt, they can extrude fully plasticized polymers almost entirely by the screw action.

These machines are generally better suited to long runs, have low wear, and offer a number of ways to introduce the reinforcing fibers. However, capital costs are high and maintenance is more complex than with single-screw machines.

Most reinforced thermoplastic used today—probably about 80%—is compounded in the extruder. Extrusion compounding give a smoother surface appearance because of better and more uniform glass dispersion. It may also produce more uniform strength from part to part.

In *direct molding,* the resin and glass are mixed as they are introduced to the molding machine. Direct molding is used primarily in high-volume applications, where capital costs for the required blending and conveying equipment, process development costs, and higher labor, handling, and maintenance costs can be offset.

During compounding, other additives are introduced to the mix to provide specific process or product properties. Depending on the application and the manufacturing process, these might include pigmentation, flame retardant, inert fillers, compounds to enhance surface finish or reduce shrinkage in the mold, release agents, curing catalysts, viscosity control materials, and weather resistance enhancers.

V. OTHER REINFORCEMENTS

Although glass fibers are the most common reinforcements for thermoplastic resins, other materials are also used. These include glass beads, various mineral fillers, asbestos fibers, carbon fibers, and organic fibers. Also, hybrids (e.g., of minerals and chopped fibers, mica and glass, glass beads and glass fibers, and carbon and glass fibers) have been used with some success.

In general, these other materials have a higher modulus than glass fibers. Whereas reinforcements are added primarily to improve strength, the fillers have other purposes—for example, increased stiffness, improved heat-deflection temperature—and may actually decrease the strength of the resin. Some property comparisons are given in Table 5. A brief discussion of characteristics and uses of the more common fillers follows.

A. Glass Beads

Solid beads, also called ballotini, made from A glass (an alkali-containing glass), offer certain processing advantages, property improvements, and cost savings when used to reinforce thermoplastics. First, the addition of these glass spheres does not increase the viscosity of the melt as much as do glass fibers or mineral particles. Thus, the production of complex or thin walled moldings is easier.

Table 5 Comparison of Mechanical Property Effects of Different Reinforcing Fillers in Some Thermoplastics

Polymer	Filler	Composition (wt%)	Tensile strength (psi)	Flexural modulus (psi)	Izod impact strength (ft-lb/in.)	Heat-deflection temperature at 264 psi (°C)
Nylon 6/6	None	0	11,800	410,000	0.9	70
	Glass fiber	40	31,000	1,600,000	2.6	260
	Glass beads	40	14,200	730,000	0.6	74
	Asbestos fiber	40	18,500	1,600,000	—	—
	Carbon fiber	40	40,000	3,400,000	1.6	260
Polypropylene	None	0	5,000	200,000	0.5	60
	Glass fiber	30	9,800	800,000	1.6	146
	Asbestos fiber	40	5,600	730,000	1.3	106
	Talc	40	4,000	500,000	0.5	115[a]

[a] At 66 psi.

Furthermore, the glass bead reinforcement reduces shrinkage, offers improved abrasion resistance, compressive strength, hardness, tensile strength, modulus, and creep of base polymers, and costs less than glass fibers.

B. Asbestos Fibers

Asbestos fibers were used primarily in propylene, but they have also been added to high-density polyethylene, polystyrene, and nylon [1]. These fibers improve stiffness and heat-deflection temperature in the components and can also reduce shrinkage. Larger fibers give better reinforcement, but also sharply decrease the melt flow index. Therefore, shorter-grade fibers are generally considered more suitable for compounding thermoplastics. A major drawback of asbestos fibers is, of course, the health hazard. Except for use in phenolics for brake linings and clutch facings, there is essentially no asbestos used today in plastics in North America.

C. Carbon Fibers

Carbon fibers for reinforcement are supplied as a continuous filament tow containing either 5000 or 10,000 parallel filaments of 8 or 9 μm diameter. Three types of fiber are provided: high modulus, high strength, and high strain [7]. Thermoplastics with carbon-fiber reinforcement show considerably higher strength and about twice the stiffness and long-term flexural strength of comparable glass-fiber-reinforced compounds [8]. They also have two to three times the thermal conductivity, reduced coefficient of friction, and reduced wear. At higher loading levels, carbon-reinforced thermoplastics are electrically conductive. The major limitations are high cost and brittleness; the impact strength usually is very low. Hybrid carbon-glass-reinforced compounds have been introduced to preserve impact strength while obtaining some of the advantages of carbon reinforcement, such as in fishing rods and poles for vaulting. The most serious problem holding back the increased use of of carbon fibers is the inertness of the surface, preventing bonding with the coupling agents that are successful with glass fibers in improving impact strength. This area is under much study and solution of the problem will lead to greatly broader usage, despite the cost. At present, there is substantial use in wound parts, such as in aerospace applications, where the nature of the construction overcomes the brittleness problem. Carbon fibers are produced from controlled pyrolysis of pitch or polyacrylonitrile (PAN).

D. Organic and Other Fibers

Excellent impact resistance is obtained with aramid (aromatic polyamide) fibers in plastics at relatively high cost. High-temperature resistance and toughness is also obtained. Surface bonding (use of coupling agents) is also difficult, but the fibers are inherently more ductile than carbon. As the price drops, one may expect replacement of carbon fibers—for example, in sporting goods.

Ultrahigh-molecular-weight polyethylene (UHMWPE) can be spun into fiber from low solids gel in solvent such as decahydronaphthalene. The low density leads to very high free-breaking length (the length of fiber that will resist breaking under its own weight). The best UHMWPE fibers have a free-breaking length of 335 km, as compared to 193 km for aramid, 171 km for carbon fiber, and 37 km for steel. The high strength-to-weight ratio must be balanced against the low melting point (145°C) and the unreactive surface. Current applications include body armor.

Short stainless-steel (SS) fibers are finding increasing use in conductive applications, such as for shielding from radio frequency (RFI) and electromagnetic fields (EMI). Such shielding has become increasingly important in computer applications, particularly as chip sizes decrease. Fine-diameter short SS fibers can often be incorporated with far less build in process viscosity than is the case with conductive carbon black.

Beyond the reinforcements and fillers discussed here, almost anything can be compounded into a thermoplastic resin, which can then be injection molded into a functional part, provided the following conditions obtain [9]:

The additive is chemically compatible with the resin up to the processing temperature of the melt.

The products of decomposition generated during compounding or molding are not excessively corrosive or toxic (assuming that proper equipment and handling procedures are used).

The additive is not harder than the metals used in the screws, barrels, and molds of the extruder.

The additive is in a form that lends itself to compounding and dispersing.

REFERENCES

1. WJ Titow, BJ Lanham, Reinforced Thermoplastics. New York: Wiley, 1975.
2. Engineering Design Handbook: Discontinuous Fiberglass Reinforced Thermoplastics, DARCOM Pamphlet, DARCOM P-706-314, U.S. Army Material Development and Readiness Command, April 1981.
3. MJ Balow, DC Fucella. Plastics 96–97, May/June 1982.
4. KC Lowenstein. The Manufacturing Technology of Continuous Glass Fibers. New York: Elsevier, 1973.
5. MC Bader, WH Bowyer. Composites 4(4):150–151, 1973.

6. RA Schweizer. Glass fiber length degradation in thermoplastics processing, Proceedings of the 36th Annual Conference, Reinforced Plastics/Composite Institute, 1981.
7. L Holloway. Glass Reinforced Plastics in Construction: Engineering Aspects. New York: Wiley, 1978.
8. RW Richards, D Sims. Composites 2(4):214, 1971.
9. PJ Cloud, RE Schulz. The Use of Extenders, Fillers and Reinforcements in Plastic Resins—An Introduction. Malvern, PA: LNP Engineering Plastics.

5
Fillers: Types, Properties, and Performance

K. K. Mathur† and D. B. Vanderheiden
Specialty Minerals, Easton, Pennsylvania

I. INTRODUCTION

The concept of cost reduction by use of filling materials has been known throughout the ages. Since the earliest days, very significant advances have been made in this area in terms of fine-particle size technology, tailoring particle morphology, beneficiation of natural materials to attain high purity, surface treatments for improved matrix compatibility, and the development of coupling agents to achieve polymer-to-filler bonding for improved mechanical properties.

In general, fillers are defined as materials that are added to the formulation to lower the compound cost. Such materials can be in the form of solid, liquid, or gas. By the appropriate selection and optimization of such materials, not only the economics but other properties such as processing and mechanical behavior can be improved. This chapter primarily covers solid fillers. Liquid fillers such as extender plasticizers and gaseous fillers such as blowing agents are covered in separate chapters.

For effective utilization of fillers, a complete understanding of individual characteristics is essential. Each class of fillers appears to exhibit specific characteristics which make them especially suited for the given application; for instance, ultrafine precipitated calcium carbonates for improved notched Izod impact strength and calcined clays for electrical properties. Although these fillers retain their inherent characteristics, very significant differences are often seen, depending on the molecular weight, compounding technique, and the presence of other

---

† Deceased.

additives in the formulation. Therefore, once the basic property requirements are established, the optimum filler type and loading for cost/performance balance must be determined. Frequently, blends of two different fillers are used to attain a balance of properties. Special consideration must also be given to determining acceptability of the processing properties. Such information will lead one to a complete understanding of the benefits to be derived from the use of filler in a given compound.

For effective utilization of fillers, it is critical that they be well dispersed in the polymer system. Some coarser fillers can be dry blended and injection molded at 1–10 phr without a major sacrifice in surface appearance and physical properties. Melt compounding is essential to disperse fine fillers for maximum impact strength improvement as well as retention of impact strength at high loadings. In a highly agglomerated state, small particle size fillers may behave like coarse fillers, causing undue weakness in the matrix which will be related to loss in notched Izod impact and low-temperature impact properties.

The addition of filler also requires a balance of formulation for optimum processing properties. For example, the use of stearate-coated calcium carbonates will require a reduction in lubricant level and maximum extruder output. The stearate-treated fillers sometimes also act as costabilizers, which should be taken into consideration for formulation economics.

Therefore, before making a final decision on a filled compound, it is critical to establish the following:

1. Optimum loading level for property/benefit
2. Optimum formulation for processing/production output
3. Economics of filled formulations

II. FILLER CLASSIFICATION

Fillers have been classified in many different ways, ranging from their shapes (e.g., spheres, rods, ribbons, flakes) to specific characteristics (e.g., conductivity, fire retardancy). For simplicity, fillers can be classified in two categories: (1) extenders and (2) functional materials.

Although practically all of the fillers exhibit some functional property, the above classification is tied to the primary reason for using a filler. By definition, an extender filler primarily occupies space and is mainly used to lower the formulation cost. A functional filler, however, has a definite and required function in the formulation apart from cost; examples are antimony oxide for fire retardancy and pyrogenic silicas for rheology modification.

As with all attempts to classify, there are gray areas where overlap and poor definition exist. For example, some of the extender fillers, when reduced

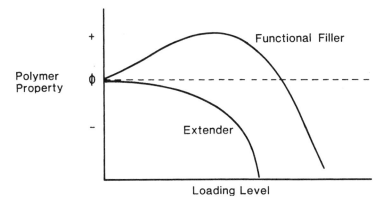

Figure 1   Characteristics of functional and extender fillers.

to a finer particle size and/or surface treated, would be reclassified as functional fillers. The development of new surface-treatment technologies has further broadened the ability to graft functional characteristics to extender fillers. Fillers that are functional in one polymer may be merely extenders in another. Such factors have seriously complicated the task of establishing sharp boundary lines between extenders and functional fillers in terms of their generic composition; however, on a performance basis, the two can be separated as shown in Fig. 1.

Therefore, the extender fillers basically lower the formulation cost and increase flexural modulus, whereas the functional fillers provide at least one specifically required function in the formulation, such as thixotropy, fire retardancy, opacity, color, or impact modification.

III.   CHEMISTRY AND PROPERTIES

A.   Extender Fillers

In general, an ideal extender filler should (1) be spherical to permit retention of anisotropic properties, (2) have an appropriate particle size distribution for particle packing, (3) cause no chemical reactivity with the polymer or additives, (4) have a low specific gravity, (5) have a desirable refractive index and color, and (6) be low in cost. Thus far, no single product stands out in all of these specifications. Ground limestones come closest to the specifications and are most commonly used as extenders. Ground talcs and clays, even though they improve some electrical properties, are also classified as extenders because they allow filling to high loadings without adversely affecting the physical properties, and they meet other extender qualifications.

1. Limestone

Limestone is a naturally occurring mineral. Chemically, it is $CaCO_3$ and may contain a small amount of $MgCO_3$ and possibly traces of other impurities such as SiO_2, Al_2O_3, and Fe_2O_3. In mineralogical terms, it is classified as a trimorphous mineral because it exists in three distinct crystal structures, namely calcite, aragonite, and vaterite.

Calcite, in its various geological forms, is one of the most widely occurring minerals. It is the chief constituent in all limestones and marbles and is found in sedimentary and metamorphic rocks. The fossiliferous form of limestone exists in oyster shells, which find special applications. Its crystal habits are shown in Figure 2.

Limestone is a moderately soft mineral. On the Mohs hardness scale, it is rated at 3. Because of this property, limestone can be ground to a very fine particle size. Depending on the ore quality, dry or wet beneficiation is often necessary to improve the color. The milling can be conducted in dry or wet form, often being dictated by the particular wet or dry beneficiation requirements. The milled

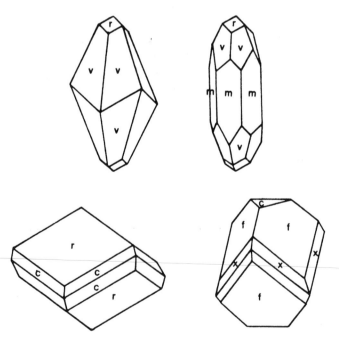

Figure 2   Crystal habits of calcite.

product is then classified through a particle classifier to achieve the desired top size.

In amorphous polymers smaller top size ground limestone is preferred for several reasons: (1) it permits greater impact strength retention up to higher loadings, (2) it minimizes abrasive wear on processing equipment, particularly extruders, and (3) it provides excellent surface appearance of parts.

Stearate surface-treated ground limestones offer many advantages over uncoated products, such as hydrophobicity, better powder flow properties (causing less hang-up in automated systems), the possibility of formulation cost reduction by lowering lubricant and stabilizer level, significant improvements in low-temperature impact strength, and reduced abrasivity. The stearate treatment chemically reacts with the limestone surface and is usually applied in sufficient quantity to form a monomolecular layer. The limestone and stearic acid reaction takes place as follows:

$$2C_{17}H_{35}COOH + CaCO_3 \rightarrow (C_{17}H_{35}COO)_2 Ca + CO_2\uparrow + H_2O$$

In general, limestones of relatively fine particle size are used in polyolefins and rigid poly(vinyl chloride) (PVC) as compared to plasticized PVC. In plastisols and organosols, where thixotropy and plastic viscosity are critical, finer-particle-size ground limestones may be preferred. With the availability of a wide variety of particle sizes and suitable morphology, these materials are frequently blended in plastisols for improved particle packing.

In rigid PVC formulations, relatively fine-particle-size stearate-coated ground limestone fillers are most commonly used. Because of the fine particle size and stearate coating, they offer improved processing as well as physical property advantages in the form of improved notched Izod and drop weight impact strength (Table 1).

Table 1   Effect of 20% Loading of Stearate-Coated 3-μm Ground Limestone on Physical Properties of PVC

Property	Unfilled rigid PVC	20% Stearate-coated 3-μm ground limestone
Flexural modulus (psi)	510,000	650,000
Tensile yield stress (psi)	7,800	6,700
Elongation at yield (%)	5.0	4.8
Notched Izod impact strength (ft-lb/in.)	0.86	1.70
Drop weight impact (ft-lb)	2.3	4.1

Note: Formulation of PVC (in phr): PVC K-62, 100.0; organotin mercaptide, 1.6; acrylic process aid, 1.5; lubricant A, 1.2; lubricant B, 0.5; and TiO$_2$, 1.0.

Table 2 Effect of Ground Limestone Particle Size on Properties of Plasticized PVC at 12-phr Filler Level

Limestone particle size (μm)		BET surface area [N$_2$ (m^2/g)]	Ultimate tensile strength (psi)	Tensile modulus at 100% elongation (psi)	Percent ultimate elongation (%)
Top	Average				
15	3	4.3	3230	3210	110
25	5	3.8	2070	1880	140
41	7	2.7	1960	1760	150
31	6	1.9	1960	1610	150

Formulation of PVC (in phr): PVC MHMW, 100.0; Sb$_2$O$_3$, 1.4; N-octyl-N-decyl phthalate, 67.0; octyl epoxytallate plasticizer, 5.5; Ba/Cd/Zn stabilizer, 2.2; lubricant, 0.2; and limestone filler, 12.0.

In plasticized PVC, the finer ground limestones impart higher tensile strength and tensile modulus, with some reduction in ultimate elongation, as compared to the coarser limestone extenders (Table 2).

The stearate-coated ground limestones in wire and cable formulations demonstrate increased volume resistivity, insulation resistance, and significantly reduced moisture sensitivity.

In addition, a series of wet ground limestone fillers of very small particle size (1.0 μm) have been introduced by a few manufacturers under the name of ultrafine ground limestone (UFGL). These products are sold commercially with and without the stearate treatment and provide improved properties over the coarser ground limestones. Key improvements are seen in impact strength retention and reduced abrasivity.

In addition to the ground limestone, there are numerous manufactured (precipitated) grades of CaCO$_3$ which are also widely used; these are discussed further in Section III.

2. Kaolin Clay

The mineral commercially known as kaolin is basically a hydrous aluminosilicate mineral with the chemical formula Al$_2$O$_3$ · 2SiO$_2$ · 2H$_2$O. Thus, a theoretical kaolin contains 46.5% silica, 39.5% alumina, and 14.0% water. The kaolin family covers three different species, namely nacrite, dickite, and kaolinite. The U.S. deposits are primarily of the kaolinite type.

Kaolinite is a platy material. The crystal elongates parallel to the c axis. The alumina octahedral sheet shares the oxygen with silica tetrahedral sheets,

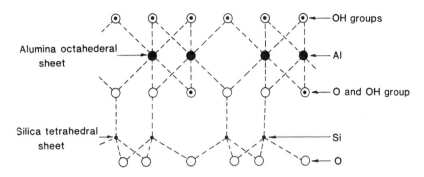

Figure 3 Idealized structure for the alumina/silica layers of hydrated aluminum silicate (kaolin).

and the OH and O form the external layers (see Fig. 3). The basal spacing is perfect for cleavage. Therefore, by selection of the appropriate milling equipment, kaolin can be delaminated to fine sheets.

The molecular weight of kaolin is 258.09. The specific gravity of the uncalcined product is 2.58, which increases to 2.63 on calcination.

The bound water begins to be liberated at 330°C. Calcination at 650°C produces partial dehydration of the kaolinite lattice, resulting in its conversion to pseudocrystalline "meta-kaolinite." Complete dehydration is accomplished at 900°C. At this stage, a significant change in density and light refractivity takes place.

Kaolin clays are commercially sold in air-floated, water-washed, calcined and silane-treated forms. The color and purity of the products depend on the beneficiation technique and posttreatments. The calcined clays are primarily used in wire and cable jacketing and insulation formulations at 5–15-phr levels in PVC and polyolefins but often higher with elastomer-modified formulations. Much higher levels cause increased stiffening and lowering of elongation. Because of the alignment of platelets in the extruder direction, the kaolin-filled wire and cables possess even further improved electrical properties.

The platelet alignment increases the electron flow path considerably (see Fig. 4). The critical electrical properties of kaolin and other fillers are given in Table 3 [1]. The low-cost air-floated, and water-washed clays are commonly used in flooring, film, toys, and upholstery applications.

3. Talc

Chemically, talc is a hydrated magnesium silicate. The chemical formula for pure talc is $3MgO \cdot 4SiO_2 \cdot H_2O$. In actuality, the ratio of MgO to SiO_2 may vary from $1:2$ to $1:1$. The structure of talc is shown in Fig. 5.

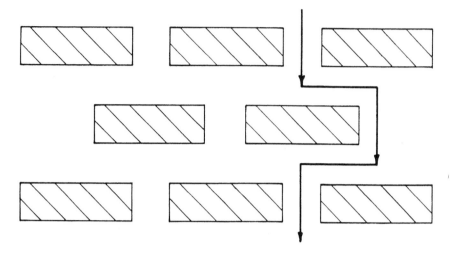

Figure 4 Flow of electrons in kaolin-filled polymer.

Table 3 Dielectric Properties of Selected Fillers

Typical fillers	Dry resistivity (Ω-cm)	Humid resistivity (Ω-cm)	Dielectric strength (V/mil)	Dielectric constant[a] K_e (at 1 mc)	Dielectric loss (%)
CaCO$_3$	10^{11}	10^7	60–80	6.1	0.05
Kaolin	10^{13}	10^6	70–120	2.6	0.1
Kaolin (calcined)	10^{13}	10^8	60–100	1.3	0.06
Kaolin (calcined and surface treated)	10^{13}	10^{12}	80–150	1.3	0.003
Kaolin, partially calcined	10^{13}	10^5	70–100	1.3	0.01
Talc	10^{14}	10^9	—	5.5–7.5	0.001

[a] Measured at 1 MHz.
1000 cycles per second = 1 MHz.

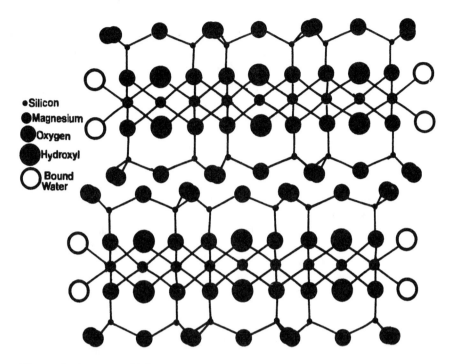

Figure 5   Structure of talc.

Talc as found in nature can be associated with a wide variety of other materials. Montana talc is an exceptionally pure talc ore that very nearly approximates the theoretical formula of talc. New York talcs generally contain only 40–60% talc. Texas talc contains high amounts of dolomite ($CaCO_3 \cdot MgCo_3$), whereas Vermont talc has varying levels of magnesite. California and Nevada talcs vary in impurities with the particular ore source.

Talc is the softest of known minerals. On the Mohs hardness scale, it is rated 1. The specific gravity of talc is 2.7. Mining of talc is carried out by classical mining techniques. Although most of the mines are open pit, there are some active underground mines.

In a typical talc operation, the ore is crushed into $\frac{1}{2}$–3-in. pieces with a jaw crusher, then water washed and sorted to remove impurities. It is then cone crushed and milled to varying particle sizes in a closed loop with pneumatic classifiers to obtain a product with desired particle sizes and distributions.

Talcs with high impurities require beneficiation by flotation, magnetic separation, bleaching, and so forth to attain a brighter and chemically purer product.

Because of the perfect basal spacing, talc can be exfoliated into perfect platelets when appropriate milling techniques are used.

In plastic processing, the platy particles align in the machine direction. In general, platy talcs improve hot strength, increased modulus, and tensile strength and are generally used in polyolefin and plasticized PVC applications, especially vinyl flooring. The fibrous New York talcs have also been used in this application in the past. A scanning electron micrograph of talc-filled PVC is shown in Fig. 6.

In addition to their use as fillers, fine-ground talcs are also used as antiblocking agents in thin films. The delaminated calcined talcs in general have properties similar to those of kaolin clay, as shown in Table 3.

Because of its high oil absorption, talc is also frequently used to increase plastisol thixotropy and plastic viscosity.

The talc surface is highly active. Talc can be surface treated to achieve a wide variety of useful properties. The most common surface-treating agents are polyethers and polyols, which give the talc improved properties for a variety of

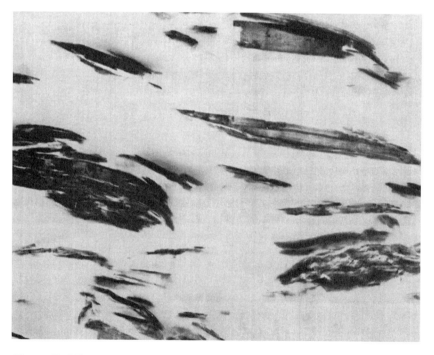

Figure 6   Microtomed section of talc-filled PVC under scanning electron microscope at ×35,000.

applications [2]. Because of the silicate surface, talc also reacts with silanes. Such products show improved dispersibility in most polymers.

B. Functional Fillers

1. Precipitated Calcium Carbonate

Precipitated calcium carbonates are among the most versatile functional fillers used. They can be manufactured from lime by three basic technologies as follows [3]:

Lime/CO$_2$:

$$Ca(OH)_2 + CO_2 \rightarrow \underline{CaCO_3} + H_2O$$

Lime/soda:

$$Ca(OH)_2 + Na_2CO_3 \rightarrow \underline{CaCO_3} + 2NaOH$$

Solvay process:

$$NH_3 + H_2O + CO_2 + NaCl \rightarrow \underline{NaHCO_3} + NH_4Cl$$

$$Ca(OH)_2 + 2NH_4Cl \rightarrow CaCl_2 + 2NH_4OH$$

$$CaCl_2 + Na_2CO_3 \rightarrow \underline{CaCO_3} + 2NaCl$$

As a result of the chemical processing, precipitated calcium carbonates have a chemical purity far beyond that normally found in the ground limestones. The chemistry and varied crystallography of calcium carbonate also permit the manufacture of products displaying a wide range of particle size, particle shape, and particle size distribution, as shown in Table 4.

Probably the most important value of precipitated calcium carbonate in amorphous polymers today stems from the ability of surface-treated products in the particle size range 0.07–0.5 μm to increase impact strength (Figs. 7 and 8) [4,5]. This is accomplished with relatively low loading levels in the 5–30-phr range and is not accompanied by a loss in the heat stability of the compound.

Improvements in ultraviolet (UV) light stability and exterior weatherability of halogenated polymers have also been reported, hypothesized to result from the HCl-scavenging ability of the precipitated calcium carbonate [4,6]. Calcium stearate-coated precipitated calcium carbonate is also being used in some plasticized applications as a thixotrope and a secondary heat stabilizer.

Precipitated calcium carbonates also provide functionality in fire resistance. Precipitated calcium carbonates of very small size (0.05–0.10 μm) and high surface area (~20 m^2/g) effectively scavenge HCl gas liberated from burning halogenated polymers, thereby reducing the corrosiveness of the generated smoke:

Table 4 Identification and Physical Properties of Surface-Treated Calcium Carbonate Products

Material	Product identification	Particle size via TEM[a] analysis (μm)		BET surface area (m²/g)	Oil absorption ASTM D281-31 (lb oil/100 lb filler)	Dry brightness (%)	Percent wet out in water (%)
		Average	Top size				
Surface-treated fine natural ground calcium carbonate	3 μ GL	3	15	3.6	14	94	0
Surface-treated very fine precipitated calcium carbonate	0.5 μ PCC	0.5	<1.5	6.0	26	94	0
Surface-treated ultrafine precipitated calcium carbonate	0.07 μ PCC	0.07	<0.2	20	35	94	0

[a] TEM: transmission electron microscopy.

$$CaCO_3 + 2HCl \rightarrow CaCl_2 + CO_2 + H_2O$$

At the same time, the carbon dioxide and water generated in the neutralization reaction serve to keep the oxygen index low enough for self-extinguishing properties. Applications of calcium carbonate in PVC jacketing compounds for electrical cables have been promoted based on the HCl-scavenging functionality of the filler [7].

2. Precipitated Silicas and Silica Gels

Precipitated silicas and silica gels are classified as wet process silicas to differentiate them from the thermally processed pyrogenic silicas. They are produced by the reaction of an aqueous alkali silicate solution with a mineral acid solution; for example [8].

$$Na_2SiO_3 + H_2SO_4 \rightarrow Na_2SO_4 + SiO_2\downarrow + H_2O$$

Precipitation can be conducted in three general ways, based on the order of addition of the reactants:

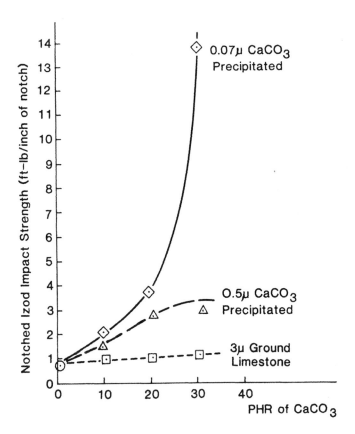

Figure 7 Notched Izod impact strength of surface-treated calcium carbonates in a rigid PVC compound.

1. Alkali silicate solution added to acid solution
2. Acid solution added to alkali silicate solution
3. Simultaneous addition of alkali silicate and acid solutions into water or neutral salt solution

The first of these processes is the preferred method for producing silica gels. These products are characterized by very small primary particle sizes and high surface areas in the range 200–800 m^2/g. These physical properties and their relatively high cost promote their use in many applications also served by the pyrogenic silicas. These applications include (1) prevention of plateout, (2) anti-blocking of film, (3) flow control aid for resin powders, (4) viscosity control

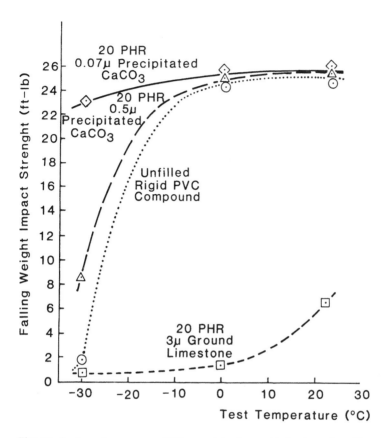

Figure 8 Low-temperature falling weight impact behavior of unfilled and 20-phr CaCO$_3$-filled rigid PVC.

agent, (5) processing aid, and (6) selective adsorbent and moisture removal agent [9].

Precipitated silicas are generally produced by precipitation procedure 2 or 3 above [8]. Simultaneous addition is generally the preferred technique, allowing variation and control of pH, precipitation time, temperature, and electrolyte concentration to produce a range of precipitated silica products.

Precipitated silicas offer a range of property improvements in thermoplastic polymers, including (1) increased hardness, (2) improved elasticity, (3) increased resistance to heat distortion, and (4) improved scratch resistance [10]. Where transparency or translucency is desirable, large additions of synthetic silicas can be made with minimal effect. In plastisols used to coat textiles, precipitated silicas can provide improved mar resistance and flatting if desired. They are also used as release agents to smooth out calendered sheets.

3. Precipitated Metallic Silicates

Precipitated metallic silicates are produced by a process very similar to that just described for precipitated silicas. In the case of metallic silicates, the aqueous alkali silicate is reacted with a solution of a metal salt instead of a mineral acid; for example,

$$Na_2SiO_3 + CaCl_2 \rightarrow CaSiO_3 + 2NaCl$$

The most important metallic silicates still in use are calcium silicate, basic lead silicate, aluminum silicate, and barium silicate. Typical physical properties are reported in Table 5. Although the importance of the metallic silicates has waned significantly in recent years in favor of the synthetic silicas, they do complement the pigment group in that they have a significantly alkaline pH of 9–12 compared to pH 6–9 for the precipitated silicas and a highly acid pH (3.6–4.3) for the pyrogenic silicas.

Metallic silicates have been added to formulations for a wide variety of reasons. Advantages reported for their use include (1) improved electrical resistivity, (2) modified rheological properties, (3) improved hand and drape characteristics in fabrics, (4) prevention of plateout, (5) optical extension of TiO_2 in opaque formulations, and (6) processing improvements in dry-blend extrusion operations.

4. Pyrogenic Silica

Pyrogenic silicas, also referred to as fumed silicas, comprise a very special class of functional silica fillers. The special properties of pyrogenic silicas result pri-

Table 5 Physical Property Comparison of Silica-Based Pigments

Physical property	Pyrogenic silica	Precipitated silica	Precipitated metal silicate
Discrete particle size (nm)	7–16	15–100	20–50
Surface area (m²/g)	100–500	40–250	35–180
Drying loss (%)	<1.5	4–7	5–9
SiO_2 (dry) (%)	>98.5	83–90	63–80
DBP absorption (g/100 g)	—	175–285	165–220
pH (5% in water)	3.6–4.3[a]	6–9	9–12
Compacted apparent density (g/L)	50–100	150–250	100–250

[a] 4% in water.
Source: Refs. 9 and 10.

marily from the unique manufacturing process, first practiced by Degussa in 1942, in which silicon tetrachloride is hydrolyzed in an oxygen–hydrogen flame [11]:

$$2H_2 + O_2 \rightarrow 2H_2O$$

$$SiCl_4 + 2H_2O \rightarrow SiO_2 + 4HCl$$

$$2H_2 + O_2 + SiCl_4 \xrightarrow{\quad 1000°C \quad} SiO_2 + 4HCl$$

In terms of physical properties, pyrogenic silicas are characterized by (1) exceptionally small discrete particle sizes ranging from 7 to 16 nm (0.007 to 0.016 μm) with attendant surface areas of 120–380 m^2/g, (2) low moisture levels, generally less than 1.5%, (3) acidic pH values in the range 3.6–4.3, (4) extremely high chemical purity, generally >99.8% SiO_2 on an ignited basis, and (5) very low bulk densities, ranging from 50 to 100 g/L (specially densified).

The most important application of pyrogenic silicas is related to their use as rheology modifiers. Very small amounts of fumed silicas provide increased viscosity [12]. This thixotropic flow behavior is essential where processing or application requires a high degree of plasticity.

Many other applications are served by pyrogenic silicas [10]. As a result of their purity and low moisture content, fumed silicas are very good insulators and highly suitable for improving the electrical properties of cable compounds. Pyrogenic silicas are also used to reduce plateout in processing. Surface-modified (hydrophobic) pyrogenic silicas are recommended to improve the dry flow (anticaking) properties of pure powders, whereas the standard grades are used to correct the dry flow deficiencies of dry blends.

5. Antimony Oxide

Antimony oxide (Sb_4O_6) is a very effective fire-retardant additive [13–15]. The pigment is produced by roasting the naturally occurring sulfide ore (Sb_4S_6) and is then purified by distillation [16]. The chemistry of antimony is much like that of its more familiar group V family member, phosphorus, and similar fire-retardant materials have been developed around each element as a base. A review prepared in 1979 [17] lists over 200 references on the uses of antimony compounds as fire retardants.

With the use of flame-retarding phosphate esters, flame resistance can be significantly increased by the addition of as little as 1–5 phr antimony oxide [18]. The combination of phosphate and antimony oxide pigment generally produces acceptable retardancy while retaining good low-temperature properties but with a sacrifice in clarity of the resulting compound [19].

6. Zinc Oxide

Interest in zinc oxide revolves around its pigmentary properties as well as heat- and light-stabilization properties. Zinc oxide pigment exists in many grades, the

Table 6 Typical Properties of Titanium Dioxide and Zinc Oxide

Property	Titanium dioxide		Zinc oxide
	Anatase	Rutile	
Average particle size (μm)	0.3	0.2–0.3	0.2
Density (g/cm^3)	3.9	4.1	5.6
Refractive index	2.55	2.76	2.01
Tinting strength	1200	1600	210
Oil absorption (lb/100 lb)	18–30	16–48	10–25
Hardness (Mohs)	5–6	6–7	4+

basic properties of which are indicated in Table 6 [20]. Two primary pyrometallurgical processes are used for its manufacture: the French or indirect process, and the American or direct method [21]. Zinc sulfide ore, the common starting raw material, is roasted, purified, and the zinc metal vapor burned directly to zinc oxide in the American process, whereas the French process contains an intermediate condensation step to zinc metal.

Based on its primary pigment properties (white color, fine particle size, high opacity/UV absorption), zinc oxide has found uses in exterior applications, although its use is small compared to that of TiO$_2$. In vinyl flooring, zinc oxide has been used often as a heat and light stabilizer, especially in formulations involving large amounts of limestone filler. Stabilization mechanisms involving synergism with alkaline earth salts have been proposed. Many proprietary stabilizer systems for vinyl floor tile are known to contain zinc oxide as an important ingredient.

7. Titanium Dioxide

The functional value of TiO$_2$ lies in its unsurpassed opacifying (hiding) power among white pigments. It exists in two primary crystal forms, anatase and rutile, the basic properties of which are given in Table 6 [20]. Two basic manufacturing processes are presently in use for TiO$_2$. The sulfate process operates from an ilmenite or iron titanate raw material, a hydrous titanium dioxide being extracted with sulfuric acid, purified, and calcined to either anatase or rutile pigment grades. The newer chloride process starts with a natural rutile ore that is converted to titanium tetrachloride, purified, and then reacted with oxygen at about 1500°C to reform the rutile pigment. Both processes include numerous finishing steps including very important inorganic surface treatments plus milling, drying, and classification to produce the final pigment products.

The pigmentary properties of whiteness, brightness, and opacity are the prime reasons for the incorporation of TiO_2. Although anatase was used in some early formulations where a blue tone whiteness was required, finer particle sized rutile specialty grades have now effectively replaced and eliminated anatase in most polymer systems. The blue tone rutile pigments develop much higher brightness and opacity and impart superior weathering characteristics compared to anatase. One area where the coarser anatase grades continue to be the product of choice is in plastisols [22].

Rutile TiO_2 is also an effective ultraviolet light absorber and is thus capable of protecting polymers in exterior applications. Numerous studies paralleling those in the paint area have shown that the inorganic-coated, nonchalking grades of rutile TiO_2 (rutile-exterior grades) provide the greatest degree of protection [23]. TiO_2 levels of 12 phr in exterior compounds appear to provide the best practical protection. It is certainly no coincidence that the 8–12-phr level is also the area of the optimum cost/performance ratio on the basis of optical properties.

8. Iron Oxide Pigments

Natural and synthetic iron oxides are frequently used as pigments [24] where the processing temperatures are relatively low. Iron oxide pigments, in addition to excellent ultraviolet stability, offer good chemical and mildew resistance. The chroma and hue of these pigments frequently require toning with brilliant organic pigments for brighter shades [25]. The blending of inorganic and organic pigments usually results in the most economical pigmentation package.

Where processing temperatures may be as high as 195–200°C, the iron oxide yellow and iron oxide black pigments are not recommended. At these processing temperatures, the yellow iron oxide ($Fe_2O_3 \cdot H_2O$) will partially dehydrate to α-Fe_2O_3, giving a reddish yellow color. Similarly, the iron oxide black begins to oxidize to γ-Fe_2O_3, giving a reddish black shade. Several patents have shown the use of phosphate doping to increase the processibility of yellow iron oxide pigment [26]. In Europe, the iron oxide pigments are very frequently used in rigid PVC. This is accomplished by selecting appropriate stabilizers and lubricants and by maintaining lower processing temperatures.

Because iron oxide pigments absorb low amounts of infrared, they are frequently used in products designed for exterior applications. Because of low infrared absorption, they exhibit low heat buildup on exposure to sunlight and could be especially suited for applications such as siding, profiles, outdoor posts, tubular patio chairs, and so forth [27].

The iron oxide pigments used for audio, video, instrumentation, and recording tapes require very special properties. The slow recording speed and high fidelity require ultrafine gamma ferrites. The magnetic inks for credit cards and railway tickets use synthetic magnetite, which permits a high concentration in the polymer system.

9. Organic Pigments

A wide variety of ultraviolet-stable organic pigments are being used for exterior applications. The key requirements for such pigments are (1) thermal stability, (2) ultraviolet stability, (3) environmental fading stability, and (4) acid rain stability.

10. Carbon Black

Carbon black pigments can provide a variety of special property advantages in plastics, even though they do not provide the reinforcing characteristics commonly associated with their primary use in rubber. Carbon blacks are produced by three quite different processes which lead to substantial differences in final product properties [28]. By far the most important process is the furnace black process, in which a highly aromatic oil feedstock is decomposed in a furnace chamber containing the combustion products of an oil or gas flame. Over 90% of carbon blacks produced and sold are made by the furnace black process. The remaining products result from use of the far less important channel black and thermal black processes. General property ranges of carbon blacks made by the three processes are compared in Table 7 [28–30].

Functionality of carbon blacks can be exceptionally broad [20,29]. For pigmentation purposes, the relatively coarse furnace or thermal blacks are commonly used because of their relatively lower cost and ease of dispersion. Carbon black pigmentation also provides significant protection from UV and thermal degradation. Carbon black is capable of absorbing harmful ultraviolet radiation and, at the same time, scavenging the free-radical degradation products capable of catalyzing further degradation. Thus, carbon blacks are well-known stabilizers and screening agents for compounds for exterior applications.

Carbon blacks are also used to control electrical properties. Depending on the carbon black chosen, high electrical conductivity or high electrical resistivity

Table 7 Carbon Black Properties Related to Production Methods

Property	Furnace black	Channel black	Thermal black
Average particle size (nm)	13–75	10–30	150–500
Surface area (N_2 BET) (m^2/g)	25–560	100–1125	6–15
Oil absorption (cm^3/g)	0.70–1.85	1.0–5.7	0.3–0.5
pH	3.3–9.0	3.0–6.0	7.0–8.0
Percent volatiles	1.0–9.5	3.0–17.0	0.1–0.5

can be imparted. In applications where high conductivity is required (e.g., to reduce static charge buildup on molded parts), a carbon black with a fine particle size, high structure, and low volatile content is required. Where high electrical resistance is desired, a carbon black with a coarser particle size, low structure, and high volatile content will be most effective. The coarse size and low structure decrease the number of particle contacts through which electrons can flow, and the surface volatiles act as insulators, further inhibiting conductivity in this case.

11. Asbestos

In terms of volume usage, asbestos was a very large volume filler for floor tiles in years past. More recent health and safety studies have shown that inhalation of fibrous asbestos over a period of time can cause asbestosis, a nonmalignant fibrotic lung condition, bronchogenic carcinoma, and mesothelioma, a rare cancer of the chest lining and abdominal cavity. Because of these health risks, the floor tile industry has in recent years reformulated their products [31].

In nature, asbestos occurs in two basic forms: chrysotile, also known as serpentine asbestos, and the amphiboles, which include tremolite, amosite, crocidolite, actinolite, and anthophyllite. These varieties differ chemically. The chemical composition from deposits varies depending on the associated impurities. The color varies from white to gray. The specific gravity varies from 2.48 to 2.56.

High-aspect-ratio asbestos exhibits excellent reinforcement and contributes to dimensional stability and impact strength, thereby allowing a greater total volume loading of filler for improved economics. Other high-aspect-ratio organic fibers lack the basic characteristics of asbestos. Organic fibers suffer from poor wetting and relatively lower hot strength, which is critical for continuous processing operations.

Both of these problems have been overcome to some degree by the industry. Overall, the industry has been quite secretive about the new formulations, but the recent literature and patents [32] from major manufacturers shed some light in this area.

The most direct patents have discussed the use of synthetic mineral wool, polyester, and cellulosic fibers to replace asbestos. A host of wetting agents, mixing equipment, and incorporation techniques have been tried and tailored to achieve maximum wetting of fibers.

IV. FILLER PROPERTIES AFFECTING PERFORMANCE

A. Color

To the formulator, the color of the filler (or, more precisely lack of color) is generally second only to cost in the ranking of filler selection criteria. Cost and color are inexorably related, however, because whiteness is a key factor on which filler value is established by the manufacturer, whether it is a product of natural origin or chemical synthesis.

Color is established by the light absorption of the pigment over the visible region. The ideal white pigment will show no absorption. The ideal is rarely if ever achieved, however, because very low levels of chemical impurities and other structural defects in the filler crystals can lead to extraordinary absorption and color effects in white fillers [33]. If the impurity exists as a separate phase or mixture in the filler, several tenths of a percent may be required to produce a significant color effect. If the impurity is well dispersed in the crystal lattice, however, only a few parts per million are needed to produce a major effect. For example, although the absolute iron content often provides a poor correlation with the color of kaolin, as little as 0.003% Fe_2O_3 in some clays can produce an intense color ranging from yellowish to reddish [34]. In another study of barium sulfate, calcium carbonate, and dolomite fillers, the brightness and color varied inversely with impurity level, particularly that of iron [35,36]. There is certainly a general trend for the brightness to fall and the color intensity to increase as the iron content and other impurities increase, regardless of the type of filler.

In examining the filler's color in a polymer, it is important that the color be measured in a fully compounded sheet. In the dry state, the actual color of the white filler is difficult to distinguish because of the enhanced scattering of the pigment in air. Visual screening in the dry state is wholly inadequate, and even instrumental measurements have questionable value. In a paste or compounded formulation, light scattering is minimized or eliminated and color differentiation is significantly enhanced. In this case, visual or instrumental assessments can easily be made and comparisons drawn among fillers of the same or closely similar refractive index.

Where the color of fillers of substantially different refractive indices must be compared, it is best to do the comparison in a formulation designed to eliminate the inherent scattering power differences of the fillers. This can be accomplished by using relatively high concentrations of the fillers (approximately 100 phr) to enhance their color contributions while adding a small amount (approximately 0.2 phr) of titanium dioxide to completely opacify the formulation and mask any filler scattering power differences. Such a technique can prove useful, for example, in comparing a silica (refractive index = 1.5) or talc (1.57) filler with a ground limestone (1.62), where the goal of the measurement is to distinguish the true color differences of the fillers apart from scattering. In nonopaque films, the color intensity is diluted by the whiteness effect of scattered light compared to the transparent film.

B. Refractive Index

From the historical perspective of unfilled systems, the present-day formulator is often looking for filled systems that perform and look as if they were unfilled.

To achieve this goal, the filler must be as transparent as possible in the polymer and display no inherent color. Both of these properties can be well characterized according to the Kubelka–Munk theory [37,38], which relates all optical properties of pigments to the fundamental processes of absorption (K) and scattering (S) of light.

The transparency of an insoluble filler in a homogeneous binder depends on the complete absence of light scattering in the heterogeneous filled system. The ability of a white pigment to scatter light (M) depends on several factors, but, by, far the most important is the index of refraction, as indicated in the following Lorentz–Lorenz equation [39,40]:

$$M = \frac{(n_p/n_0)^2 - 1}{(n_p/n_0)^2 + 1}$$

where n_p is the index of refraction of the pigment and n_0 is the index of refraction of the surrounding medium. When $n_p = n_0$ and $M = 0$, there is no scattering and a perfectly transparent compound results. With the refractive index of PVC, for example, at 1.55 and most phthalate plasticizers in the range 1.48–1.50, a typical formulation with 100 phr DOP would have a refractive index of about 1.53. Most silica fillers with refractive indices ranging from 1.48 to 1.55 would be quite transparent in such a system.

As the refractive index of the filler varies progressively from 1.53, the optical transparency of such a filled system progressively decreases. Fillers with refractive indices in the 1.57–1.65 range, including talc (1.57), clay (1.57), mica (1.59), calcium silicate (1.59), calcined clay (1.62) and calcium carbonate (1.65), generally produce translucent PVC compositions ($M = 0.0174$ to 0.0255). Fillers with refractive indices above 1.7 are generally considered as hiding pigments and are added specifically for their opacifying power. Totally opaque vinyl systems will generally include from 0.5 to 10 phr of ZnO (R.I.) $= 2.01$ or rutile TiO_2 (R.I. $= 2.76$). As with any such system, the degree of transparency or opacity will also be affected by the concentration of the filler and the film thickness. With knowledge of the refractive index, the above can be generalized to all polymers.

In addition to the index of refraction, particle size is also an important variable affecting the scattering power of a filler. Mitton [39,40] has offered the following empirical relationship of scattering power (S) to particle size (d) for white pigments, where λ is the wavelength of light and M and n_0 are as defined earlier:

$$S = \frac{\alpha M^3 \sqrt{\lambda}}{(\lambda^2/2d) + n_0^2 \pi^2 M^2 d}$$

Plots of S versus d for TiO_2 and $CaCO_3$ are given in Fig. 9 for $\lambda = 560$

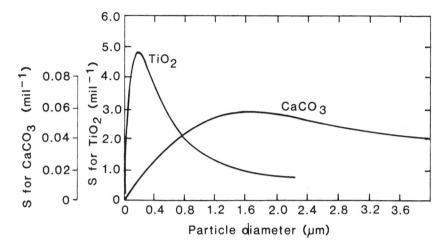

Figure 9 Empirical plot of scattering coefficient (S) versus particle size (PS) using a green filter (560 nm) for TiO_2 and $CaCO_3$ pigments.

nm. Note the differing optimum particle sizes depending on the different refractive indices for the two pigments. The optimum size is also affected by changes in the refractive index of the surrounding medium (n_0). Finally, the scattering power of all pigments progressively decreases as the particle size falls below approximately 0.1 μm. This size is so much smaller than the wavelength range of visible light that the scattering interaction is essentially lost. Thus, whereas a compound filled with a 1-μm calcium carbonate might be translucent, the same formulation filled with a 0.07-μm calcium carbonate could be quite transparent.

When a ductile plastic is bent or otherwise stressed, an obvious whitening generally occurs along the fold. This stress whitening results from a discontinuity in refractive index. The mechanical stressing force produces a partial phase separation, producing small voids that are now filled with air having a refractive index of 1.0. These air–resin and air–filler interfaces refract and scatter light very efficiently because of the large differences in refractive index at the interface. The ease and extent to which stress whitening occurs generally increase in filled systems and are related to poor filler wetting and/or weak filler bonding to the polymer–plasticizer network.

C. Particle Size, Size Distribution, and Shape

Particle size, size distribution, and shape of fillers are key variables affecting performance [41–44]. They are also key specification properties in the manufacture of most fillers because of their important contribution to overall production costs. These filler properties strongly influence the mechanical properties (tensile strength, impact strength, modulus, and hardness), dimensional properties (shrinkage and creep), surface gloss, fire resistance, permeability, and air entrapment. Rheology and resulting processibility are also primarily controlled by these variables, whether the system is a plastisol or melt processed via calendering, extrusion, or injection molding. These properties play such a central role in the ultimate properties of the formulation that most filler producers will offer the same filler chemistry in several particle sizes, size distributions, and even particle shapes to give the formulator the greatest latitude in developing the desired properties in the ultimate product.

Although the selection of fillers is quite large in terms of these properties, some generalizations can be drawn to help the formulator. For mineral fillers of natural origin, particle shape is largely predetermined. Limestone, silicas, and other minerals ground to a specific size display an irregular particle shape characterized by the fracture surfaces resulting from the grinding process. Particle size distributions are largely determined by the milling and classification equipment used and are thus relatively similar from product to product. Clay and talc products are generally platy in shape with greater variations in shape and distribution based on ore source and process variations. In general, particle size distributions are broad, a characteristic that typifies the natural products.

For the synthetic minerals and pigments that have been made available, much wider varieties of particle shapes and sizes are possible. In general, the synthetic products display smoother, more regular crystal surfaces and much narrower particle size distributions. These characteristics generally offer property advantages over the less costly natural products. Synthesis technology must generally allow optimization of particle shape, size, and distribution for a particular filler function (opacity, impact strength, gloss, etc.) in order to provide a property advantage or benefit commensurate with the added cost of the filler.

The importance of size, shape, and size distribution measurements to the filler manufacturer and user has resulted in the development of numerous measurement techniques and equipment. A general list of these techniques with appropriate references is provided in Table 8. Excellent books and references are also available covering this important area in greater detail [45–47]. Familiarization with the general techniques employed is necessary for the filler user to develop accurate comparisons of competitive materials.

In general, sieve analyses are relied on for particle analyses down to 44 μm (325 mesh). The methods are simple, fast, and convenient. From 44 to about 1 μm, sedimentation methods are most commonly employed with some use of microscopy, often as a reference technique. From 1 to about 0.1 μm, centrifugal sedimentation and electron microscopy techniques are commonly used. Below 0.1 μm, electron microscopy was the only available method until the recent

Table 8 Types of Particle Size Analysis Technique

	Applicable size range (μm)	Method or reference
Screen analysis	44	ASTM D-1921-63
		ASTM C-92-46
Elutriation		
Air	100–5	50, 51, ASTM B293-60
Liquid	100–5	52
Gravitational sedimentation		
Pipette	50–2	53
Balance	50–2	54
Micromerograph	250–1	55
Divers	30–2	56, ASTM D422-39
Turbidimetric (extinction)	150–2.5	57
Centrifugal sedimentation		
Beaker centrifuge	3–0.05	58
Centrifuge pipette	2–0.1	59
Microscopy		
Visible	100–0.5	60, ASTM E20-62T, D1366-65
UV	100–0.2	60
Electron	5–0.005	61

release of equipment utilizing dynamic laser-light scattering to provide particle size measurements from approximately 3 to 0.003 μm [48,49]. It is important to note that sieve and sedimentation analyses provide particle statistics based on weight distributions, whereas microscopy and the optical or laser-light-scattering techniques provide number distributions. Care must be taken that proper comparisons are made based on the technique and type of data generated.

D. Surface Area and Porosity

In addition to particle size, size distribution, and shape, surface area and porosity of the filler are key factors affecting the rheology and, ultimately, the mechanical properties and performance of the filled system. Specific surface area (SSA) is the term most commonly used for powders and is defined as the area on a molecular scale that is exposed to a liquid or a gas by 1 g of the powder (common units are meters squared per gram. Porosity in this case is defined as the collection of surface flaws that are deeper than they are wide and can be further characterized in terms of pore size, pore volume, pore area, and pore shape.

Table 9   Relation Between Specific Surface Area and Particle Diameter

Diameter	0.1 μm	1 μm	10 μm	100 μm	1 mm
SSA	60 m^2	6 m^2	6000 cm^2	600 cm^2	60 cm^2

The specific surface area is clearly related to particle size and can be calculated for a host of particle geometries, assuming nonporous surfaces and well-defined size distributions. In the simple case of monodisperse spheres.

$$\text{SSA} = \frac{6}{dP}$$

where SSA is the specific surface area, d is the particle diameter, and P is the true density of the powder. Herden [62] used this relationship to tabulate specific surface area as a function of particle diameter, as shown in Table 9.

This exercise is instructive in that it provides a general but useful frame of reference for interpreting surface area results. It also illustrates the sensitivity of surface area measurements to the fraction of fine particles in a powder sample. It should be noted that surface areas calculated from particle size data will at best establish the lower limits for the sample. True SSA values are often higher by a factor of 1000 or more based on variations in particle shape, surface irregularities, and porosity that always exist in the real filler sample.

The surface area can be determined by several methods based on adsorption [63], photoextinction analysis [64,65], and permeability [66,67]. Because of its simplicity and broad applicability, however, the Brunauer, Emmet, and Teller theory employing gas adsorption is almost universally used for surface area measurements today [68,69]. The equipment allowing full characterization of the gas desorption processes also provides the necessary data for characterization of the sample porosity in terms of pore size, pore volume, pore area, and pore shape [70].

For the formulator, surface area data on the filler(s) represent an important complement to particle size data in estimating reasonable loading levels for rheology and processibility of the filled system. Whereas particle size data on a mass basis tend to overemphasize the coarse fraction of the filler and provide no measure of porosity, surface area data compensate for these shortcomings. The entire available filler surface must be properly wet for optimum properties and performance of the filled system. Because some fillers are capable of adsorbing and inactivating many formulation additives, such as plasticizers and liquid stabilizers and lubricants, increased additive levels are generally called for, based on the total filler surface area incorporated in the formulation. In a similar way, porosity (pore volume, size, area, shape) information is an essential complement to surface area in that it describes that part of the total surface (as determined by the adsorption of nitrogen) which may not be available to the larger polymer molecules. It is surface that should not be overlooked, however, because it can provide adsorption sites for smaller organic molecules (plasticizers, stabilizers, inhibitors, antioxidants), resulting in severe performance problems in the ultimate formulation [71].

E. Surface Treatment

It is rare today to find a line of filler products that does not include several grades that are surface modified for particular end-use properties or applications. In altering the filler surface chemistry, the energy and degree of interaction at the polymer–filler interface can be changed significantly, leading to surprising enhancement of mechanical and/or dimensional properties in filled vinyl formulations.

Surface treatment of a mineral filler can provide a host of useful benefits in the ultimate formulation. The uncoated mineral surface, which is generally hydrophilic, can be made hydrophobic, thereby increasing its compatibility, dispersibility, and processibility. The coated surface is also less likely to adsorb and deactivate other formulation additives such as plasticizers, heat stabilizers, and/ or antioxidants that are required for optimum performance. Finally, the degree of filler–polymer loading or adhesion can be increased through the use of an appropriate coupling agent, thereby providing functional reinforcement from an otherwise nonfunctional filler. In this regard, silicone and silane coupling agents have been widely used on the various silica and silicate fillers with reasonable success [72,73]. Their effectiveness with calcium carbonate has been somewhat limited. Fatty acids have generally proved to be effective dispersing/coupling agents for calcium carbonate fillers [74–76]. As always, the formulator should carefully evaluate the cost/benefit performance of the coated versus uncoated filler. The high cost of many potential coupling agents has limited the commercial viability of many coated fillers.

F. Dispersibility

Proper dispersion is essential to the performance and ultimate success of the filled system (see Figs. 10 and 11). Responsibility in this important area is shared by the filler manufacturer and the formulation compounder. Lack of performance on the part of either contributor will almost surely result in inferior performance of the filled formulation.

Dispersibility of a filler is established primarily by (1) the chemical nature of the filler, (2) its surface characteristics, (3) its particle size, and (4) the process by which it is made. The chemical nature of the filler defines the bond strengths

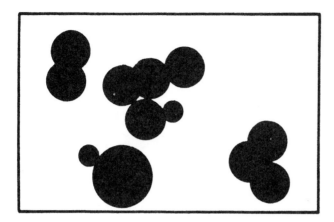

Figure 10   Agglomerated filler in matrix.

and many of the special surface characteristics that will be encountered in any given material. Filler manufacturers have generally gone to great lengths to define the chemical and surface characteristics of their materials. The physical and chemical properties of the filler surface will establish its tendency to aggregate and the strength of the clusters formed.

Particle size of the filler is also important to its ease of dispersion. In general, the surface energies of particles greater than 1 μm are such that the

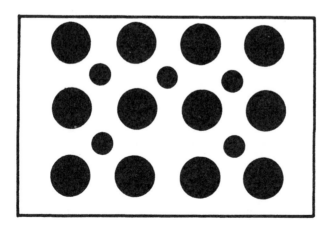

Figure 11   Filler dispersed to discrete size in matrix.

driving force for aggregation is minimal and the aggregates that do form are generally weak. As the particle size drops below 1 μm, and particularly as it enters the colloidal region below 0.1 μm, the tendency of the particles to aggregate in large clusters increases progressively and substantially. In this region, protective colloids or surfactants must also be employed to stabilize the particles from reagglomerating once the initial aggregates are broken during the dispersion process.

Finally, the closely guarded processes by which the fillers are produced have an all-important effect on the ultimate dispersibility of the product. It is in this area that substantial effort is expended to maintain the optimum filler performance at a minimum manufacturing cost. An incredibly extensive patent literature has grown out of an almost universal need by filler manufacturers to improve the ease of dispersion of their products [77]. Favorite techniques include (1) the use of various surface treatments to change the surface chemistry and reduce the forces of particle–particle attraction and bonding and (2) methods to circumvent the drying process or reduce the forces of agglomeration that occur during drying. Clearly, techniques to improve pigment dispersibility will continue to receive a great deal of attention from filler manufacturers and users.

In compounding, three basic dispersion processes are commonly used: dry blending, melt shear, and liquid dispersion [78,79]. In practice, a combination of these techniques is often employed, although there are few hard and fast rules. The experience of the compounder is still the primary factor determining the dispersion processes used. The dry-blending technique achieves dispersion primarily by impact and attrition grinding. Unless followed by another dispersion process, there is little opportunity for effective wetting of the pigments by the resin. Melt shear dispersion is most commonly employed in thermoplastics compounding and works well in follow-up combination with dry blending. The effectiveness of melt shear depends on the ease of wetting of the filler by the polymer and a high enough viscosity to allow high shear forces to be effectively transmitted to the filler aggregates. Liquid dispersion processes for plastisols involve predispersion of the filler in the plasticizer. Equipment type can vary, but the type of dispersion forces is generally determined by viscosity, which is most often controlled by the filler concentration. The choice of dispersion technique usually is dictated by cost, the ultimate property requirements of the filled composite, and, finally, the dispersibility of the fillers required to provide those properties.

G. Abrasion and Hardness

The abrasivity of mineral fillers is generally recognized to depend on three key factors: (1) filler hardness, (2) particle size, and (3) particle shape.

When the filler particle contacts a softer surface (the surface being abraded), the extent of the damage is controlled by the difference in hardness, the energy of the interaction (which involves the mass and acceleration of the particle), and, finally, the efficiency of the interaction (which involves the shape of the particle, including the existence of sharp points and edges).

The hardness of mineral fillers is established by their comparative ratings on the Mohs scale. The higher the Mohs hardness value, the harder the mineral. The Mohs hardness scale is defined as follows:

1. Talc	6. Feldspar
2. Gypsum	7. Quartz
3. Calcite	8. Topaz
4. Fluorite	9. Corundum
5. Apatite	10. Diamond

The Mohs hardness is useful in establishing a general expectation of abrasivity for a synthetic or natural mineral product of high purity. When dealing with ground natural fillers, however, it is important to assess the mineral purity. For soft minerals like talc, the abrasivity of the product may be established primarily by the level of impurities of higher hardness. Abrasivity of mixtures increases disproportionately with the level of the more abrasive component.

Abrasion studies conducted on synthetic mineral filler samples that were essentially monodispersed but of different average size showed a generally linear increase of abrasion with size [80]. Similar studies with ground natural products have tended to confirm this abrasion–size relationship, but exceptional care must be taken to control the particle size distribution and purity of each sample [80]. The coarse particle fraction of the total size distribution disproportionately controls the abrasivity of the sample. In addition, any harder mineral impurities tend to be more difficult to grind and thus are concentrated in the coarse fraction.

Particle shape also tends to affect abrasivity, although the effect appears to be somewhat smaller than that of hardness or particle size. Clearly, spherical shapes and relatively smooth surfaces decrease the overall abrasivity of the product. Particles with sharp crystal points or edges appear capable or more abrasion damage per impact than other less aggressive shapes. Once again, the irregular particle shapes created in grinding a natural product tend to be more abrasive than particles of the same mineral composition and size produced synthetically. It is admittedly difficult, however, to eliminate the effects of purity and particle size distribution in these comparative assessments.

To the compounder, abrasion has two important effects. First, excessive wear on equipment, such as processing rolls, screws, cylinder walls, dies, molds, and mixers, has a decidedly negative economic impact on operations. Costs of equipment repair and replacement as well as lost production time should be considered when filler evaluation and choices are made. Unfortunately, wear is not related to fillers alone, and abrasion and wear are difficult to assess in tests of short duration. Second, excessive wear leads to fine metal contamination in the compound. This can be manifested as a discoloration or a more insidious premature heat-stability failure resulting from catalytic degradation influenced by the metallic contamination [81]. Clearly, abrasivity should be assessed early in the development of filled formulations.

V. PROCESSING WITH FILLERS

Filler materials, as they are purchased, are generally found in an agglomerated state. The degree of agglomeration is dependent on processing steps, filler size, surface energy, surface coating, and so forth. The energy required to deagglomerate these particles is directly proportional to the energies holding the particle agglomerates together.

In general, the small particle size fillers (below 1 μm average size), because of their high surface area, contribute to multiplied surface and electrostatic charges, causing high amounts of agglomeration. The presence of a monomolecular layer of a surface coating such as a fatty acid satisfies some of the surface energetics, resulting in soft agglomerates that are easier to break down by low-level mechanical energy.

The filler aggregates form weak contact points in the matrix. Therefore, a dispersion to discrete particle size resulting in complete wetting of filler with the polymer is essential. Usually, the presence of 2–3% agglomerates will cause measurable changes in impact strength of a filled PVC. This property change is further magnified on low-temperature impact strength testing.

The plastics industry uses single-screw, twin-screw, and internal batch mixers to incorporate fillers. Melt-mix compounding prior to final shaping becomes highly desirable when the particle size of filler is very small and the loading level is above 5 phr. Some recent work has shown that a medium-particle-size precipitated $CaCO_3$ (0.5 μm) can be dry blended up to 10 phr on a high-intensity mixer and injection molded without sacrificing the impact strength property [82]. In fact, injection-molding equipment manufacturers are currently working on screw designs for better compounding on the molder to eliminate added melt compounding cost.

Both the single-screw and twin-screw extruders are excellent means of compounding and simultaneously forming materials. In general, when varying the filler size, surface treatment, and loading levels, a study to balance the lubricant package and stabilizer level is essential. Overall, the fine-particle-size fillers raise torque, promoting early fusion, with some reduction in compound stability. Such studies can be made on a torque rheometer and can be magnified by an

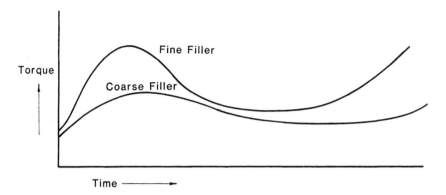

Figure 12   Effect of filler size on processing properties.

increase in filler loading as well as the rotor speed for each formulation before use (Fig. 12).

The calcium stearate coating formed on the $CaCO_3$ basically acts as an external lubricant and a costabilizer, requiring an overall reduction in external lubricant for optimum processing. For large runs, the production output of formulations containing coated (lubricating) and uncoated fillers can be optimized by designing a 2^3 factorial experiment to study the effect of key parameters (see Fig. 13).

The extrusion index for single- or twin-screw extruders can be calculated as follows:

$$\text{Single screw} = \frac{(\text{Output})^2}{(\text{Amps})(\text{Melt pressure})} \times 100$$

$$\text{Twin screw} = \frac{(\text{Output})^2}{(\text{Torque})(\text{Thrust})} \times 100$$

Several other studies [83,84] have examined the processing efficiency of filled systems which become necessary from an economic point of view.

VI.   MECHANICS OF FILLED POLYMERS

The composite properties of filled compounds are primarily governed by the following filler factors: (1) particle size, (2) morphology, and (3) interfacial adhesion with polymer.

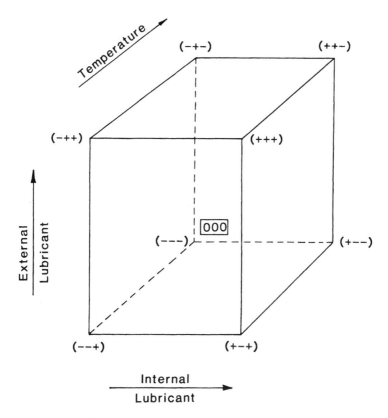

Figure 13 2^3 factorial design for extrusion performance study.

Reduction in filler size leads to a decrease in the thickness of the polymer interlayer and, consequently, an increase in the proportion of polymer at the filler boundary. This has a definite effect on mechanical properties because of the increased surface contact with the polymer. The particle morphology—platy, fibrous, spherical, and so forth—has a major effect on the processing and reinforcement of composites. Interfacial adhesion affects stress–strain behavior, thus affecting tensile properties.

When properly selected and compounded, there are significant benefits to be obtained by using particulate fillers. The key advantages are as follows:

1. Rheology and particle packing
2. Improved dimensional stability
3. Increased stiffness (modulus)

4. Increased toughness (impact strength)
5. Improved electrical properties
6. Reduced cost

Most of these factors have been studied in detail in special polymer composites. A discussion of some mathematical treatments of these factors now follows.

A. Plastisol Rheology

Besides improving key polymer properties such as tear resistance, hardness, and creep resistance, the addition of appropriate filler material to plastisols and organosols improves thixotropy, with the added possibility of cost reduction.

1. Thixotropy

Many studies have been done to relate the theoretical description of the rheology of dispersed systems, beginning with Einstein's equation [85]

$$\eta_s = \eta_0 (1 + 2.5\phi)$$

In this equation, η_s is the viscosity of the suspension, η_0 is the viscosity of the liquid medium, and ϕ is the volume fraction of suspended particles. This equation is based on the assumptions that (1) all the filler particles are smooth-surfaced spheres with low concentration and (2) there is no particle–particle interaction. Unfortunately, most plastisol and organosol filler materials have nonuniform shapes, sizes, and surface characteristics and the tendency is to load these fillers much above the range of Einstein's equation. The complexities of real-world systems notwithstanding, significant work has been reported explaining the rheological behavior of particle suspensions since Einstein.

A generalized Mooney equation [86] appears to give an excellent fit:

$$\ln(\eta_r) = \frac{K\phi}{1 - s\phi}$$

In this equation $\eta_r = \eta_s/\eta_0$ is the relative viscosity, ϕ is the volume fraction of filler, K is an adjustable parameter related to the size and shape of particles, and s is an adjustable parameter related to the space-filling properties of the particles. Thus, a practical user of filler would like to have ϕ as high as possible and η_r as low as possible [87].

2. Particle Packing

The definitive analysis of the packing of monodisperse spheres was done by Graton and Fraser [88]. They found that there are six possible ordered packing

Table 10 Packing of Spheres

	Points of contact	Volume of unit cell	Volume of unit void	Voids (%)
Cubic	6	$8.00r^3$	$3.81r^3$	47.6
Orthorhombic (two orientations)	8	$6.93r^3$	$2.74r^3$	39.5
Tetragonal	10	$6.00r^3$	$1.81r^3$	30.2
Rhombohedral (two orientations)	12	$5.66r^3$	$1.47r^3$	26.0

arrangements and, associated with each, a characteristic void content. These results are summarized in Table 10 [88].

A classical empirical study on the packing of spheres was done by McGeary [89]. He found that in random packing, a roughly orthorhombic arrangement having a void content of 37.5% predominates. By studying binary, ternary, and quarternary mixtures, he was able to derive relationships related to relative sizes and amounts of components and the observed packing density. His results are shown in Table 11.

For binary systems (see Fig. 14), McGeary found that the packing density improved as the ratio of d_1 to d_s tended to infinity. The maximum density achievable was 86% of the theoretical maximum. Packing density improved rapidly as d_1/d_s increased from 1 to about 7 and then changed abruptly and increased gradually as d_1/d_s ranged from 7 to 100.

For three spheres in contact, as shown below, the largest diameter sphere which will just fit through the triangular pore is $0.154d$ or $d_1/d_s = 6.5$. Thus, if d_1/d_s is less than 6.5, the smaller spheres will be able to penetrate the interstitial area of the packing, but if d_1/d_s is greater than 6.5, infiltration of the smaller spheres may occur. McGeary's results indicate that very efficient packing can be obtained by a mixture of only four or five components.

Table 11 Packing Density for Mixtures of Spheres

Component	d Ratio	X_1	X_2	X_3	X_4	% Theoretical
1	1	1.00	—	—	—	60.5
2	7	0.726	0.274	—	—	86
3	38	0.647	0.244	0.109	—	94
4	316	0.607	0.230	0.102	0.061	97.5

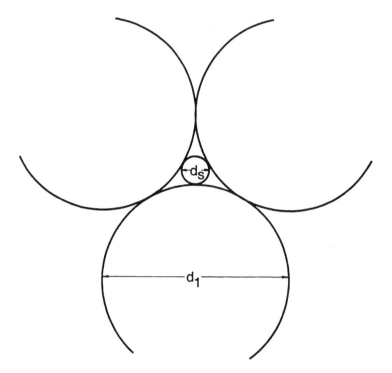

Figure 14 Binary particle packing.

For nonspherical fillers, the following equation has been used:

$$\eta = \eta_0 \exp\left(\frac{\alpha F}{1/f - KF}\right)$$

where α and K are shape factors of particles ($\alpha = 10.5$ to 24.8 and $K = 1.35$ to 1.90) and f is a magnitude determined by the ratio of thickness of the surface layer on the particle and its size.

Milewski [90] has studied the particle packing of more complex systems such as spheres–flakes, spheres–fibers, and so forth. Because of the excessive costs involved in selecting appropriate lots for blending of materials for optimum packing, such systems have been mostly justified only in specialized applications thus far.

B. Moduli of Filled Systems

One of the most dramatic effects of adding a filler to the polymer matrix is seen by an increase in stiffness (modulus) of the composite. Smallwood [91], on the basis of a hydrodynamic concept, proposed the following equation for calculation of the elastic modulus of filled composites, which is analogous to Einstein's equation for viscosity:

$$E_{\text{fill}} = E_{\text{unfill}} (1 + 2.5F)$$

where E_{fill} and E_{unfill} are the elastic moduli of filled and unfilled composites, respectively, and F is the volume proportion of the filler. This equation has been found suitable only for low concentrations of filler in the composite.

Kerner's [92] equation, given as follows, is well recognized for calculation of this property in filled polymers:

$$\frac{E_{\text{fill}}}{E_{\text{unfill}}} = 1 + \left(\frac{ABF}{1 - BF} \right)$$

where $A = (7 - 5v)/(8 - 10v)$, $B = [(E_{\text{fill}}/E_{\text{unfill}}) - (E_{\text{fill}}/E_{\text{unfill}} + A)$, and E_{fill} is the elastic modulus of the filler. A further modification of this equation is

$$\frac{E_{\text{fill}}}{E_{\text{unfill}}} = \frac{1 + ABF}{1 - B\psi F}$$

where ψ is a function depending on the maximum degree of filling F_m:

$$\psi F = \left(1 + \frac{1 - F_m}{F_m^2} F \right) F$$

or

$$\psi F = 1 - \exp\left(- \frac{F}{1 - F/F_m} \right)$$

The versatile Halpin–Tsai equation for predicting the composite modulus is based on micromechanics theory and is fully discussed by others [93–95]. For a composite containing a flake or platelet oriented parallel to the stress direction, the modulus enhancement ratio is given by

$$\frac{\overline{E}}{E_m} = \frac{1 + \zeta \eta v_f}{1 - \eta v_f} \tag{1}$$

where

$$\eta = \frac{E_f/e_m - 1}{E_f/E_m + \zeta} \tag{2}$$

and $\zeta = 2(d/t)$ or $2 \times$ (aspect ratio).

The quantities are identified as follows:

$\bar{E}$ tensile modulus of composite
E_m corresponding tensile modulus of unfilled polymer (matrix)
E_f corresponding tensile modulus of filler
v_f volume fraction of filler in composite
ζ geometric factor which takes into account the filler aspect ratio
d average equivalent diameter of platelets
t average thickness of platelets

For a composite containing spherical particles (a particulate-filled system). Eqs. (1) and (2) are still applicable, where the aspect ratio now becomes equal to 1 and $\zeta = 2$.

1. Prediction of Composite Moduli for Spherical Fillers

Work reported by Radosta [96] basically shows that all that is needed to predict the composite modulus from Eqs. (1) and (2) for a calcium carbonate-filled polymer are the corresponding moduli for the unfilled polymer and the filler and the volume fraction of filler used. The actual polymer used for the matrix phase and the test temperature at which the values were obtained do not enter into the calculations but are taken into account by the value of the modulus used for the unfilled polymer. (Note: It is assumed that the modulus of a mineral filler does not change appreciably over the useful temperature range of most thermoplastics.) Therefore, if one were to vary the polymer matrix modulus either by using different polymers as the matrix phase or by varying the test temperature, and if the data were plotted as the modulus enhancement ratio versus the matrix modulus, then the data points should lie on one continuous curve. This is indeed the case, as seen in Fig. 15. By using the data from the previous section, the matrix modulus has been varied by changing both the test temperature and the matrix polymer (polypropylene and PVC). The solid line in Fig. 15 represents the curve calculated using Eqs. (1) and (2) and a value of 5×10^6 psi for the elastic modulus of calcium carbonate. The agreement between the experimental data and the calculated curve is excellent. (Note that although the Halpin–Tsai equations were derived for tensile modulus, it has been shown that the modulus enhancement ratios for flexural modulus are equivalent to the ratios for tensile modulus, and either can be used as long as one is consistent.)

2. Prediction of Composite Moduli for Platy Fillers

Because of the platy nature of some fillers, the additional parameter of aspect ratio is needed to predict the composite modulus of these materials. If all the platelets are aligned parallel to the stress direction, we can use Eqs. (1) and (2). However, examinations of sections obtained from injection-molded specimens

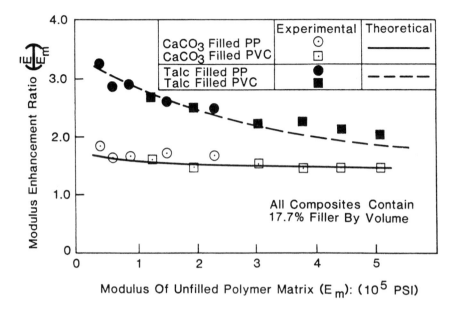

Figure 15 Comparison of experimental modulus data with the theoretical curves from the Halpin–Tsai equations.

show that the platelets are not perfectly aligned parallel to the mold surfaces, but are more randomly arranged throughout the specimen. At present, a good predictive equation for random three-dimensional orientation of platelets has not been established.

For the purposes of this discussion, a constant K is introduced into Eq. (1) as a scale factor to adjust for the decrease in modulus due to deviations of the platelets from perfect parallel alignment. The constant K in Eq. (3) below is expected to vary between 1 (for perfect parallel alignment) and 3 (for a completely random arrangement).

$$\frac{\overline{E}}{E_m} = \frac{1 + \zeta\eta v_f}{K(1 - \eta v_f)} + \frac{K - 1}{K} \tag{3}$$

As seen in Fig. 15, the data points for the platy composites all lie on one continuous curve. As with the spherical filler data, the matrix modulus has been varied by changing both the test temperature and the matrix polymer. The dashed curve in Fig. 15 represents the curve calculated by using Eqs. (3) and (2) with values of 2.5×10^7 psi for the elastic modulus of the platy filler, 20 for the average aspect ratio of the platy filler, and 3.0 for the scale factor K. Although there is good agreement between the experimental data and the calculated curve, a theoretical equation derived to accommodate three-dimensional orientation of platelets would give a better fit and would not require the empirical scale factor K used here.

C. Stress–Strain Properties

The addition of fillers raises the flexural modulus, as discussed previously. Generally, such an increase in modulus is associated with some reduction in tensile strength and elongation. The addition of finer spherical fillers tends to contribute good flexural modulus and improved tensile strength, with some reduction in elongation. This effect is much more dominant when platy materials such as talc and clays are used.

The stearate-treated small-particle-size spherical fillers (e.g., precipitated calcium carbonate), which exhibit low adhesion (coupling) with the polymer matrix, enable the compound to retain greater tensile strength and elongation up to much higher loading levels. The filled composites show a clear yield point. This is believed to be due to destruction in adhesion of filler with the polymer as well as filler–filler agglomerates. This is described as "dewetting."

The elongation of a filled composite can be described by

$$\epsilon_B = \epsilon_B{}^0 (1 - \phi^{1/3})$$

where ϵ_B is elongation at break for a filled compound, $\epsilon_B{}^0$ is elongation at break for an unfilled polymer, and ϕ is the volume fraction of the filler. The change in elongation and cross-sectional area has been expressed in terms of Poisson's ratio:

$$V = \frac{C/C_0}{L/L_0}$$

where V is Poisson's ratio, C_0 the compound initial cross-sectional area, and L_0 the original length. Because the filled polymer increases in cross-sectional area, Poisson's ratio is less than 0.5.

The measurement of impact strength at various strain rates also provides very useful performance information. For a filler exhibiting perfect adhesion with the polymer matrix and following Hookean behavior, the impact strength is described as

$$I_s = \frac{E\epsilon_B^2}{2}$$

where I_S is the impact strength, E is Young's modulus, and ϵ_B is the elongation at break.

Work reported by Mathur and Driscoll [4] and Radosta [5] has clearly demonstrated that fillers with smaller particle sizes improve impact resistance dramatically. The surface treatment with fatty acid further improves the ductile–brittle transition.

The optimum particle size range appears to be about 0.05 μm. Fillers much smaller than 0.05 μm form tightly held agglomerates and are very difficult to disperse, thus offering no further advantage in terms of impact strength. It is hypothesized that high-surface-area coated fillers, when subjected to impact, go through "dewetting." exhibiting ductility up to much higher loadings and lower temperatures.

D. Diffusion in Filled Polymers

The addition of filler materials affects the permeability and diffusion in compounds. Nielsen's [97] basic model essentially assumes that diffusing materials have to go around the filler particle. Therefore, platy fillers aligned with the surface render a more lengthy path. The addition of platy talc and clays to wire and cable compounds essentially improves the insulating efficiency due to these principles.

In general form, the diffusion of materials through a filled compound can be described by

$$\frac{P_{\text{fill}}}{P_{\text{unfill}}} = \frac{V_{\text{pol}}}{t}$$

where P_{fill} and P_{unfill} are permeability of filled and unfilled compounds, respectively, V_{pol} is the volume fraction of the polymer, and t is tortuosity defined as the extended distance material has to travel to go through the film [92]. This equation assumes that all the filler platelets are oriented parallel to the surface.

Nielsen has further modified the equation to include the particle shape and orientation in the matrix:

$$\frac{P_{\text{fill}}}{P_{\text{unfill}}} = \frac{P_M}{P_{\text{unfill}}F^{\eta} + P_M(1 - F^{\eta})} \frac{V_1}{t} + \frac{V_{\text{pol}} + V_2}{t}$$

where η is a constant varying from 0 to 1, characterizing the average specific diffusion path. This constant depends on the particle shape and orientation.

1. Cost Reduction

In addition to modifying various properties as previously discussed, the selection of appropriate fillers is expected to lower the formulation cost.

Table 12   Calculation of Pound-Volume Cost of a CaCO₃-Filled Compound[a]

Ingredients	Formula weight	Specific gravity	Volume	Hypothetical cost/lb	Pound cost
Resin	100.0	1.40	71.43	$0.35	$35.00
Lubricant A	1.2	0.85	1.41	$0.98	1.18
Lubricant B	0.4	0.82	0.48	$0.80	0.32
Stabilizer	2.0	1.02	1.96	$1.85	3.70
Stearic acid	0.5	0.85	0.58	$0.44	0.22
CaCO₃	12.0	2.71	4.43	$0.07	0.84
	116.1		80.29		$41.26

[a] Calculation:

$$\text{Cost per pound} = \frac{41.26}{116.1} = \$0.355$$

$$\text{Compound specific gravity} = \frac{116.1}{80.29} = 1.446$$

$$\text{Pound-volume cost} = \frac{\$0.355}{1.446} = \$0.245$$

Because fillers are sold on a per-pound basis, their economic value comes from the volume of resin they replace. Therefore, it is essential to compare them on the basis of unit volume of resin they will replace (see Table 12). The normal cost units are cents per cubic inch. In order to calculate this, the cost of each material in cents per pound and the specific gravity (sp.gr.) of fillers under consideration is necessary. The cost per cubic inch can then be calculated from the following relationship, where 0.0361 is the specific gravity conversion factor:

$$\text{in.}^3 \text{ Cents} = \text{lb cents (sp. gr.) (0.0361)}$$

VII.   ANALYTICAL METHODS AND FILLER QUALITY CONTROL

A.   Chemical Properties

The key chemical properties of interest to a compounder include the following:

Material purit
Percent acid insolubles
Iron or other metallic impurities
Water solubility
pH (water slurry)

The mineral purity will vary from mine to mine. The presence of high amounts of foreign materials in the ore can cause multiple problems. High silica levels have been related to abrasivity of filler. High iron and other metallic impurities affect heat stability of the compound. The typical chemical analysis of a commercial-grade ground limestone is reported by its producer as follows:

$CaCO_3$	96.00%
$MgCO_3$	1.50%
SiO_2	1.20%
Al_2O_3	0.30%
Fe_2O_3	0.08%
H_2O	0.25%

B. Physical Properties

The physical properties of prime interest to compounders are as follows:

Particle size and distribution
Surface area
Dispersibility
Specific gravity
Bulk and tap densities
Dry brightness
Wet color
Oil absorption
Refractive index
Moisture content
Mohs hardness

These properties are tested by the ASTM tests indicated in Table 13.

Table 13 Physical Property Tests

Property	Test method
Particle size	ASTM D1366-55T
Surface area	ASTM E20-62F
Specific gravity	ASTM D153-54
Bulk density	ASTM D1895-65T
Oil absorption	ASTM D1483-60
Brightness—color	ASTM E313-67
Moisture content	ASTM C25-58

VIII. CURRENT USAGE

The following is a summary of currently reported usage of common fillers:

Alumina trihydrate	ABS, polyethylene (PE), PVC, EVA, epoxy, phenolic, polyester, urethanes
Barium sulfate	Polyesters, urethanes
Calcium carbonate	ABS, PE, polypropylene (PP), polystyrene (PS), PVC, polyamides, polyesters, epoxy, phenolics, urethanes
Calcium sulfate	PE, PVC, polyamides, polyesters
Carbon black	ABS, PE, PP, PS, PVC
Feldspar	Acrylics, cellulosics, PP, PS, epoxy, polyesters
Kaolin (clay)	PE, PP, PVC, polyamides, polyesters, urethanes
Metal fillers	ABS, PS, PVC, polyamides, epoxy, phenolics
Mica	ABS, PE, fluoroplastics, polyamides, PC, epoxy, polyesters, phenolics, polysulfone, urethanes
Organic fillers	Cellulosics, PVC, polyesters, epoxy, phenolics
Silica	Acrylics, PE, PS, PVC, polyesters, urethanes
Talc	PC, PE, PP, PS, PVC, phenolics, urethanes, polyamides
Wollastonite	PC, PE, PP, PS, polyamides, polyesters, epoxy, urethanes

REFERENCES

1. RF Conley. PM&AD—RETEC Proceedings, 1985, p 144.
2. L Dientenfass. Colloid Z 155:121–130, 1957.
3. A Standen (ed). Kirk-Othmer Encyclopedia of Chemical Technology. 2nd ed. New York: Interscience, 1964, pp 7–11.
4. KK Mathur, SB Driscoll. SPE Annual Technical Conference, pp 912–917.
5. JA Radosta. SPE Annual Technical Conference, 1979, pp 593–595.
6. KK Mathur, DB Vanderheiden. In: Proceedings of the ACS International Conference on Polymer Additives, (HF Mark, ed.). New York: Plenum, 1982, pp 371–389.
7. O Leuchs. 19th International Wire and Cable Symposium, 1970 p 239.
8. What Are "White Reinforcing Fillers"? NJ: Teterboro, Degussa Corp., 1980.
9. D Marvel. Silica Gel. Baltimore: W. R. Grace & Co., 1980.
10. Synthetic Silicas and Silicates for PVC. Tech. Bull. Pigment No. 51. Teterboro, NJ: Degussa Corp., 1978.
11. Basic Characteristics and Applications of Aerosil®. Tech. Bull. Pigment No. 11. Teterboro, NJ: Degussa Corp., 1980.

12. Aerosil® for PVC Plastisols. Tech. Bull. Pigment No. 41. Teterboro, NJ: Degussa Corporation, 1983.
13. JC Furnivall, AD Kupfer, JL Irvine. SPE Annu Tech Conf Proc 27:541–544, 1981.
14. EJ Augustyn, JM Schwarcz. SPE Annu Tech Conf Proc 23:202–207, 1977.
15. ED Nelson, SJ Kaufman. J Fire Flammabil 13:79–103, April 1982.
16. CY Wang. Antimony—Its Geology, Metallurgy, Industrial Uses and Economics. London: Charles Griffin & Co., 1952.
17. Use of Antimony Compounds as Fire and Flame Retardants. Rahway, NJ: M&T Corp., 1979.
18. VM Bhatnagar. In: Proceedings of 1975 International Symposium on Flammability and Fire Retardants, (VM Bhatnagar, ed.). Westport, CT: Technomic Press. 1975, pp 238–248.
19. R Hindersinn, GM Wagner. In: Encyclopedia of Polymer Science and Technology, (HF Mark and NG Gaylord, eds.). New York: Wiley—Interscience, 1967, p 7.
20. TC Patton. In: Pigment Handbook, Vol. I, (TC Patton, ed.). New York: Wiley–Interscience, 1973.
21. CH Mathewson. Zinc: The Metal and Its Alloys and Compounds. ACS Monogr. 142. New York: Reinhold, 1959.
22. LC Komar. Titanium dioxide in polymers. 26th SPE Annual Technology Conference 1968, p 303.
23. JB DeCoste, VT Wallder. Weathering of Polyvinyl Chloride-Farbe Lack, 77: 325–330, 1971.
24. E Herman. Kunst Plast 10, 1963.
25. DB Vanderheiden. Paint Varnish Prod 19–24, September 1974.
26. U.S. Patents 3,652,334 (1972) and 4,053,325 (1984).
27. KK Mathur, K Kramer. J Vinyl Technol 5(1):32–38, 1983.
28. A Standen (ed.). Kirk-Othmer Encyclopedia of Chemical Technology. 2nd ed. Interscience, New York, 1964, pp 243–282.
29. HS Katz, JV Milewsk (eds.). Handbook of Fillers and Reinforcements for Plastics. New York: Van Nostrand Reinhold, 1978.
30. M Grayson. (Exec. ed.). Kirk-Othmer Encyclopedia of Chemical Technology. 3rd ed. New York: Wiley-Interscience, 1978, p 643.
31. Chem Week 16, May 3, 1978.
32. U.S. Patents 4,260,534 (1979), 4,250,064 (1974), 4,242,397 (1979), 4,193,841 (1977), 4,138,521 (1976), 4,097,644 (1972), and 3,962,507 (1976).
33. PB Mitton. In: Pigment Handbook, Vol. III D-C. New York: Wiley, 1973.
34. VV Leoin, DA Danilova, LV Shoets. Bum Prom 4:15–16, 1972.
35. M Cremers. Polym Paint Colour J 852–862, November 3, 1976.
36. M Cremers. Polym Paint Colour J 936–941, December 1, 1976.
37. JT Atkins, FW Billmeyer. Color Eng 6(6):40, 1968.
38. DB Judd, G Wyszecki. Color in Business Science and Industry. 3rd ed. New York: Wiley, 1975.
39. PG Mitton, LW Vejnosk, M Frederick. Off Digest Fed Soc Paint Technol 33, 1961.
40. PG Mitton, LW Vejnosk, M Frederick. Off Digest Fed Soc Paint Technol, 34(444): 73–89, 1962.

41. T Nikaido. Japan Kokai 78:82,851 (1978).
42. HG Shanks. J Appl Polym Sci. 26(9):3099–3102, 1981.
43. KK Mathur, J Greenzweig, SB Driscoll. SPE Annu Technol Conf 24:732–736, 1978.
44. KK Mathur, SB Driscoll. J Vinyl Technol 4(2):81–86, 1982.
45. T Allen. Particle Size Measurement. London: Chapman & Hall, 1974.
46. BA Jarrett, H Heywood. Br J Appl Phys 3(Suppl):S21–S26, 1954.
47. PJ Lloyd, B Scarlett, J Sinclair. Particle Size Analysis Conference, 1970.
48. B Chu. Laser Light Scattering. New York: Academic Press, 1974.
49. Instrument companies: Leeds & Northrup Instruments (Microtrac), St. Petersburg, FL; Coulter Electronics Inc. (model N4), Hialeah, FL; Nicomp Instruments, Santa Barbara, CA.
50. PS Roller. Proc ASTM 32:608, 1932.
51. JM Dalla Valle. Micromeritics. New York: Pitman, 1948, p 72.
52. L Andrews. Industrial Engineering Symposium, 1947, p 114.
53. AH Andreasen. Kolloid. Beith 27:349–358, 1928.
54. S Odean. Soil Sci 19:1, 1925.
55. F Eadie, R Payne. Iron Age 174:99, 1954.
56. S Berg. Ingenioervidensk Skr 2, 1940.
57. E Sharatt, E Van Somersen, E Rollenson. J Soc Chem Ind 64:63, 1945.
58. S Martin. Ind Eng Chem Anal Ed 11:47, 1939.
59. HJ Komack. Anal Chem. 23:844, 1951.
60. J Green. Industrial Rheology and Rheological Structures. New York: Wiley, 1949.
61. G Riedel, H Ruska. Kolloid Z 96:86, 1941.
62. G Herden. Small Particle Statistics. 2nd ed. London: Butterworths, 1960.
63. S Brunauer, P Emmet, E Teller. J Am Chem Soc 60:309, 1938.
64. HE Rose. J Appl Chem 2:80, 1952.
65. ASTM Book of Standards, Part 2. New York: ASTM, 1944, pp. 47, C115–C142.
66. FM Lea, RW Nurse. J Soc Chem Ind 58:278, 1939.
67. PC Carman. J Soc Chem Ind 57:225, 1938.
68. S Lowell. Introduction to Powder Surface Area. New York: Wiley, 1979.
69. SJ Gregg, KSW Sing. Adsorption Surface Area and Porosity. 2nd ed. London: Academic Press, 1982.
70. Gas Adsorption Equipment. Norcross, GA: Micromeritics Instrument Corp., 1976.
71. KK Mathur, FE Witherell. SPE Annual Technology Conference 1980.
72. JG Marsden. J Appl Polym Sci 14:107–120, 1970.
73. LJ Friedman. SPE Tech. Paper 15:287, 1969.
74. A McCord. U.S. Patent 3,333,980 (1967).
75. Solvay Cie, Fr. Demande, 2,231,695 (1974).
76. T Shikata. Japan Kokai 75,02,754 (1975).
77. MH Gutcho (ed.). Inorganic Pigment Manufacturing Processes. Park Ridge, NJ: Noyes Data Corp., 1980.
78. DB Miller. Color Eng 6(4):46–51, 1968.
79. TB Reeve, WL Dills. SPE Annual Technology Conference, 1970; Preprint 16, pp 574–576.
80. KK Mathur, Pfizer Inc., unpublished work.

81. GL Levy. Wire J 8:39, 1971.
82. KK Mathur. Pfizer Inc., unpublished work.
83. GS Baronin, EV Minkin, TG Artemova. Deposited Doc. SPSTL 663, Khp-D81, 1981; available from SPSTL.
84. VYa Laukhin. Khim Neft Mashinostr 11:14–17, 1982.
85. A Einstein. Ann Phys 17:549, 1905; 18:289, 1906; 34:591, 1911.
86. M Mooney. J Colloid Sci 6:162, 1951.
87. R Simha, HL Frisch. In: FR Eirich, ed. Rheology, Vol 1. New York: Academic Press, 1956.
88. LC Graton, HJ Fraser. J Geol 43:785, 1935.
89. RK McGeary. J Am Ceram Soc 44(10):513, 1961.
90. JV Milewski. RP/C 1974, 29th Annual Conference, 1974.
91. H Smallwood. J Appl Phys 15:758, 1944.
92. RM Barrer, et al. J Polym Sci Part A 1:2565, 1963.
93. JE Aston, JC Halpin, PH Petit. Primer on Composite Materials: Analysis. Stamford, CT: Technomic Publications, 1969.
94. JL Kardos. Crit Rev Solid State Sci 3:419–450, August 1973.
95. JC Halpin, JL Kardos. J Appl Phys 43:2235, 1972.
96. JA Radosta. SPE Tech Paper 21:526, 1975.
97. LE Nielsen. J Macromol Sci Part A 1:929, 1967.

6
Flame Retardants and Smoke Suppressants

Joseph Green
Joseph Green Associates, Monroe Township, New Jersey

I. INTRODUCTION

The objective of flame-retarding thermoplastics is to increase the resistance of a material to ignition and, once ignited, to reduce the rate of flame spread (slow burning). The product does not become noncombustible, but use of a flame-retardant* additive may prevent a small fire from becoming a major catastrophe. The primary flame retardants used to accomplish this objective are halogen- and phosphorus-containing organic compounds. Antimony oxide is generally required as a synergist for halogen compounds. Inorganic compounds containing high concentrations of water of hydration, such as alumina trihydrate and magnesium hydroxide, are also used. The type of flame retardant and the quantity needed to meet specific objectives depend on the specific polymer. Additive as well as reactive flame retardants are available commercially. The latter type are used mainly in thermoset resins.

The addition of large quantities of a flame retardant may severely degrade the properties of the thermoplastic and also may present processing problems. Flame retardants that plasticize the polymer reduce thermal properties such as heat distortion temperature, whereas nonmelting solid additives may severely

---

* The terms "flame retardant," "self-extinguishing," and similar expressions, and the numerical flame-retardant ratings in this chapter are not intended to reflect hazards under actual fire conditions. They are descriptive of the specific small-scale laboratory tests only. Data presented should not be used to predict performance or to assess hazard under actual fire conditions.

173

degrade impact properties. Furthermore, many flame retardants have limited thermal stability and may place restrictions on processing temperature. Polymers containing high concentrations of filler may be difficult to process, and good dispersion may require special handling. The use of flame retardants in thermoplastics is therefore a compromise. The desired properties of a flame-retardant additive are high decomposition temperature (to avoid formation of corrosive gases), low volatility, minimal effect on mechanical properties such as resistance to impact, minimal effect on heat-distortion temperature, good ultraviolet-light resistance, absence of toxicity, and nontoxicity of combustion products.

This chapter reviews flame-retardant additives and flame-retardant technology for specific plastics, with emphasis on the effect of these additives on the mechanical properties and processibility of the plastic. The chemistry of commercial halogen, phosphorus, and inorganic flame retardants is discussed. Smoke suppressants are discussed. The mechanism of combustion and flame retardancy is discussed, and there is a brief description of laboratory flammability tests used by the industry to determine the relative effectiveness of various flame retardants.

II. FLAME-RETARDANT MECHANISM

The performance of halogens as flame retardants is rated as:

I > Br > Cl > F

aliphatic > alicyclic > aromatic

Iodine compounds, apparently the most effective, are not used in plastics because of their poor thermal stability. Fluorocarbons are inherently nonburning, but they do not impart flame retardancy to other plastics because either the carbon–fluorine bond is too stable or the fluoride radical that may form is highly reactive, so that hydrogen fluoride forms preferentially, reacts in the solid state and does not enter into the flame-quenching mechanism. Commercial halogen-containing flame retardants include aliphatic, alicyclic, and aromatic organic compounds containing chlorine or bromine. The choice of the type of bonding in the molecule will depend in large measure on the thermal stability required for processing of the polymer; aromatic halogen compounds are usually the most stable. Both additive and reactive compounds are available, but only the additive type is discussed.

It is generally agreed that the combustion of gaseous fuels proceeds via a free-radical mechanism [1,2]. A number of propagating and chain branching reactions are critical for maintaining the combustion process. Some of these reactions are as follows:

$$CH_4 + O_2 \rightarrow CH_3 + H + O_2$$
$$H + O_2 \rightleftharpoons OH + O$$
$$CH_4 + OH \rightarrow CH_3 + H_2O$$
$$CH_3 + O \rightarrow CH_2O + H$$
$$CH_2O + OH \rightarrow CHO + H_2O$$
$$CHO + O_2 \rightarrow H + CO + O_2$$
$$CO + OH \rightarrow CO_2 + H$$

Here, H, OH, and O radicals are chain carriers, and the reaction of H with O_2 is an example of chain branching in which the number of carriers is increased. The reaction of CO with OH converting CO to CO_2 is a particularly exothermic reaction.

In the radical trap theory of flame inhibition, it is believed that HBr competes in the above reactions for the radical species that are critical for flame propagation:

$$CH_4 + Br \rightarrow HBr + CH_3$$
$$H + HBr \rightarrow H_2 + Br$$
$$OH + HBr \rightarrow H_2O + Br$$
$$O + HBr \rightarrow OH + Br$$

The active chain carriers are replaced with the much less active Br radical, and, concomitantly, the burning velocity or flame strength is reduced.

Antimony oxide is known as a flame-retardant synergist when used in combination with halogen compounds. Volatile antimony oxychloride (SbOX) and/or antimony trihalide (SbX_3) are formed in the condensed phase and are transported into the gas phase (Table 1). [3]. It has been suggested that the antimony is also a highly active radical trap [4]. Halogen transports the otherwise nonvolatile

Table 1 Reactions of Antimony Oxide

$RCl + SB_2O_3$	$\longrightarrow$	$SbOCl(s)$
$5SbOCl(s)$	$\xrightarrow{245-280°C}$	$Sb_4O_5Cl_2(s) + SbCl_3(g)$
$4Sb_4O_4Cl_2(s)$	$\xrightarrow{410-475°C}$	$5Sb_3O_4Cl(s) + SbCl_3(g)$
$3Sb_3O_4Cl(s)$	$\xrightarrow{475-565°C}$	$4Sb_2O_3(s) + SbCl_3(g)$
$Sb_2O_3(s)$	$\xrightarrow{658°C}$	$Sb_2O_3(l)$

Source: Ref. 3.

antimony to the gas phase. Although both the halogen as HX and the antimony as SbO inhibit the flame, inhibition is due in larger part to the antimony, particularly when the halogen is chlorine [4]. It has also been suggested that the antimony oxide particulates function as a third body for radical termination.

Antimony oxide itself imparts no flame inhibition to plastics, but when used with halogen compounds, it significantly reduces the halogen level required to obtain a given degree of flame retardancy. This reduction is desirable and/or necessary because the required halogen concentration for many polymers is so high that the mechanical properties of the plastic would be drastically affected. Laboratory tests indicate the optimum halogen/antimony atom ratio to be about $3:1$, although it may vary with the polymer system. Other synergists include iron, boron, and tin compounds.

Interference with the antimony–halogen reaction will affect the flame-retardant properties of the polymer. For example, metal cations from color pigments or seemingly inert fillers such as calcium carbonate may lead to the formation of stable metal halides, rendering the halogen unavailable for reaction with the antimony oxide. The result is that neither the halogen nor the antimony is transported into the flame zone [5]. Silicone mold release has also been known to interfere with the flame-retardant mechanism. As a result, the total plastic composition must be considered in producing a flame-retardant plastic product.

Phosphorus-containing flame retardants include phosphate esters, phosphonates, phosphine oxides, chlorophosphates, chlorophosphonates, and inorganic phosphates. The commercial chloroaliphatic compounds are not sufficiently stable for thermoplastic processing (but used commercially in castable polyurethanes), and many of the inorganic compounds are too hygroscopic for thermoplastic application.

The flame-retardant mechanism for phosphorus compounds varies with the phosphorus compound, the polymer, and the combustion conditions [5]. For example, some phosphorus compounds decompose to phosphoric acids and polyphosphates. A surface glass or viscous melt forms and shields the polymer from the flame. This coating serves as a physical barrier to heat transfer from the flame to the polymer and to diffusion of gases (e.g., oxygen in and combustible pyrolysis gases out).

Triaryl phosphate esters are thermally stable, high-boiling ($>350°C$) materials and effective flame retardants for poly(vinyl chloride) (PVC), modified polyphenylene oxide (PPO), and cellulose acetate. In PVC, the phosphate ester replaces the more flammable organic esters such as dioctyl phthalate (DOP). In modified PPO, the preponderance of evidence indicates that the phosphate ester functions mainly in the gas phase [6] as a radical trap to help quench the flame. The phosphate ester volatilizes without significant decomposition into the flame zone, where it decomposes. Flame-inhibition reactions, similar to the halogen radical trap theory, have been proposed [7]:

$$H_3PO_4 \rightarrow HPO_2 + PO + \text{etc.}$$
$$H + PO \rightarrow HPO$$
$$H + HPO \rightarrow H_2 + PO$$
$$OH + PO \rightarrow HPO + O$$

The literature shows phosphorus to be three to eight times more efficient as a flame retardant than bromine, depending on the polymer and the specific flame retardant. In a 2/1 polycarbonate/polyethylene terephthalate blend, phosphorus was shown to be 10 times more effective than aromatic bromine [8].

The literature contains many claims for flame-retardant synergy for bromine and phosphorus. Many of these claims appear to be based on a nonlinear response–concentration relationship. Bromine–phosphorus synergy was convincingly demonstrated for a polycarbonate/polyethylene terephthalate polymer blend [8]. Brominated phosphates, where both bromine and phosphorus are in the same molecule, show enhanced synergy. Other references confirm this observation.

The formation of phosphorus halides and oxyhalides has been postulated to support a synergistic mechanism by analogy with antimony/halogen. Compounds such as PCl_3 and $POCl_3$ are known commercial compounds and could function as free-radical quenchers [9,10]. However, there is no evidence for the formation of such compounds during combustion. Bond energies appear unfavorable for the formation of phosphorus/halogen compounds from phosphate esters and halocarbons [11].

A generally positive effect is observed with the joint use of phosphorus- and halogen-containing flame retardants. It has been proposed that this is due largely to a change in direction of the chemical reactions in the condensed phase [12]. This increase in effectiveness of aromatic phosphates when combined with halogenated derivatives could be caused by the conversion of triaryl phosphate into acidic derivatives of phosphorus:

$$
\begin{array}{cc}
O & O \\
\parallel & \parallel \\
(RO)_2P\!-\!OR + HX \rightarrow & (RP)_2P\!-\!OH + RX
\end{array}
$$

This reaction may inhibit volatilization of the triaryl phosphate. As a result, the halogen would function in the gaseous phase and the phosphorus in the condensed phase.

Scattered reports in the literature claim less than additive results, or even antagonistic results, where phosphorus and antimony oxide are used together. For example, the flame retardancy of polyethylene containing both halogen and phosphorus was not increased by adding antimony oxide [13]. It has been speculated that the antimony oxide is converted to the nonvolatile antimony phosphate [14].

Metal cations have been shown to have an antagonistic effect on phosphorus-based flame retardants in cellulosics [15]. It has been postulated that stable metal phosphates (e.g., calcium phosphate) form, interfering with the flame-retarding mechanism of the phosphorus or orthophosphoric acid.

Intumescent flame-retardant systems for polypropylene are commercially available. Decomposition during combustion yields a thick char that insulates the substrate from the flame. The system contains no halogen or antimony. The mechanism depends on a char former such as a polyol, an acid source such as a phosphate, and a nitrogen blowing agent such as melamine. Combustion gives low smoke and nonacidic gases of low toxicity.

Inorganic flame retardants such as alumina trihydrate and magnesium hydroxide decompose endothermically and release large amounts (30%) of water of hydration to cool the system. The result is that flaming ignition of the material is delayed and may actually be inhibited in some cases. High concentrations of additive are required, with anticipated severe degradation of mechanical properties. These hydrates are available in coated form for better dispersion.

In burning polymers, melt flow may carry away a significant portion of the thermal energy from the flame zone. Flow of molten polymer from the burning zone prevents heat-up of the polymer adjacent to the flaming zone and significantly reduces combustible pyrolysis products. One can conceive of a mechanism of increasing melt flow to pass a specific flammability test. This has been demonstrated with polystyrene containing very low levels of flame retardant or with plasticizing compounds. Radical-generating peroxides or azo compounds have been added which decompose the polystyrene, thereby increasing dripping [16]. The flammability test results suggest, perhaps erroneously, that less flame retardant need be used.

The flame retardancy of polymers decreases (as measured by the oxygen index) as the temperature of the polymer increases [17]. For example, increasing the temperature of polypropylene to about 100°C can lower its oxygen index substantially. This finding is significant because in actual fires, combustible polymers are subjected to radiant heat and may be more flammable than otherwise expected from small-scale laboratory tests.

III. FLAME-RETARDANT TESTS

Flame-retardant plastic systems are frequently designed to meet specific flammability tests. The laboratory tests used most frequently for thermoplastics are described here.

A. Ease of Ignition: Oxygen Index

Ease of ignition may be defined as the facility with which a material or its pyrolysis products can be ignited under given conditions of temperature and oxygen concentration. This characteristic provides a measure of fire hazard. One of the tests used in the laboratory is ASTM D 2863-77 of the American Society for Testing and Materials for measuring the oxygen index [18].

The oxygen index test employs a vertical glass tube 60 cm high and 8.4 cm in diameter, in which a rod or strip specimen is held vertically by a clamp at its bottom end. A mixture of oxygen and nitrogen is metered into the bottom of the tube, passing through a bed of glass beads at the bottom to smooth the

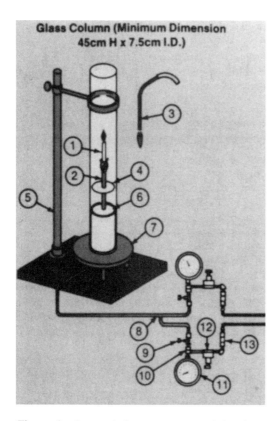

Figure 1 Oxygen index test apparatus. 1: burning specimen; 2: clamp with rod support; 3: igniter; 4: wire screen; 5: ring stand; 6: glass beads in a bed; 7: brass base; 8: tee; 9: cutoff valve; 10: orifice in holder; 11: pressure gauge; 12: precision pressure regulator; 13: filter.

Table 2 Oxygen Indices of Non–Flame-Retardant Polymers

Polymer	Oxygen index
Polyacetal	14.9
Poly(methyl methacrylate)	17.3
Polyethylene	17.4
Polypropylene	17.4
Polystyrene	17.8
Cellulose acetate (4.9% water)	18.1
Impact polystyrene	18.2
Acrylonitrile–butadiene–styrene	18.8
Styrene–acrylonitrile copolymer	19.1
Cellulose butyrate (2.8% water)	19.9
Polyethylene terephthalate	20.6
Noryl 731 (modified polyphenylene oxide)	24.3
Nylon 6/6	24.3
Polycarbonate 141	24.9
Polysulfone	30.4
Poly(vinylidene fluoride)	43.7
Poly(vinyl chloride)	45
Poly(vinylidene chloride)	60
Polytetrafluoroethylene	95

Source: Data from Refs. 19 and 20.

flow of gas, providing a specific environment for the sample. The sample is ignited at its upper end with a hydrogen flame, which is then withdrawn; the sample burns like a candle from the top down (Fig. 1). The atmosphere that permits steady burning is determined. The limiting oxygen index or simply the oxygen index is the minimum percent of oxygen in an oxygen–nitrogen mixture that will just sustain burning for 2 in. or 3 min, whichever comes first.

Oxygen index values for non–flame-retardant polymers are given in Table 2 [19,20].

B. Flammability Test for Electrical and Electronic Materials: UL-94

Laboratory tests for determining the flammability of materials intended for electrical and electronic applications are described in Underwriters Laboratories (UL) standards. A very popular test known as the UL-94 standard measures the vertical burning of self-extinguishing plastic materials [21].

In the Underwriters Laboratories Subject 94 vertical burn test, a sample 127 mm (5.0 in.) long and 12.7 mm (0.5 in.) wide is exposed vertically to a Bunsen burner blue flame 18.75 mm (0.75 in.) high for 10 s. The sample is ignited at the bottom and burns upward (Fig. 2). If the specimen self-extinguishes within 30 s, another 10-s application is made. Flaming droplets are allowed to fall on dry absorbent surgical cotton located 12 in. below the bottom edge of the sample. If the average burning time is less than 5 s (no single burn more than 10 s) and the drips to not ignite the cotton, the material is classified 94 V-0. If the average time is more than 5 s but less than 25 s (no single burn greater than 30 s) and the drips do not ignite the cotton, the material is classified 94 V-1. If

Figure 2   Underwriters Laboratories Subject 94 vertical burn test.

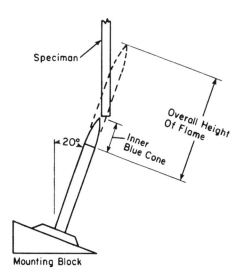

Figure 3   Vertical burn test for UL-94-5V classification.

the sample is self-extinguishing but the cotton is ignited, the material is classified as 94 V-2. The rating is obtained on a specific sample thickness up to 12.5 mm (0.5 in.)

The vertical burn test for classifying materials as 94-5V is a more stringent test. The burner is positioned 20° from the vertical and the overall height of the flame is adjusted to give a 127-mm (5-in.) flame with a 38-mm (1.5-in.) inner blue cone (Fig. 3). The flame is applied for 5 s and removed for 5 s. This is repeated until the 127 × 12.5-mm (5.0 × 0.50-in.) specimen has been subjected to five applications. If any specimens are observed to undergo shrinkage, elongation, melting, and so on, additional testing is carried out with test plaques of approximately 152 × 152 mm (6 × 6 in.). This method calls for testing plaques in various positions: vertical lower corner, vertical lower edge, vertical center of one side, horizontal center of bottom surface, and horizontal with flame directed downward to the top surface. Materials are classified 94-5V when no specimens burn with flaming and/or glowing combustion for more than 60 s after the fifth flame and no specimens drip any particles.

C. Flame Spread

Flame spread, or the rate of travel of a flame front under given conditions of burning, is a measure of fire hazard. The spread of flame along the surface of a material can transmit fire to more flammable materials in the vicinity.

1. Tunnel Test

The Underwriters Laboratories 25-ft tunnel test developed by Steiner (also ASTM E-84) is perhaps the most widely accepted test for surface flame spread. It requires a specimen 24 ft long and 20 in. wide, mounted face down to form the roof of a 25-ft-long tunnel 17.5 in. wide and 12 in. high. The fire source, two gas burners 1 ft from the fire end of the sample of select-grade red oak flooring would spread flame 19.5 ft from the end of the igniting fire in 5.5 min $\pm$ 15 s. The end of the igniting fire is considered as being 4.5 ft from the burners, this flame length being due to an average air velocity of 240 $\pm$ 5 ft/min. Flame spread classification is determined on a scale on which an asbestos–cement board is zero and select-grade red oak flooring is 100. Fuel contribution and smoke density are determined on a similar scale.

2. Radiant Panel Test

The radiant panel test ASTM E162-83 provides a laboratory test procedure for measuring and comparing the surface flammability of materials when exposed to a prescribed level of radiant heat energy. A radiant heat source consists of a panel measuring 12 $\times$ 18 in. (300 $\times$ 460 mm). A specimen 6 $\times$ 18 in. (150 $\times$ 460 mm) is supported in front of it with the 18-in. dimension inclined 30° from the vertical. The orientation of the sample is such that ignition is forced near its upper edge and the flame front progresses downward. The temperature rise recorded by stack thermocouples is used as a measure of heat evolution.

A factor derived from the rate of progress of the flame front and another relating to the rate of heat liberated by the material are combined to give a flame spread index. Smoke evolved during the test can also be measured.

D. DOT302 MVSS Motor Vehicle Safety Standard

The U.S. Department of Transportation Motor Vehicle Safety Standard known as DOT302 MVSS measures the flame spread rate of all materials used for automobile interiors. A 14 $\times$ 4-in. specimen is mounted horizontally in a test chamber. One end is ignited with a 1.5-in. vertical flame from a Bunsen burner. The flame is allowed to progress along the sample until it extinguishes itself or consumes the sample. The rate of burning must not exceed 4 in./min.

E. Federal Test Method Standard CCCT 191B, Method 5903

The federal test identified as Standard CCCT 191B, Method 5903 measures the rate of flame spread of plastic film and coated fabrics. Test specimens, 2.75 $\times$

12 in., are placed in a frame and suspended vertically in a test chamber. The lower end is exposed to a 1.5-in. flame from a burner for 12 s. The "afterflame" and "afterglow" times required for the flame to extinguish and the char length or the distance the flame travels up the sample are reported.

F. Ohio State University Heat Release Apparatus

No one test that describes a specific behavior in a fire is an adequate measure of fire hazard. A material may be hazardous because it is easily ignited, because flame travels rapidly across its surface, because it burns rapidly (i.e., has a high rate of heat release), or because it produces a large quantity of smoke at a rapid state.

The rate of heat release is believed to be most important during the "steady" burning period following flame spread. It is most important to consider in the stage of fire growth preceding flashover. In a typical heat-release-rate calorimeter, a sample of material is exposed to a controlled airflow and an external radiant heat flux simultaneously. When the specimen ignites, heat is released as a function of time. The back surface of the specimen is either insulated or water cooled. In the latter case, the temperature rise of the water stream is used to calculate the heat released through the back surface of the specimen.

The Ohio State University (OSU) heat-release-rate apparatus consists of a chamber 35 in. high × 16 in. wide × 8 in. deep (Fig. 4). An electrically heated ceramic radiant panel is located in this chamber. The test panel dimensions are 13.8 × 18.1 in., with an exposed surface area of 10 × 10 in. The test sample is positioned 3 in. from the radiant panel, and the distance between the sample and radiant panel varies from 0 to 7 in. Calculations of the heat-release rate are made in terms of watts per unit of surface area exposed. The Ohio State University release-rate apparatus has been used to measure the rate of heat release, smoke, toxic gases (e.g., CO, CO_2, nitrogen oxides, HCN, and HBr), and consumption of oxygen [22,23].

G. Cone Calorimeter

Attempts to improve on the OSU radiant panel heat-release test resulted in the development of the cone calorimeter based on the principle of oxygen consumption (i.e., the relationship between the mass of oxygen consumed during combustion and the heat released).

The oxygen consumption calorimeter is more commonly called the cone calorimeter because its radiant electric heater is wrapped inside a truncated cone-shaped hood. The device was developed at the U.S. National Bureau of Standards and is a test method for heat- and visible smoke-release rates for materials and products (ASTM E-1354-94).

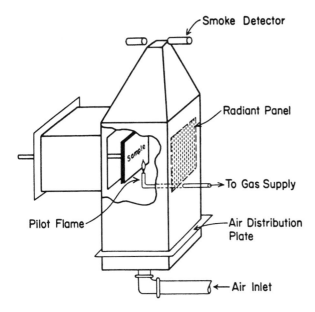

Figure 4   Ohio State University rate of heat release apparatus.

The equipment makes two primary measurements: oxygen concentration and exhaust gas flow rate. Additional measurements include the mass-loss rate of the specimen, time to ignition, and smoke obscuration. This method is used to determine the ignitability, heat-release rate, mass-loss rate, effective heat of combustion, and visible smoke development of materials and products. The sample is a square about 100 cm (4 in.) per side and up to 5 cm (2 in.) thick.

H. Smoke Measurements

Smoke or smoke density is defined as the degree of light obscuration produced from the burning of a material under a given set of combustion conditions. This characteristic provides a measure of fire hazard in that occupants have a better chance of escaping from a burning structure if they can see their way. One of the important laboratory tests is the National Bureau of Standards (NBS) smoke density test. ASTM E662-83 ''Specific Optical Density of Smoke Generated by Solid Materials'' is also known as the NBS smoke chamber test. This method is identical to the National Fire Protection Association (NFPA) test 258-1976. It is used solely to measure the smoke density characteristics of a material under controlled laboratory conditions.

The smoke density chamber test is used to determine the specific optical density of smoke generated within a closed chamber due to nonflamming pyrolytic decomposition and/or flaming combustion. The nonflaming mode employs an electrically heated radiant energy heat source with an irradiance level of 2.5 W/cm^2. For the flaming mode, a six-tube burner, fueled with a mixture of propane and air, is used in combination with the radiant heat to apply a row of equidistant flamelets across the lower edge of the specimen and into the specimen trough.

A vertical photometer path for measuring light absorption is employed to minimize measurement differences due to smoke stratification, which could occur with a horizontal photometer path at a fixed height. The full 3-ft height of the chamber is used to provide an overall average for the entire chamber.

Light transmission measurements are used to calculate the specific optical density, which is derived from a geometrical factor associated with the dimensions of the test chamber and specimen, and the measured optical density, a measurement characteristic of the concentration of smoke. The photometric scale used to measure the smoke generated in this test is similar to the optical density scale for human vision.

Specimens measuring 3 in. square are taken from the specimen to be evaluated. Size can be varied only in terms of specimen thickness, the exposed surface area being a constant defined by the specimen holder. Specific gravity differences are a factor. Specimens are backed with heavy-duty aluminum foil and mounted in troughed specimen holders backed with a 0.5-in. inorganic millboard block. The samples are preconditioned for 24 h at 140°F, then conditioned to equilibrium at 70°F and 50% relative humidity.

The test method consists of three exposures in the flaming mode and three exposures in the nonflaming mode. In many instances, testing is done in only one mode and reported.

The test chamber is preconditioned to a ready state and the irradiance level of the radiant heat energy source is verified. The 100% and 0% transmittances of the photometric system are calibrated. The specimens mounted in the troughed specimen holders are tested in the appropriate mode (i.e., flaming or nonflaming). The test period is 20 min or until minimum light transmittance is obtained, whichever occurs first.

The test chamber is then evacuated of the accumulated smoke. Another light transmittance reading is taken and is recorded as the clear beam reading. This represents the accumulation of soot and other effluent on the optical system, and it is converted to specific optical density and used as a correction factor in calculating final test results.

The recorded data for light transmittance are used to calculate the specific optical density of the smoke generated during the test for each of the specimens tested. The results are then averaged and reported as average specific optical density.

IV. FLAME-RETARDANT ADDITIVES

A. Bromine-Containing Flame Retardants

No single organobromine compound is used to impart flame retardancy to all thermoplastics. There are more than 25 commercial bromine-containing flame retardant additives, many used in only a single thermoplastic application. Table 3 lists the more common additives by trade name and, in some cases, by chemical name. The bromine content listed is for aromatic bromine unless noted otherwise (e.g., aliphatic and alicyclic). The melting point, molecular weight, thermal stability, and manufacturer are also listed. Many flame retardants are based on the same organic compound, the derivatives being significantly different from each other in properties and in applications.

Figure 5 shows the chemical structure of four brominated flame retardants based on diphenyl oxide. The fully brominated compound decabromodiphenyl oxide is a high-melting solid (>300°C) and is the major flame retardant used for high-impact polystyrene (HIPS). Octabromodiphenyl oxide is a blend of brominated diphenyl oxides with an average bromine content of eight bromines. This product melts over a range of 70–150°C and had been used to flame retard acrylonitrile–butadiene–styrene (ABS). Because it can form toxic brominated dioxins and furans, it has been replaced by brominated epoxy oligomers in ABS. Pentabromodiphenyl oxide is also a blend and a very high-viscosity liquid. It is used commercially in a blend with phosphate ester to flame retard flexible polyurethane foam; no antimony oxide is used. Saytex 120, a fully brominated diphenoxybenzene, has the highest melting and decomposition temperature of the series and is recommended for use in nylon and other engineering plastics. To avoid the diphenyl oxide structure, decabromodiphenyl ethane is available commercially.

Figure 6 shows five organobromine flame retardants based on brominated phenols. FF-680 is a coupled tribromophenol produced from tribromophenol and ethylene dibromide. It has a melting point of 223°C and is melt blended into ABS. Pyro-Chek 77B is a coupled pentabromophenol prepared by brominating diphenoxyethane. It melts at 322°C and is a filler-type flame retardant. This product was withdrawn from the market because of its potential to form toxic brominated dioxins and furans. Tribromophenylallyl ether has been used to flame retard expandable polystyrene bead board. PO-64P is a polymeric flame retardant that does not "plateout" during injection molding, like FF-680. Tris-dibromophenyl phosphate contains both bromine and phosphorus in the same molecule. Bro-

Table 3 Commercial Bromine-Containing Flame Retardants

Flame retardant	Chemical/trade name	Bromine[a] (%)	Melting point (°C)	Molecular weight	TGA (wt loss/°C)	Producer
BA-43	BA-50-BIS (acrylate)	43.2	125–128	740.1	5/329	Great Lakes
BA-50	Tetrabromobisphenol A-bis(2-hydroxyethyl ether)	50.6	113–119	632	5/322	Great Lakes
BC-52	Tetrabromobisphenol A carbonate oligomer	52	210–230	2,494	5/430	Great Lakes
BC-58	Tetrabromobisphenol A carbonate oligomer	58	230–260	3537	5/430	Great Lakes
BCL-462	Dibromoethyldibromocyclohexane	74 aliphatic/ alicyclic	65–80	428	1/135	Albemarle
BE-51	Tetrabromobisphenol A-bis(allyl ether)	51.2	118–120	624	5/230	Great Lakes
BE-62	Bis ether of FF-680	63.6	156–160	1,258	5/360	Great Lakes
BN–451	Ethylene bis(dibromonorbornane) dicarboximide	45 alicyclic	294	672	—	Albemarle
BT-93(w)	Ethylene bistetrabromophthalimide (white grade)	66	446	952	dec. 450	Albemarle
CD-75P	Hexabromocyclododecane	74.7 alicyclic	185–195	641.7	5/265	Great Lakes
DE-60F Special	Pentabromodiphenyl oxide/triaryl phosphate blend	50 Br/2.5 P	Liquid	—	5/245	Great Lakes
DE-71	Pentabromodiphenyl oxide	70.8	Very viscous liquid	564.7	5/243	Great Lakes
DE-79	Octabromodiphenyl oxide	79.8	70–150	801	5/304	Great Lakes
DE-83	Decabromodiphenyl oxide	83.3	300–315	959.2	5/357	Great Lakes
Decabromodiphenyl oxide	DE-83; Satex 102; FR-1210	83.2	300–315	959.2	3/357	
DP-45	Tetrabromophthalate ester	45.3	Liquid	706.1	5/320	Great Lakes
F-2000 Series	Brominated epoxy oligomers	46–54	50–155 (softening point)	720–50,000	—	Dead Sea Bromine
FF-680	Bis(tribromophenoxy) ethane	70	223–225	655.5	5/290	Great Lakes
FM-836	Halogenated phosphate ester	46.9/7.5 P	Liquid	416.5	5/224	Great Lakes
FM-935	See PO-64P					
FR-651-P	Pentabromochlorocyclohexane	77.9 Br/6.9 Cl alicyclic	>180	513.5	1/175 dec 230	Albemarle
FR-804	Tetrabromoxylene	75	250	422	2/230	Ameribrom
FR-913	Tribromophenyl allyl ether	62	75–76.5	371	1/50 (sublimes)	Ameribrom
FR-1025	Poly(pentabromobenzyl) acrylate	70–71	205–215	80,000	—	Ameribrom
FR-1033	Tribromophenylmaleimide	58	142	410	1/182	Amerib rom
FR-1201	Hexabromocyclododecane	73.8 alicyclic	180–190	641.7	1/195	Ameribrom
FR-1205	Pentabromodiphenyl oxide	69	2000 cPs/50°C	564.7	1/195	Ameribrom
FR-1208	Octabromodiphenyl oxide	78	127–160	801	2/330	Ameribrom
FR-1210	Decabromodiphenyl oxide	83	300–305	959.2	1/343	Ameribrom
FR-1525	Tetrabromobisphenol A ethoxylate	51	116–120	632	—	Ameribrom
HBCD	Hexabromocyclododecane	74 alicyclic	185–195	641.7	—	Albemarle
Hexabromo- cyclododecane	CD-75P; FR-1206; HBCD	74.6 alicyclic	185–195	641.7	5/265; stable to <304°C	Albemarle
Octabromodiphenyl oxide	DE-79; FR-1208; Saytex 111	79.8	70–150	801	5/304	Albemarle
PE-68	Tetrabromobisphenol A bis(2,3-dibromopropyl ether)	67.7 aromatic aliphatic	90–100	943.6	5/302	Great Lakes
Pentabromo- chlorocyclohexane	FR-651-P	77.9 Br/6.9 Cl aliphatic	>180	513.5	1/175; dec 230	Albemarle
Pentabromodiphenyl oxide	DE-71; FR-1205; Saytex 115	70.8	Very viscous liquid	564.7	5/243	
Pentabromomethyl benzene	Saytex 105	79	136–138	501	10/231; dec >300	Albemarle
PHT-4	Tetrabromophthalic anhydride	68.2	270–276	6,000	5/250	Great Lakes
PHT-4 Diol	Tetrobromophthalate diol	46	Liquid	627.9	5/188	Great Lakes
PO-64P	Polydibromophenylene oxide	62	210–240	6,000	5/400	Great Lakes
Pyro-Chek 60PB	Brominated polystyrene	58–61	200	>200,000		Ferro
Pyro-Chek 68PB	Brominated polystyrene	68	220	200–300,000	Initial 230; 1/340	Ferro

(continued)

Table 3 Continued

Flame retardant	Chemical/trade name	Bromine[a] (%)	Melting point (°C)	Molecular weight	TGA (wt loss/°C)	Producer
Pyro-Chek 77B	Bis(pentabromophenoxy) ethane	77	322	1,004	dec >310	Ferro
RB-49	Tetrabromophthalic anhydride	68	270	463.7	—	Albemarle
Saytex 102	Decabromodiphenyl oxide	83	300–310	959.2	330	Albemarle
Saytex 105	Pentabromoethylbenzene	79	136–138	501	10/231; dec >300	Albemarle
Saytex 111	Octabromodiphenyl oxide	79	70–150	801	5/305	Albemarle
Saytex 115	Pentabromodiphenyl oxide	69–72	Liquid	564.7	5/258	Albemarle
Saytex 120	Tetradecabromodiphenoxybenzene	82	370	1,368	dec 380	Albemarle
Saytex 8010		82				Albemarle
Tetrabromophthalic anhydride	PHT-4; RB-49	68.2	270–276	463.7	5/250	Albemarle/ Great Lakes
Tetrabromoxylene	FR-804	75	250	422	2/230	Ameribrom

[a] All aromatic bromine except as noted.

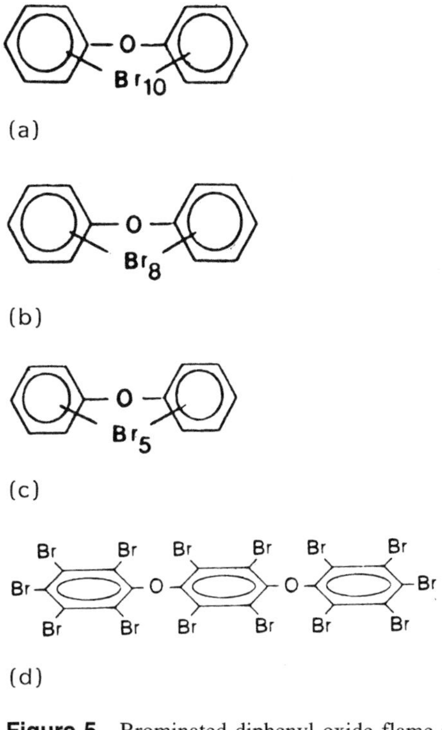

(a)

(b)

(c)

(d)

Figure 5 Brominated diphenyl oxide flame retardants: (a) decarbromodiphenyl oxide [melting point (m.p.) 300°C], (b) octabromodiphenyl oxide (m.p. 70–150°C), (c) pentabromodiphenyl ether (very viscous liquid), and (d) tetradecabromodiphenoxybenzene (m.p. 370°C).

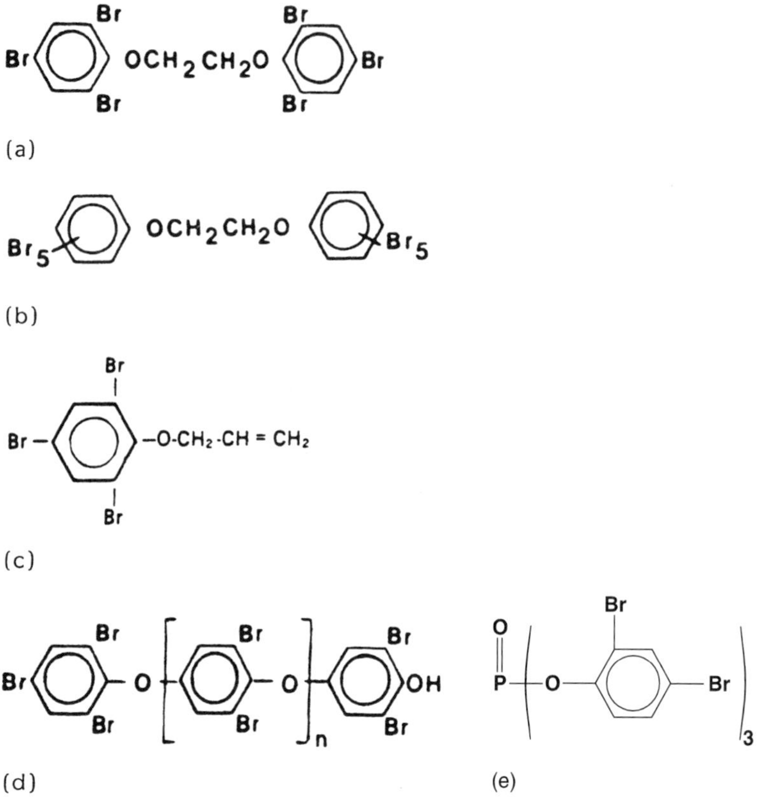

(a)

(b)

(c)

(d)   (e)

Figure 6 Flame retardants based on brominated phenols: (a) 1,2-bis-(2,4,6-tribromophe-noxy)ethylene, FF-680; (b) bis(pentabromophenoxy) ethane, Pyro-Chek 77B; (c) tribromo-phenylallyl ether, FR-913; (d) polydibromophenylene oxide, PO-64P; and (e) tris-dibromo-phenyl phosphate.

mine and phosphorus has been shown to be synergistic in polymers containing oxygen, such as polyethylene terephthalate (PET).

BT-93 is a yellow bisimide prepared by coupling tetrabromophthalic anhy-dride (Fig. 7) with ethylene diamine. It has excellent thermal stability and good ultraviolet stability. It is used in cross-linked polyethylene wire and cable. The good UV stability may be of interest in impact polystyrene for television and computer housing, except for the relatively high cost compared to decabromodi-phenyl oxide. A white version is available as BT-93W.

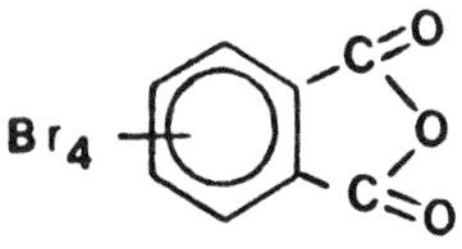

(a)

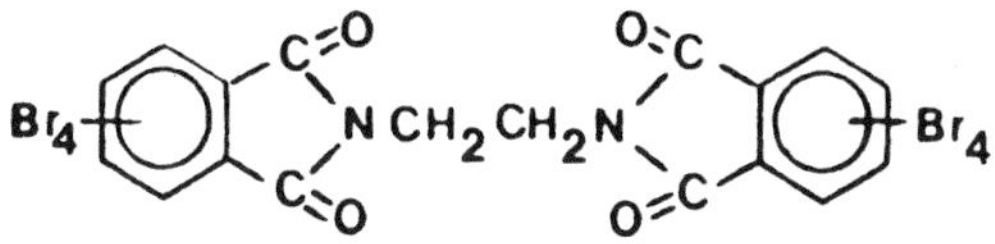

(b)

Figure 7 Flame retardant based on (a) tetrabromophthalic anhydride and (b) 1,2-bis(tetrabromophthalimido)ethane, BT-93.

BN-451, another bisimide (Fig. 8), is used commercially as a flame retardant for polypropylene with an UL-94 V-2 rating. This bromine is of the alicyclic type.

Tris-dibromophenyl phosphate containing both bromine and phosphorus in the same molecule has been recommended for use in engineering thermoplastics. The bromine and phosphorus show flame-retardant synergy in these polymers.

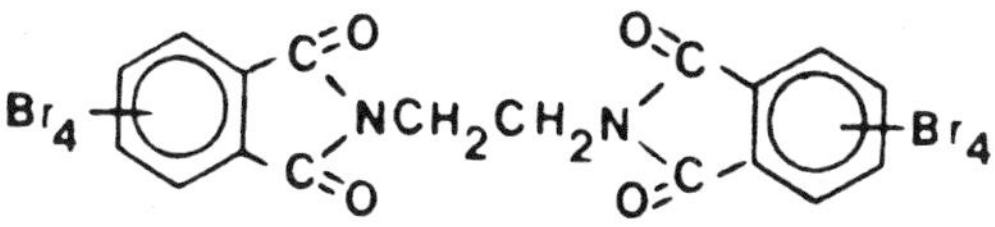

(a)

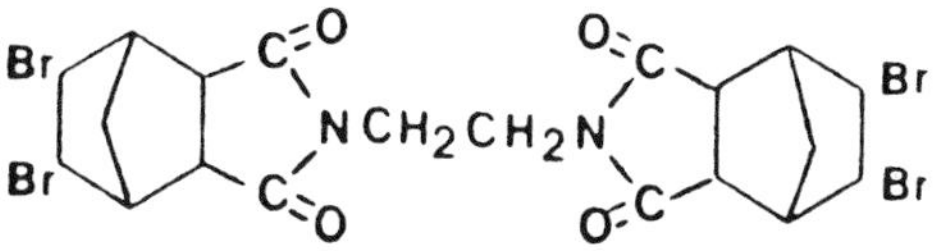

(b)

Figure 8 Bisimide flame retardants: (a) 1,2-bis(tetrabromophthalimido)ethane, BT-93, and (b) ethylene bis(dibromonorbornane)-dicarboximide, BN-451.

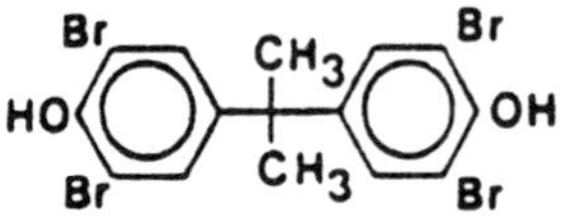

(a)

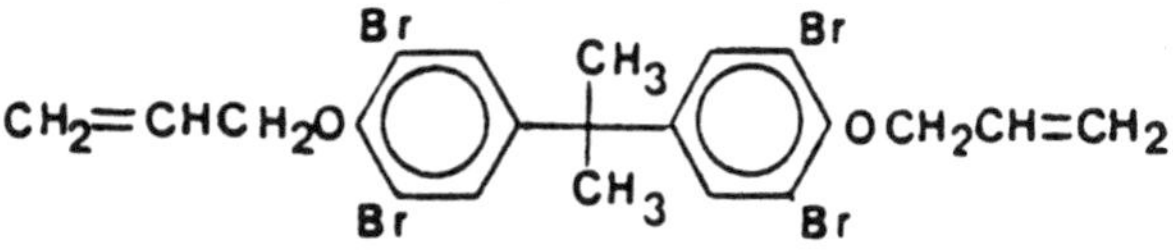

(b)

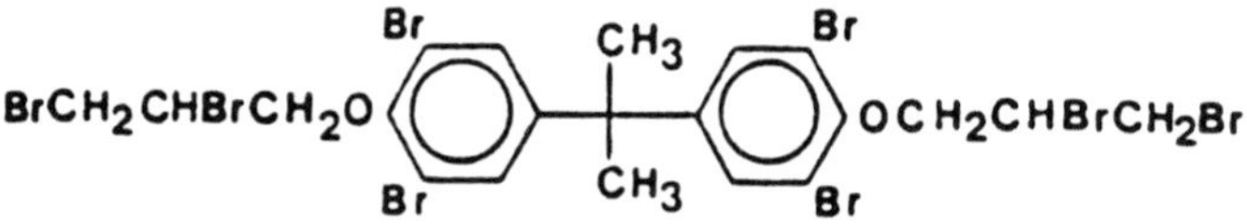

(c)

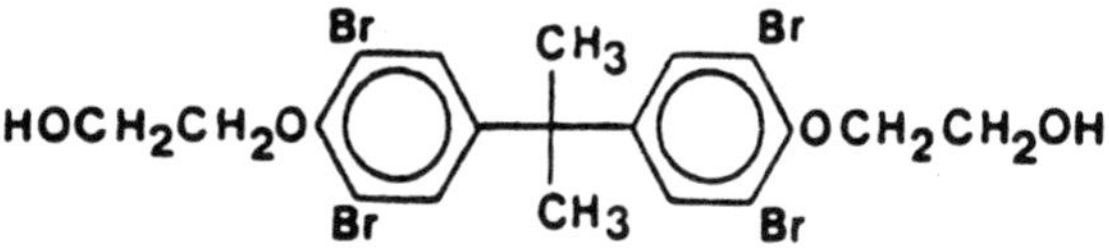

(d)

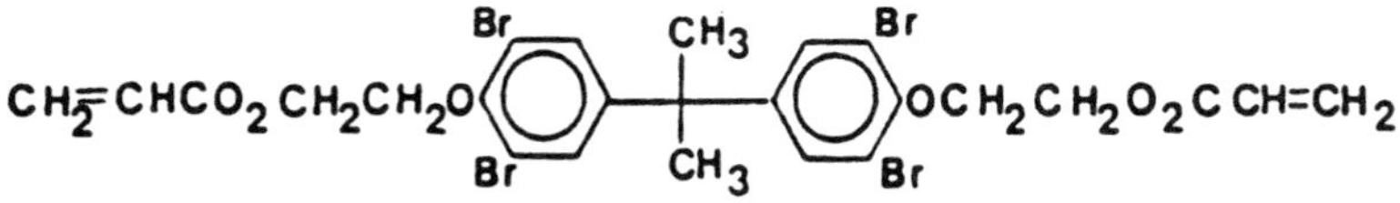

(e)

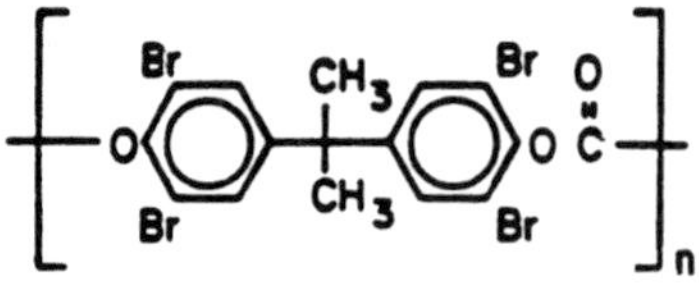

(f)

Figure 9 Five flame retardants based on tetrabromobisphenol A (a) are as follows: (b) tetrabromobisphenol A, bis(allyl ether), BE-51; (c) bis(dibromopropyl) ether of tetrabromobisphenol A, PE-68; (d) ethylene oxide adduct of tetrabromobisphenol A, BA-50; (e) bis(acryloxyethyl) ether of tetrabromobisphenol A, BA-43; (f) tetrabromobisphenol A polycarbonate oligomer, BC-52/BC-58.

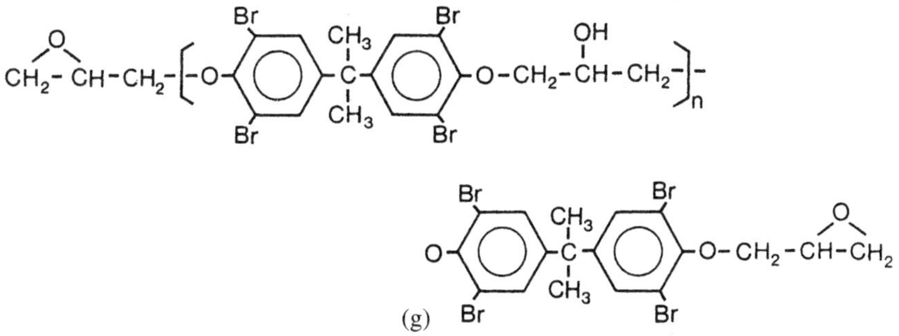

Figure 9 Continued. (g) brominated epoxy oligomer/polymer.

It melts at 110°C and is melt blendable aiding processing. It is not available commercially.

Flame retardants based on tetrabromobisphenol A (TBBP-A) are shown in Fig. 9. TBBP-A is the largest volume flame retardant. It is used as a reactive in epoxy resins for printed circuit boards. Many of these products can be used as reactive flame retardants. PE-68 is of particular interest as an additive because it contains both aromatic bromine and the more effective aliphatic form. It is very effective in flame retarding polypropylene. Its major deficiency is blooming. BC-52 and BC-58 differ mainly in the chain cap: The former is chain capped with phenol and the latter with tribromophenol.

Alicyclic bromine containing hydrocarbons (Fig. 10) generally have poor thermal stability (<200°C). They are used commercially to flame retard expandable and extruded polystyrene. Expandable polystyrene bead board is produced using pressurized steam. Hexabromocyclododecane (HBCD) is the preferred additive and "stabilized" or more stable isomers are offered. Tris-tribromoneopentyl phosphate is very effective in flame retarding polypropylene molding compounds and polypropylene fiber.

Figure 11 shows several polymers containing high concentrations of bromine. These products were developed specifically as flame-retardant additives. Polymeric materials would not volatilize during high-temperature processing and would not be expected to migrate, bloom, or plateout.

Figure 12 gives five reactive flame retardants, some used in condensation polymerization. Diallyl tetrabromophthalate has been used to cross-link unsaturated polyester resin and has been of particular interest in pultrusion applications. Vinyl bromide is a reactive flame retardant used in modacrylic fiber.

Table 4 lists thermoplastics and the typical flame-retardant additives used commercially for each material.

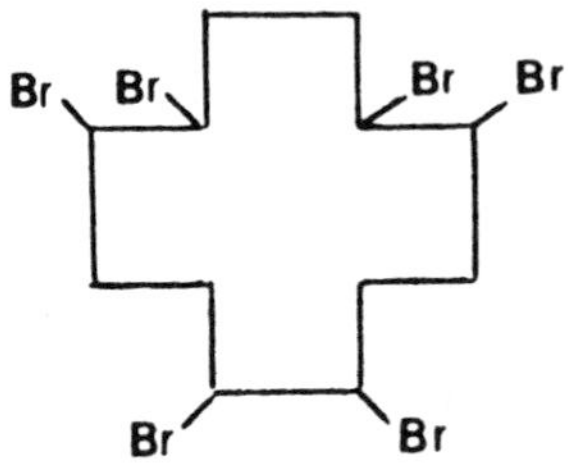

(a)

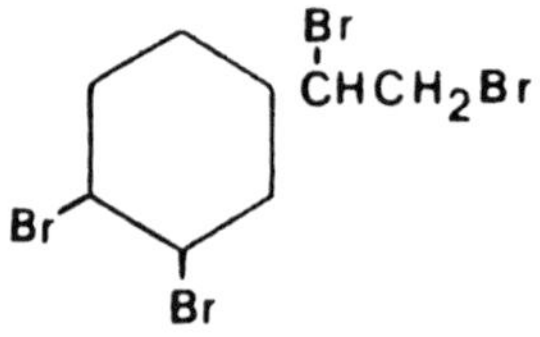

(b)

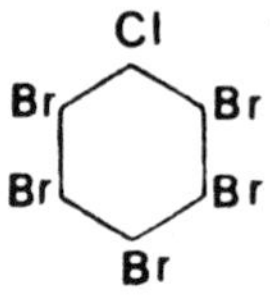

(c)

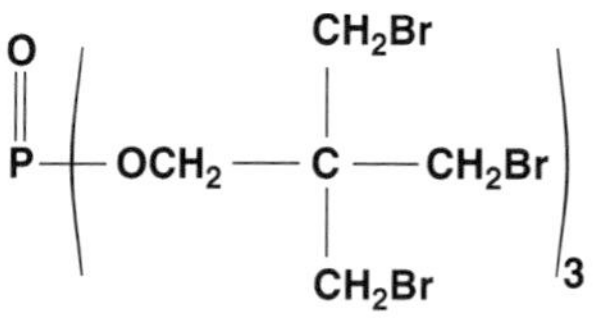

(d)

Figure 10 Alicyclic bromine-containing hydrocarbon flame retardants. (a) hexabromo-cyclododecane; (b) 1,2-dibromoethyl-3,4-dibromocyclohexane, BCL-462; (c) pentabro-mochlorocyclohexane, FR-651F; and (d) tris-tribromoneopentyl phosphate (Reoflam PB-370).

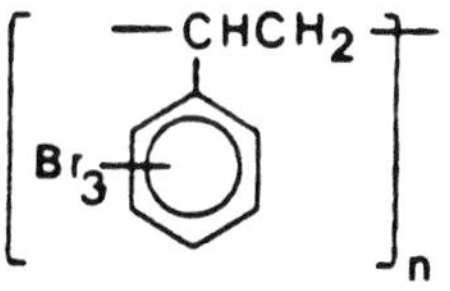

(a)

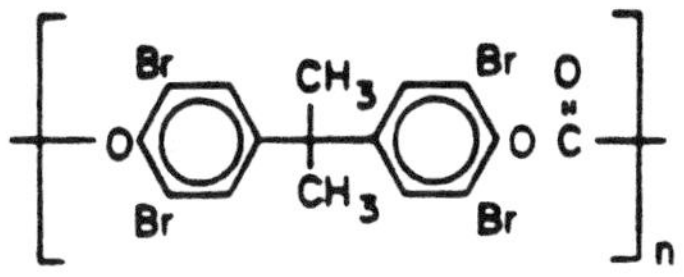

(b)

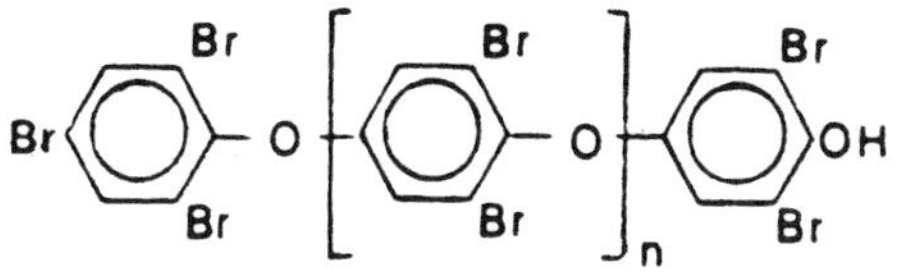

(c)

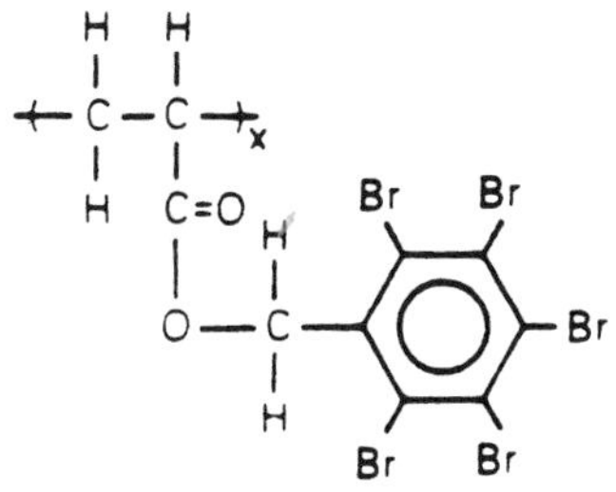

(d)

Figure 11 Polymeric flame retardants containing bromine: (a) brominated polystyrene, Pyro-Chek 68PB; (b) tetrabromobisphenol A polycarbonate oligomer, BC-52/BC-58; (c) polydibromophenylene oxide, PO-64P; and (d) poly(pentabromobenzyl) acrylate, FR-1025.

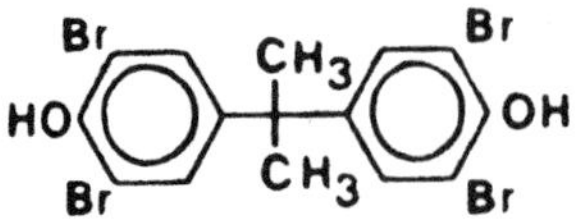

(a)

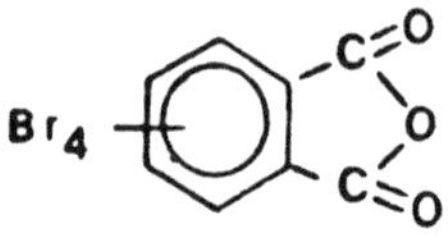

(b)

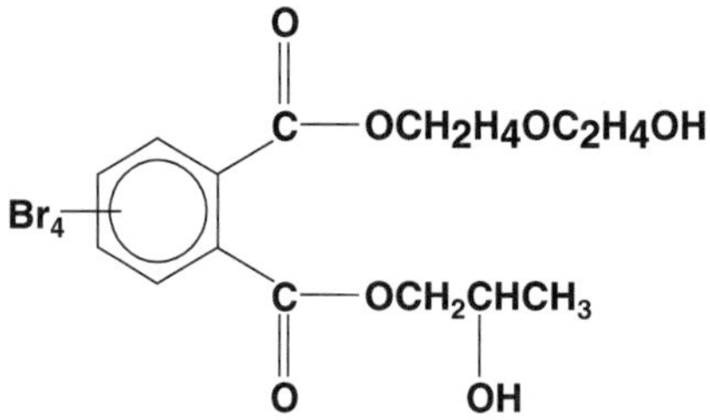

(c)

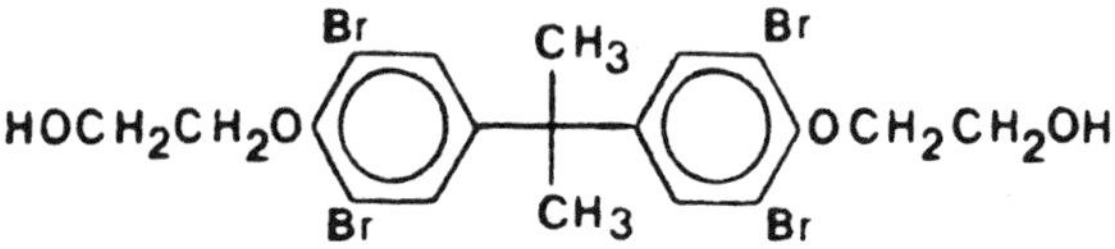

(d)

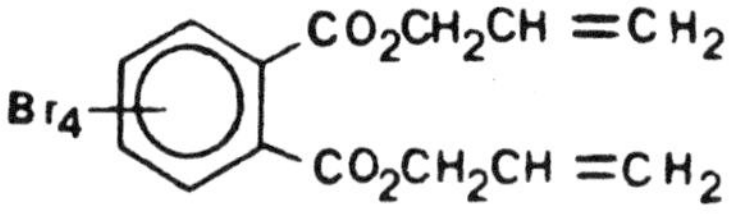

(e)

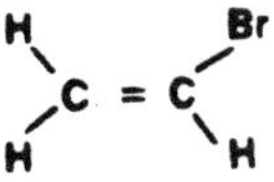

(f)

Figure 12 Reactive flame retardants containing bromine: (a) tetrabromobisphenol A; (b) tetrabromophthalic anhydride; (c) dibromoneopentyl glycol; (d) ethylene oxide adduct of tetrabromobisphenol A; (e) diallyl tetrabromophthalate; and (f) vinyl bromide.

Table 4 Typical Flame Retardants Used Commercially, by Polymer Type

Polymer	Additive
Polyethylene	Chlorinated paraffin
	Decabromodiphenyl oxide
	Dechlorane Plus
	Saytex BT-93
	Alumina trihydrate
	Magnesium hydroxide
Polypropylene	Dechlorane Plus (UL-94 V-0)
	Non-Nen 52 (UL-94 V-0)
	Saytex BN-451 (UL-94 V-2)
	PE-68
	Reoflam PB-370
Polystyrene	Hexabromocyclododecane
	Pentabromochlorocyclohexane
	Saytex BCL-462
Impact polystyrene	Decabromodiphenyl oxide
Acrylonitrile–butadiene–styrene	FF-680
	Octabromodiphenyl oxide
	Brominated epoxy oligomers
Modified polyphenylene oxide	Phosphate ester
	Resorcinol diphosphate
Polybutylene terephthalate	Decabromodiphenyl oxide
	Pyro-Chek 68PB
	Tetrabromobisphenol A carbonate oligomer
	Brominated epoxy oligomer
Polycarbonate	Tetrabromobisphenol A carbonate oligomer
	Sulfonate salt
Poly(vinyl chloride)	Phosphate ester
	Zinc borate
	Ammonium octamolybdate
	Alumina trihydrate
Nylon	Dechlorane Plus
	Pyro-Chek 68PB
	Melamine cyanitrate
	Red phosphorus

B. Chlorine-Containing Flame Retardants

1. Chlorinated Paraffin

Chlorinated paraffins are manufactured by the chlorination of a paraffin hydrocarbon to a maximum chlorine content of 75–76%. By far, they are the lowest-cost halogen-containing flame retardants. Commercial products of 40–60% chlorine are liquids with a wide range of viscosities. Paraffins of 70% chlorine are solids with typical softening points of about 100°C. Thermal degradation, which starts at about 180°C, is initially a dehydrochlorination yielding an olefinic structure. The poor thermal stability of chlorinated paraffins limits their use to PVC and polyurethanes, although some claims are made for polypropylene and polyethylene applications. The latter application is in polyethylene blown film for which processing temperatures are low. Chlorex 760 (Dover Chemical), a high-softening-point resinous chlorinated paraffin (160°C), is claimed to have considerably improved thermal stability, allowing for safe polypropylene processing and possible use in high-impact polystyrene (HIPS).

   To meet higher flame-retardant requirements, bromochlorinated paraffins have been developed. A typical product has 27% chlorine and 32% bromine.

2. Dechlorane Plus

Dechlorane Plus is a thermally stable alicyclic chlorine-containing flame retardant made by a Diels-Alder reaction of 2 mol of hexachlorocyclopentadiene and 1 mol of cyclooctadiene (Fig. 13). It is one of the earliest flame retardants developed and was used commercially in the mid-1960s. The product contains 65% alicyclic chlorine and is stable up to its melting point of about 350°C. It is produced in three particle size ranges for various applications.

   Major uses for Dechlorane Plus are in UL-94 V-0 polypropylene, where it is used with antimony oxide and zinc borate, cross-linked polyethylene wire and cable (UL-83 VW-1 vertical wire flame test), and nylon using either antimony oxide or ferric oxide as a synergist. Table 5 gives the properties of Dechlorane Plus, which is a product of Occidental Chemical Corporation (Niagara Falls, NY).

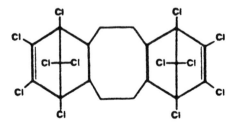

Figure 13   Chlorine-containing flame retardant: Dechlorane Plus.

Table 5 Properties of Dechlorane Plus

Appearance	White crystalline solid
Chlorine content (%)	65; alicyclic
Melting point (°C)	350°C dec
Specific gravity (g/cm^3)	1.8–20

	Maximum size (μm)	Retained on 325 mesh screen (%)
Dechlorane 515	15	0.15
Dechlorane 25	5	0.15
Dechlorane 2520	5	0.05

Source: Occidental Chemical, Niagara Falls, NY.

Table 6 Types of Phosphate Ester Flame Retardants

Phosphates	Flame retardants	Sold as
Triaryl		
Natural esters	Tricresyl phosphate	TCP
	Cresyl diphenyl phosphate	CDP
	Trixylenyl phosphate	TXP
Synthetic esters	Isopropylphenyl	Reofos 50[a]
		Reofos 100[a]
		Phosflex 41-P[b]
	t-Butylphenyl	Reofos B80[a]
		Phosflex 71-B[b]
		Santicizer 154[c]
	Triphenyl	TPP
Diphosphates	Resorcinol diphosphate RDP[a,b]	Resorcinol diphosphate RDP[a,b]
	Bis-phenol 4 diphosphate[a,b]	Bis-phenol A diphosphate[a,b]
Alkyl diaryl	Isodecyl diphenyl	Santicizer 148[c]
	Ethylhexyl diphenyl	Santicizer 141[c]
		Santicizer 2148[c]
Trialkyl	Trioctyl	TOF
	Tributyl	TBP
	Tris-butoxyethyl	KP-140[a]

[a] FMC.
[b] Akzo Nobel.
[c] Monsanto.

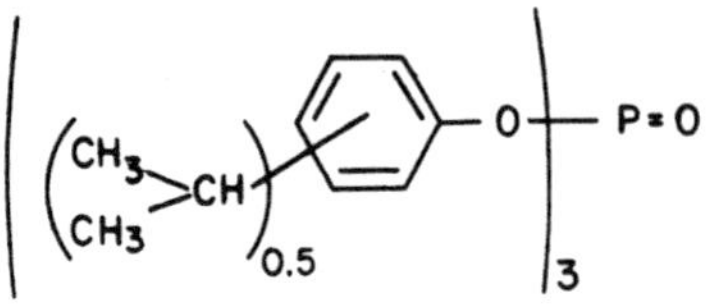

(a)

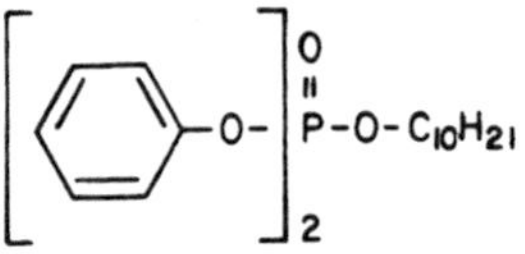

(b)

$(C_4H_9OC_2H_4O)_3P = O$

(c)

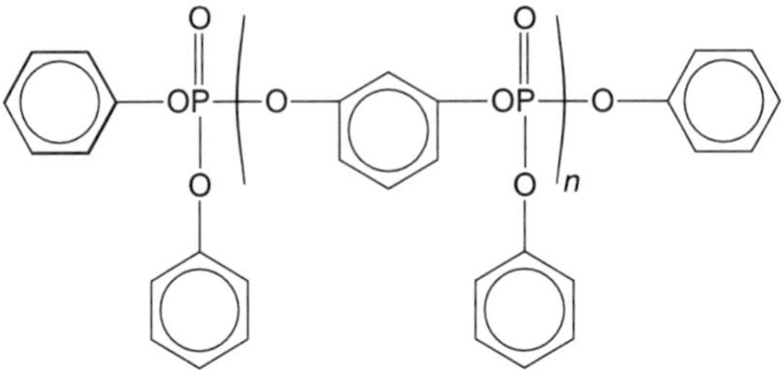

(d)

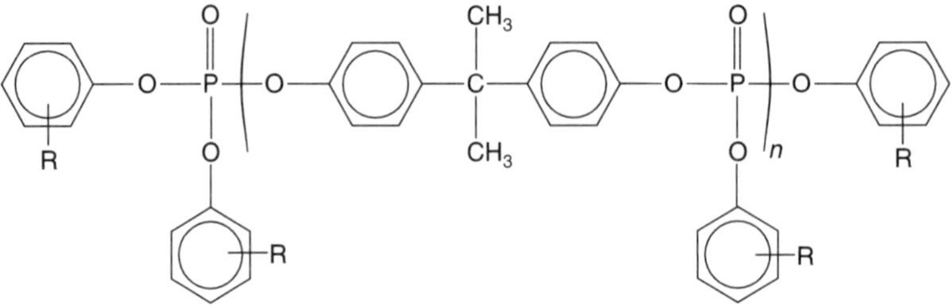

(e)

Table 7 Properties of Phosphate Ester Flame Retardants

Property	Phosphate			Diphosphate (RDP)
	Triaryl (Kronitex 100)	Alkyl diaryl (Santicizer 148)	Trialkyl (KP-140)	
Molecular weight (average)	390	390.5	398	574
Boiling point (°C/mm)	220–270/4	245 dec/10	215–228/4	—
Viscosity of 20°C (cPs)	90	25	12.2	400–800
Specific gravity (g/cm^3)	1.150–1.165	1.070	1.018	1.294–1.318

C. Phosphorus-Containing Flame Retardants

Phosphate esters are classified into three types, namely triaryl, alkyldiaryl, and trialkyl phosphates. Triaryl phosphates can be either ''natural'' esters produced from naturally derived raw materials (cresols and xylenols) or synthetic esters made from synthetically produced feedstocks (isopropylphenol and t-butylphenol) (Table 6). The ''synthetic'' phosphates are the major commercial products. Resorcinol diphosphate (RDP) is less volatile than the monophosphates and does not ''juice'' or fume-off during polymer processing. Chemical structures are shown in Fig. 14 and properties in Table 7. Phosphate esters are used commercially to flame PVC, modified polyphenylene oxide polycarbonate/ABS blends and cellulose acetate.

Phosphine oxides are hydrolytically more stable than phosphate esters. Additive and reactive compounds are known. Structures of some phosphine oxides are shown in Fig. 15. Properties are given in Table 8 [24,25].

Phosphine oxide diols can be polymerized into polyesters [polybutylene terephthalate (PBT), PET], polycarbonates, epoxy resins, and polyurethanes.

An alkyl bishydroxymethyl phosphine oxide may be in commercial use. The application is epoxy printed circuit boards and the volume is small; phosphine oxides are expensive.

Figure 14 Phosphate ester flame retardants: (a) tiaryl phosphate, Kronitex 100; (b) alkyl diaryl phosphate, Santicizer 148; (c) trialkyl pyosphate, tris-(butoxyethyl)-phosphate, KP-140; (d) resorcinol diphosphate (ADP); and (e) bis-phenol A diphosphate.

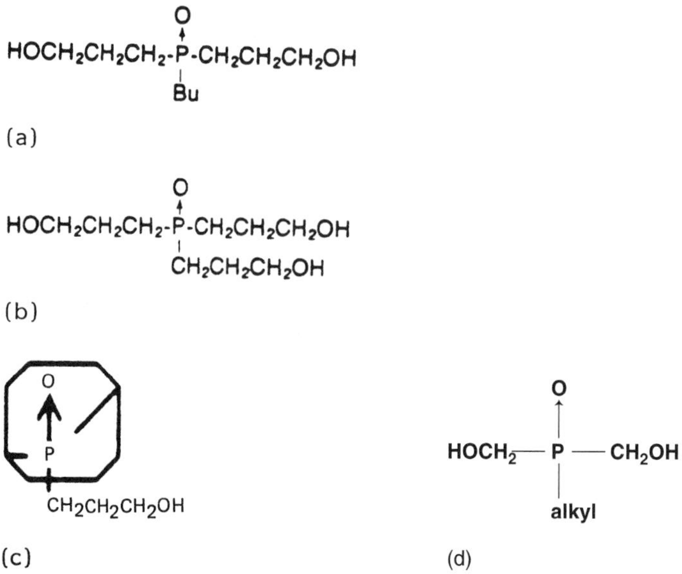

(a)

(b)

(c)   (d)

Figure 15   Phosphine oxide flame retardants: (a) *n*-butyl bis(hydroxypropyl) phosphine oxide, FR-D; (b) tris(hydroxypropyl) phosphine, FR-T; (c) cyclooctyl hydroxypropyl phosphine oxide; and (d) alkyl dihydroxymethyl phosphine oxide.

Table 8   Properties of Phosphine Oxide Flame Retardants

Property	FR-D[a]	FR-T[b]
Physical form	Liquid	White solid
Phosphorus content (%)	14	13.8
Boiling point (°C)	235/0.4 mm	—
Melting point (°C)	—	112–115
Viscosity at 77°F (cPs)	12,000	—
Specific gravity (g/cm^3)	1.12	—
Hydrolytic stability	Excellent	Excellent

[a] *n*-Butyl bis(3-hydroxypropyl) phosphine oxide.
[b] Tris(3-hydroxypropyl) phosphine oxide.
Source: FMC, Philadelphia, PA.

Table 9 shows structures and properties of cyclic phosphate compounds. These have been recommended for use in poly(methyl methacrylate), cellulose acetate, low-density polyethylene film, and viscose rayon [26].

D. Antimony Oxide

Antimony oxide is a fine white powder with a melting point of 656°C and a specific gravity of 5.7. A typical chemical analysis of a commercial grade gives 99.5% Sb_2O_3 (83.0% Sb). A superfine powder having an average particle size of 1.3 μm is especially recommended for use in thermoplastics. It has a high tinctorial strength and is satisfactory for white/opaque and pastel stock. A fine powder of 3.0 μm average particle size has low tinctorial strength and is used in poly(vinyl chloride) and polyethylene film, sheet, and molded products where deep color tones and low opacity are desired.

The use of high concentrations of antimony oxide in thermoplastics frequently leads to "afterglow" after the flame has been extinguished. In these cases, the addition of zinc borate or phosphorus compounds will reduce glow.

The fine powder will dust readily during handling. It is recommended that breathing the dust be avoided. Antimony oxide has caused lung tumors in laboratory inhalation studies with animals [27]. Low-dusting, surface-treated powder grades are available. Light coatings of wetting agents such as mineral oil, diisodecyl phthalate, chlorinated paraffin, or other liquids are added to substantially reduce the dusting. Plastic concentrates are also available to eliminate powder dust. These products are antimony oxide dispersed in polymers such as polyethylene, polystyrene, ABS, and ethylene vinyl acetate (EVA) and are available in pelletized form.

Colloidal antimony pentoxide is available in aqueous dispersion and in powder form. It is less tinting than the trioxide. Claims that it is more effective because of its finer particle size are contradictory, presumably due to agglomeration of the powder. The pentoxide is considerably more expensive.

Sodium antimonate is used in polyethylene terephthalate (PET), because antimony oxide may act as a depolymerization catalyst. Particle size and color are closely controlled for compatibility with thermoplastic resins.

E. Alumina Trihydrate

Hydrated minerals such as alumina trihydrate (ATH) contain high concentrations of water and function as flame retardants by liberating their chemically combined water (34.6%) (34.6% for ATH), thereby delaying or inhibiting ignition of a polymer. The endothermic decomposition absorbs heat from the plastic and retards the rate of thermal degradation. The accompanying release of water dilutes the combustible concentration; the lower the fuel contribution, the less heat and

Table 9 Cyclic Phosphate Ester Flame Retardants

Product[a]	Structure

5060

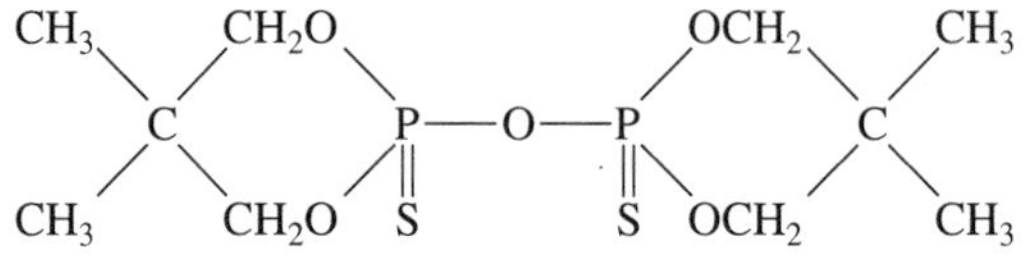

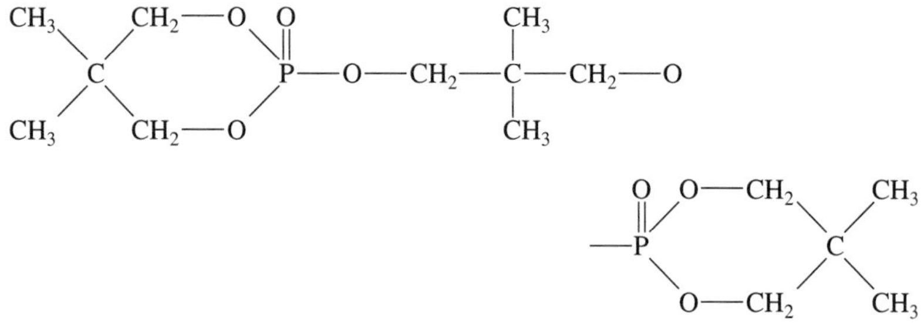

$$\begin{array}{c} CH_3 \\ CH_3 \end{array}\!\!\!>\!C\!<\!\!\!\begin{array}{c} CH_2O \\ CH_2O \end{array}\!\!\!>\!P(=S)\!-\!O\!-\!P(=S)\!<\!\!\!\begin{array}{c} OCH_2 \\ OCH_2 \end{array}\!\!\!>\!C\!<\!\!\!\begin{array}{c} CH_3 \\ CH_3 \end{array}$$

—

$$\begin{array}{c} CH_3 \\ CH_3 \end{array}\!\!\!>\!C\!<\!\!\!\begin{array}{c} CH_2\!-\!O \\ CH_2\!-\!O \end{array}\!\!\!>\!P(=O)\!-\!O\!-\!CH_2\!-\!\underset{\underset{CH_3}{|}}{\overset{\overset{CH_3}{|}}{C}}\!-\!CH_2\!-\!O$$

$$-P\!\!\!\begin{array}{c} \overset{O}{\parallel} \\ \end{array}\!\!\!<\!\!\!\begin{array}{c} O\!-\!CH_2 \\ O\!-\!CH_2 \end{array}\!\!\!>\!C\!<\!\!\!\begin{array}{c} CH_3 \\ CH_3 \end{array}$$

5085

$$\begin{array}{c} CH_3 \\ CH_3 \end{array}\!\!\!>\!C\!<\!\!\!\begin{array}{c} CH_2\!-\!O \\ CH_2\!-\!O \end{array}\!\!\!>\!P(=O)\!-\!O\!-\!CH_2\!-\!\underset{\underset{CH_2Cl}{|}}{\overset{\overset{CH_2Cl}{|}}{C}}\!-\!CH_2\!-\!O$$

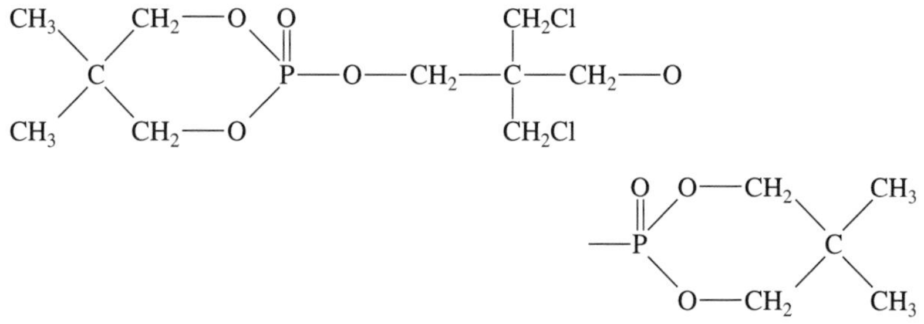

$$-P\!\!\!\begin{array}{c} \overset{O}{\parallel} \\ \end{array}\!\!\!<\!\!\!\begin{array}{c} O\!-\!CH_2 \\ O\!-\!CH_2 \end{array}\!\!\!>\!C\!<\!\!\!\begin{array}{c} CH_3 \\ CH_3 \end{array}$$

VP5086

$$\begin{array}{c} BrCH_2 \\ BrCH_2 \end{array}\!\!\!>\!C\!<\!\!\!\begin{array}{c} CH_2O \\ CH_2O \end{array}\!\!\!>\!P(=O)\!-\!OCH_3$$

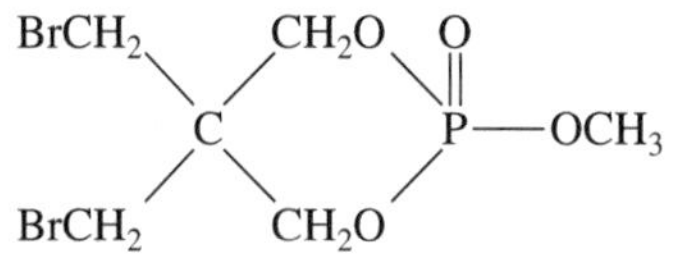

[a] Trade name: Sandoflam (Sandoz, Basel).

Phosphorus, halogen (%)	Melting point (°C)	Molecular weight	TGA (% wt loss/°C)	Application
17.9	228	346	5/212	Low-density polyethylene film; viscose rayon
15.5	127–131	400	5/255	—
P: 13.2 Cl: 15.1	186–190	469	5/270	Poly(methyl methacrylate) molding compound
P: 9.1	65–72	338	5/175	Cast poly(methyl meth-methacrylate); cellulose acetate

Table 10 Physical Properties of Alumina Trihydrate, Magnesium Hydroxide, and Basic Magnesium Carbonate

Property	Additives		
	ATH	$Mg(OH)_2$	$Mg_4(CO_3)_3(OH)_2$ $\cdot 4H_2O$
Bound water (%)	34.6	31.0	20.2
Carbon dioxide (%)	—	—	36.1
Specific gravity (g/cm^3)	2.42	2.36	2.24
Minimum decomposition temperature (°C)	205	320	230
Enthalpy of decomposition (cal/g)	-280	-382	-295

smoke. The water content and temperature of water release for a number of hydrated minerals are shown in Table 10 [28]. For ATH, slow decomposition begins as low as 205°C.

Alumina trihydrate was first used in polyester resins as a low-cost resin extender, and use in thermoset systems with low processing temperatures is still the major application. ATH is used in thermoplastics such as wire and cable polyethylene compounds and PVC.

High concentrations of ATH are required to be effective, and there is a deleterious effect on mechanical properties. Large selections of particle sizes and surface coatings are commercially available to aid in processing and to improve mechanical properties. Surface modification aids in the dispersion, allows for better and more complete filler wet-out, and offers higher filler loading. Tensile properties, elongation, and impact strength are improved.

F. Magnesium Hydroxide and Magnesium Carbonate

Magnesium hydroxide differs from alumina trihydrate in decomposition temperature (320°C versus 205°C). The former will release about 31% water on ignition, slightly less than ATH (Table 10). Plastic compositions containing magnesium hydroxide can be processed at much higher temperatures.

Basic magnesium carbonate differs in that it decomposes over a broader temperature range and releases CO_2 in addition to water. Properties are shown in Table 10.

V. FLAME-RETARDANT THERMOPLASTICS

A. Flame-Retardant Polypropylene

1. Halogen-Containing Flame Retardants

Polypropylene is flame retarded commercially to meet either the Underwriters Laboratories (UL-94) V-0 or V-2 specification [29]. Both products are self-extinguishing by the definition discussed in Section III, but the V-0 product does not drip, whereas the V-2 product drips and ignites the cotton. A V-0 product that drips but does not ignite the cotton also is commercially available.

Nondripping flame-retardant polypropylene requires about 40% of additives. A V-2 product can be prepared with only 5–6% of total flame retardants. Flame-retarding copolymer for a V-2 rating may require slightly more flame retardant [29].

Dechlorane Plus, an alicyclic chlorine-containing compound with 65% chlorine, is used for flame-retarding polypropylene to a V-0 specification. A basic composition contains 27% Dechlorane Plus and 13% antimony oxide as a synergist. At this high antimony oxide level, the product exhibits long "afterglow" time and may fail the UL-94 test. Zinc borate has been used to replace some of the antimony oxide to give a composition with reduced glow and a UL-94 rating of V-0.

The V-0 product containing Dechlorane Plus described in Table 11 has 48% filler. At this high filler loading, the polypropylene no longer exhibits the resiliency of virgin polypropylene, nor does it retain the hinge property. The specific gravity of the compound is increased from 0.91 for the virgin resin to 1.25, because of the high specific gravity of Dechlorane Plus (1.8–2.0) and antimony oxide (5.7). Properties of flame-retardant polypropylene are compared to virgin resin in the table.

Decabromodiphenyl oxide, 83% aromatic bromine, has also been claimed as a flame-retardant additive for UL-94 V-0 polypropylene. The use of talc as a filler is recommended. Presumably, this additional filler is required to inhibit the drip, but it also reduces the cost. The properties of such a product are also shown in Table 11.

A UL-94 V-2 self-extinguishing polypropylene can be obtained with about 4% flame retardant and 2% antimony oxide. Two flame retardants can be used, Saytex BN-451, containing 45% alicyclic bromine, and PE-68, containing equal levels of aliphatic and aromatic bromine for a total bromine concentration of about 66%. Antimony oxide is required. The low add-on levels do not affect the properties of the polymer to any great extent (Table 12). The flame retardant polypropylene is still resilient, melts and drips on heating, and still exhibits hinge

Table 11   Injection-Molded Polypropylene Resin

	Formulations		
	1[a]	2[a]	3
Ingredients (%)			
Polypropylene	100	52	58
Dechlorane Plus 25	—	38	—
Decabromodiphenyl oxide	—	—	20
Antimony oxide	—	4	10
Zinc borate	—	6	—
Talc	—	—	12
Flame Retardancy			
Oxygen index	18	29	26.5
UL-94, 1/16 in.	B	V-0	V-0
Physical Properties			
Heat deflection at 66 psi (°F)	230	252	193 (264 psi)
Tensile strength (psi)	3270	2680	3550
Elongation (%)	33.4	49.0	78
Flexural strength (psi)	9180	7260	—
Flexural modulus $\times$ 10^5 psi	3.03	4.76	—
Izod impact strength, notched (ft-lb/in.)	0.47	0.38	0.55
Hardness, Rockwell, R	84	86	—
Specific gravity (g/cm^3)	0.91	1.25	1.31

[a] Data from Occidental Chemical, Niagara Falls, NY.

properties. The effect of the flame-retardant concentration on oxygen index for various types of halogen-containing compounds is shown in Figure 16.

The impact resistance of flame-retarded polypropylene can be improved by the use of polypropylene copolymers or the addition of low levels of impact modifiers such as ethylene–propylene rubber. This does not appear to interfere with the flame retardancy to a great degree. The Izod impact resistance is improved from about 0.5 to 1.9 ft-lb/in.

A dripping but UL-94 V-0 rated polypropylene is commercially available. The product drips during the UL-94 flammability test, but the polymer drops do not ignite the absorbent cotton. The flame retardant used is Non-Nen 52, containing aliphatic and aromatic bromine in equal concentrations for a total bromine content of about 66%. A flame retardant/antimony oxide concentration ratio of 7:3.5% gives a UL-94 V-0 product with an oxygen index of 31. At a 3:1.5 concentration, a V-2 composition with an oxygen index of 28.0 is obtained. Reoflam PB-370 will also give a V-0 drip product.

Table 12 Flame-Retardant Polypropylene Compounds: V-0 versus V-2

	Product and flammability (UL-94 at 1/16 in.)			
			Reoflam PB-370	
	Dechlorane plus (V-0)	BN-451[a] (V-2)	(V-2)	(V-0)
Specific gravity (g/cm^3)	1.30	0.938	—	—
Hinge properties	No	Yes	Yes	Yes
Heat deflection temperature at 66 psi (°F, annealed)	272	252	—	—
Izod impact strength, notched (ft-lb/in.)	0.6	0.5	0.7	0.7
Continuous-use temperature index (°C at 0.030 in.)	105	105	—	—
Tensile strength (psi)	3000	4400	4,780	4,600
Elongation (%)	25	19	—	—
Flexural modulus ($\times 10^{-5}$ psi)	2.7	2.15	3.4	3.2

[a] See Fig. 8b.

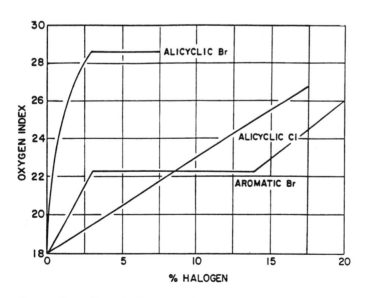

Figure 16 Effect of halogen in polypropylene (FR/Sb$_2$O$_3$ = 2 : 1 bw).

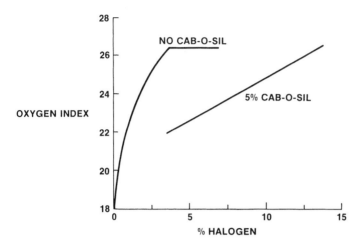

Figure 17 Effect of Cab-O-Sil on the flame retardant polypropylene.

The effect of additives in the plastic composition that can interfere with flame retardancy was discussed in Section II. Figure 17 shows that Cab-O-Sil (fumed silica) can have a dramatic, negative effect on oxygen index as well as the UL-94 rating. This interference can be due either to the formation of stable silicon bromides or to inhibition of dripping by the high-surface-area fumed silica (see pgs 176 and 178). The addition of talc also demonstrates the effect of inhibition of drip (Fig. 18). Here, a self-extinguishing compound is not obtained up to an alicyclic bromine concentration of 8%.

A number of companies supply ready-to-run compounds meeting both UL-94 V-0 and V-2 ratings. Table 13 shows the properties of typical commercial compounds available from Monmouth Plastics and A. Schulman. Note the density differences; presumably the Monmouth Plastics product is a V-0 dripping product, whereas the Schulman material contains a high filler level.

Standard grades of chlorinated paraffins have not been used successfully to flame retard polypropylene. A novel resinous chlorinated paraffin with considerably improved thermal stability has been claimed to give a nondrip UL-94 V-0 product. Chlorez 760 (ICC–Dover Chemical) has a softening point of 160°C and sufficient thermal stability for safe processing of polypropylene formulations. The low cost of this product should be of interest. Table 14 shows V-2 and V-0 compositions.

a. Flame-Retardant Concentrates. Filler-type flame retardants can be difficult to disperse in polypropylene because of the low viscosity of the polymer melt and thus the low shear during processing. It is preferable that these products first be prepared as a master batch or concentrate containing 50–60% of flame retar-

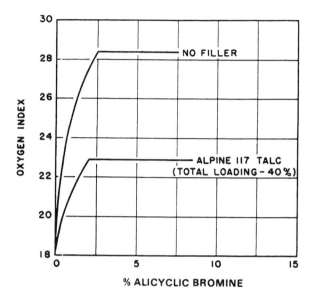

Figure 18   The Flame retardant polypropylene filled with talc (FR/Sb$_2$O$_3$ = 2 : 1). The flame retardant is the alicyclic bromine compound BN-451.

Table 13   Commercially Available Flame Retardant Polypropylene

Property	Flame retardant	
	Monmouth PP301	Schulman RPP 1174
Flammability, UL-94	V-0 at 1/16 in.	V-0 at 1/32 in.
Oxygen index	28	—
Density (g/cm^3)	0.9888	1.39
Melt index (g/10 min)	16	—
Tensile strength at yield (psi)	4900	3400
Elongation at break (%)	—	45
Izod strength, notched (ft-lb/in.)	0.44	0.5
Flexural modulus ($\times$ 10^3 psi)	238	245
Heat distortion temperature (°F)		
At 66 psi	223	240
At 264 psi	—	155
Hardness, Rockwell R	102	93
Dielectric strength (V/mil)	1000	876–950
Volume resistivity ($\times$ 10^{15} Ω-cm)	1.46	2.6–3.2
Dielectric constant	1.52	2.53
Dissipation factor at 10^6 Hz	0.001	0.0035

Table 14   Flame-Retarding Polypropylene with Chlorinated Paraffin

	Formulations		
	1	2	3
Ingredients (%)			
Polypropylene	100	84	62
Chlorez 760	—	12	27
Antimony oxide	—	4	10
Stabilizer	—	—	1
Flame Retardancy			
Oxygen index	18.0	—	27.0
UL-94 at 1/8 in.	Burns	V-2	V-0
Physical Properties			
Hardness, Shore D	64	—	71
Izod impact strength (ft-lb/in.)	10.0	—	0.82
Tensile strength (psi)	3200	—	4000
Elongation (%)	187	—	30
Specific gravity (g/cm^3)	0.920	1.00	1.165

Source: Data sheet, ICC–Dover, Dover, Ohio.

dant, which subsequently is let down into virgin polypropylene. These concentrates may use a 50 : 50 blend of polypropylene and low-density polyethylene (LDPE) or a blend of high-density polyethylene (HDPE) and ethylene–propylene rubber. When let down, the concentration of these polymers is very low in the final let-down product.

A number of companies supply polypropylene concentrates that are let down by the customer in a ratio of 6 : 1 to 12 : 1, depending on the composition of the concentrate and desired composition of the final product (Fig. 19).

b. Hydrated Fillers.   Flame-retarding polypropylene with alumina trihydrate (Table 15) or magnesium hydroxide (Table 16) requires at least a 50% concentration, generally 65% add-on.

Samples of polypropylene containing ATH, magnesium hydroxide, and decabromodiphenyl oxide/antimony oxide were examined for levels of smoke produced. The results in Table 17 show that the onset of smoke emission is significantly retarded for the samples containing the hydrated fillers, presumably due to the early decomposition of the hydrates. Furthermore, the maximum level of smoke evolved is very much lower. Although the detailed mechanism of smoke suppression is not fully understood, it has been suggested that decomposition products of the hydrated fillers function in part through modification of the pyro-

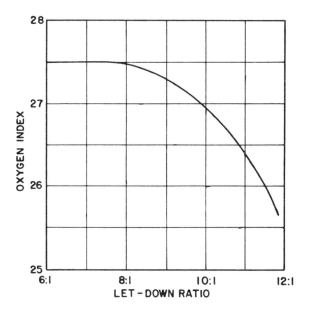

Figure 19 Oxygen index versus let-down ratio of flame-retardant concentrate into polypropylene. (From data sheet, Monmouth Plastics, Asbury Park, NJ.)

lytic process in the condensed phase [31]. These products also have significantly less fuel (40% polymer) to burn.

The use of Ucarsil organosilicone chemicals (Union Carbide) substantially improves the dispersibility of hydrated mineral fillers in polyolefins (Table 18). This allows for a high concentration of filler for greater flame retardancy. Compounding problems and the reduction in mechanical properties associated with high loadings of hydrated minerals are substantially reduced. Particularly im-

Table 15 Polypropylene[a] (35%) Flame Retarded with Alumina Trihydrate (65%)

Property	Value
ASTM radiant panel flame spread index	25
NBS smoke density, D_s (4 min)	10
UL-94 flammability	V-0

[a] Including stabilizer package.

Table 16 Flame-Retarding Polypropylene with Magnesium Hydroxide

	$Mg(OH)_2$ concentration (%)	
Property	50	60
UL-94 flammability	V-1	V-0
Oxygen index	37	34
Izod impact strength (ft-lb/in.)	4.1	2.9
Tensile modulus ($\times 10^3$ psi)	446	607
Ultimate tensile strength (psi)	5010	5220
Elongation (%)	3.74	1.66

Source: Ref. 30.

Table 17 Smoke from Flame-Retardant Polypropylene

Resin	Time to onset of smoke (s)	Maximum obscuration (%)
Polypropylene (PP)	85	34
PP + 60% ATH	220	22
PP + 60% $Mg(OH)_2$	210	24
PP + 15% decabromodiphenyl oxide + 8% Sb_2O_3	25	96

Table 18 Effect of Silicon Dispersion Aid Ucarsil FROSG on Composition Containing 35% Polypropylene and 65% Alumina Trihydrate

	Formulations	
Property	No dispersion aid	With Ucarsil FROSG, 1.3%
Tensile strength (psi)	2800	4250
Gardner impact strength (in.-lb)	16	24
Izod impact strength, unnotched (ft-lb/in.)	0.31	0.96
Tangential flexural modulus ($\times 10^3$ psi)	490	860
Flexural yield strength (psi)	4200	7800

Source: Data sheet, Ucarsil, FR, Union Carbide, Danbury, CT.

proved is the impact strength. Only about 2% of the organosilicone compound is needed.

c. *Intumescent Coatings.* Intumescent coatings function by decomposing under heat to give a thick char that insulates the substrate from flame, heat, and oxygen. The intumescent system depends on a char former, a polyol such as pentaerythritol, an acid source (catalyst or dehydrating agent) such as a phosphate, and a blowing agent or nitrogen source such as melamine. These flame retardants contain no halogen or antimony.

Several companies have developed intumescent additives for polypropylene. One of these additives is CN-329, a nitrogen–phosphorus additive recommended for polypropylene. CN-329 is stable at processing conditions, does not migrate from the polymer, and yields flame-retardant polypropylene compounds of low specific gravity. An additive level of 30% gives a UL-94 rating of V-0 at $\frac{1}{16}$ in. and an oxygen index of 34. The specific gravity is only 1.03. The product has good electrical properties and the capacity to form a "living hinge." Flame-retardant polypropylene containing CN-329 generates smoke density values that are very close to those of the non-flame-retardant resin [30,32,33]. Typical properties are given in Table 19.

Table 19 Typical Properties of Intumescent Polypropylene

	Flame retardant	
Property	None	CN-329, 30%
Specific gravity (g/cm^3)	0.92	1.03
Flammability, UL-94		
1/8 in.	Fail	V-0
1/16 in.	Fail	V-0
Oxygen index	<18	34
Melt flow index (g/10 min)	4.3	3.3
Heat distortion temperature at 66 psi (°F)	215	265
Tensile strength (psi)	5620	4930
Elongation (%)	22	15
Izod impact strength (ft-lb/in.)		
Notched	0.9	0.9
Unnotched	22.5	4.5
Dart impact strength (in.-lb)	20	20
Flexural modulus ($\times$ 10^5 psi)	1.64	3.70
Flexural fatigue (1%) ($\times$ 10^6 cycles)	9.5	4.2
NBS smoke density (D_m corr)		
Flaming mode	50	80
Nonflaming mode	660	720

Source: Data sheet, Great Lakes Chemical, West Lafayette, IN.

Table 20 Silicone Flame-Retarded Polypropylene

	Formulation			
	Base resin	Halogen + ATH	Low halogen	No halogen
Ingredients (%)				
Polypropylene A[a]	100	47.4	74.2	—
Polypropylene B[b]	—	—	—	78.5
Silicone retardant[c]	—	10.5	9.5	10.7
Magnesium stearate	—	4.7	4.4	3.6
Decabromodiphenyl oxide	—	13.5	6.9	—
Alumina trihydrate	—	23.9	—	—
Talc	—	—	5.0	7.2
Physical Properties				
UL-94 flammability at 1/16 in	—	V-0/5V	V-1	V-1
Oxygen index (%)	—	30	27	23
Specific gravity (g/cm^3)	0.9	1.2	1.01	1.06
Melt index (g/10 min)	4	10–25	10	6
Tensile strength (psi)				
At yield	5400	2730	3700	2400
At break	—	1250	—	2100
Gardner impact strength (in.-lb)				
At 23°C	20	160–200	160–200	160
At 0°C	5	80	40–60	40
At −18°C	5	35	20–30	40
Notched Izod impact strength (ft-lb/in.)	0.47	0.9	1.6	2.6
Flexural modulus (10^3 psi)	195	200	302	120
Heat distortion temperature (°F)				
At 66 psi	220	185	280	141
At 264 psi	—	130	131	109
Hardness Rockwell R	100	60	83	93
Volume resistivity (× 10^3 Ω-cm)	297	1.6	21.0	58
Dielectric strength (V/mil)	666	790	811	622
Dielectric constant	1.67	2.42	1.61	1.08
Dissipation factor at 10^6 Hz	0.0004	0.0036	0.0120	0.0008

[a] Nortech NPP 8000GK.
[b] Himont SE 191.
[c] General Electric SF100.
Source: Ref. 36.

MF-80, another intumescent additive for polypropylene, is a water-insoluble, hygroscopic, organic polycondensate containing 28% nitrogen, which is blended with 2.5 parts of ammonium polyphosphate [34]. Reference 35 is another patent on an intumescent flame retardant for polypropylene.

2. Silicone-Based Flame Retardants

Polypropylene has been flame retarded using a silicone-based system with some halogen and no antimony. The flame retardant designated SFR 100 by General Electric is a combination of reactive silicone polymers, a linear silicone fluid or gum, a silicone resin that is soluble in the fluid, and a metal soap. Magnesium stearate is the preferred soap. Silicones have also been used to flame retard cross-linked polyethylene.

During combustion, the silicone mixture functions synergistically with the magnesium stearate to produce a hard, slightly intumescent char that insulates the remaining substrate from heat and eliminates drip. Some decabromodiphenyl oxide and alumina trihydrate are used. Bromine from the decabromodiphenyl oxide is held mostly in the char and does not show up as a gaseous product.

The use of the silicone system alone raises the oxygen index of polypropylene to 23. The addition of 7% decabromodiphenyl oxide brings the oxygen to 27. The further addition of alumina trihydrate raises the oxygen index value to 30. Properties are shown in Table 20.

The impact strength of the SFR 100 polypropylene formulation, as measured by the Gardner impact test procedure, is significantly improved: Values are up to 10 times those reported for current conventional flame-retardant grades of polypropylene. The SFR 100 system also offers another nonhalogen system. The product appears to be of particular interest to the wire and cable industry.

3. Summary

Table 21 summarizes the various methods of flame retarding polypropylene discussed in this section. Table 22 gives the properties of the various halogen-containing flame retardants used in polypropylene.

B. Flame-Retardant Polyethylene

The major applications for flame-retardant low-density polyethylene include blown film, wire and cable insulation and jacketing, molded goods, and extrusion coating. Flame retardants used commercially in polyethylene include chlorinated paraffins, Dechlorane Plus (alicyclic chlorine), decabromodiphenyl oxide (aromatic bromine), and Saytex BT-93 (aromatic bromine) [29].

Although LDPE and polypropylene are both polyolefins, aromatic bromine compounds appear to be more effective in the former. The higher pyrolysis temperature for polyethylene ($600 +$ °C) compared to polypropylene (480°C) is a possible explanation for this difference [37].

Table 21   Flame Retardants for Polypropylene

Flame retardant	UL-94 rating	Active ingredient(s) (%)	Total flame retardant (%)
Dechlorane Plus	V-0	65 Cl	40+
Decabromodiphenyl oxide	V-0	83 Br	40+
Chlorinated paraffin	V-0	70 Cl	40+
Saytex BN-451	V-2	45 Br	6
PE-68	V-2	67.7 Br	4
Reoflam PB-370	V-2/V-0	70 Br/3 P	6/12
Non-Nen-52	V-0[a]	66 Br	8–10
CN-329	V-0	P and N	30
SFR 100	V-0	Several[b]	52.6

[a] Drips but does not ignite the cotton.
[b] Silicone, bromine, magnesium stearate, and alumina trihydrate.

The relatively poor thermal stability of standard grades of chlorinated paraffin (~190°C) restricts the use of this very low-cost additive to low-temperature processing such as in low-density polyethylene blown film. Another problem attributable to the low thermal stability of the additive is difficulty in reusing the scrap, which discolors upon reprocessing. However, the low cost of the chlorinated paraffins and their high flame-retardant efficiency in polyethylene make it a favorite additive whenever it can be used. Wire and cable compounds use a

Table 22   Properties of Halogen Flame Retardants Used for Polypropylene

Property	Dechlorane plus[a]	Saytex BN-451[b]	Great Lakes PE-68[c]	Non-Nen-53[d]	Reoflam PB-370
Appearance	White crystalline, solid	White, free-flowing powder	White to pale yellow powder	Light yellow powder	White powder
Halogen content	65% chlorine	45% bromine	65–67% bromine	66% bromine	70% bromine
Halogen type	Alicyclic	Alicyclic	33% aromatic, 33% aliphatic	33% aromatic, 33% aliphatic	Aliphatic
Melting point (°C)	350 (dec)	294	90–100	~50	181
Decomposition temperature (°C)	350	ca. 300	~275	~320	~230
Specific gravity (g/cm³)	1.8–2.0	2.07	~2.2	2.29	2.275
Solubility	1–2% in aromatics, 1.4% in trichloroethylene, 0.7% in methylethyl ketone, 0.1% in hexane, 0.1% in methyl alcohol	Essentially insoluble in water and common solvents	Insoluble in water, soluble in halogenated hydrocarbons, toluene, acetone		<1% in methanol, hexane, toluene, 9.75% in methylene chloride

[a] Occidental Chemical.
[b] Saytech/Ethyl.
[c] Great Lakes Chemical.
[d] Marubishi.

Table 23   Flame-Retardant Polyethylene

	Formulations		
	1	2	3
Ingredients			
Polyethylene, LDPE	100%	84%	66%
Chlorez 700	—	12	24
Antimony oxide	—	4	10
Flame Retardancy			
UL-94 at 1/8 in.	Burns	V-2	V-0
Oxygen index (%)	18.5	—	24.5
Arapahoe smoke test result (%)	2	—	11
Physical Properties			
Hardness, Shore A/D	96/45	—	98/54
Melt index (g/10 min)	2.4	—	8.8
Izod impact strength (ft-lb/in.)	6.2	—	2.0
Tensile strength (psi)	1461	—	1459
Heat-distortion temperature (°F)			
At 264 psi	95	—	104
At 66 psi	113	—	122
Specific gravity (g/cm^3)	0.920	—	1.129

Source: Data sheets, Dover Chemical, Dover, OH.

number of additives such as Dechlorane Plus, decabromodiphenyl oxide, and Saytex BT-93.

Chlorinated paraffins come in many grades, ranging in chlorine content from 40% to 70%. The lower-halogen-containing products are liquids and the 70% products are resinous solids. A concentration of 12 parts of a 70% chlorine-containing paraffin and 4 parts of antimony oxide has been used effectively to impart a UL-94 V-2 rating to most grades of low-density polyethylene. Higher levels of flame retardants (specifically, 24 parts of Chlorex 700 and 10 parts of antimony oxide) are needed for a UL-94 V-0 rating ($\frac{1}{8}$ in.), as shown in Table 23.

The chlorinated paraffins tend to bloom from polyethylene, resulting in an oily film on the surface. This may present problems in heat sealing and in surface printing.

The amount of flame retardant necessary to flame retard thermoplastic polyethylene appears to be a function of the melt index of the polyethylene; the higher the melt index (greater flow) of the base resin, the less flame retardant required, as measured by oxygen index and by UL-94 [29]. The most likely explanation involves loss of heat from the flame zone by the dripping hot polymer, as discussed earlier.

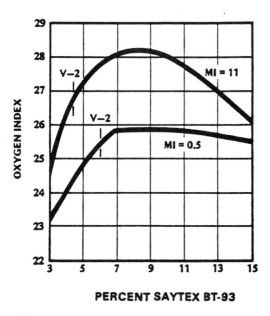

PERCENT SAYTEX BT-93

Figure 20   Effect of aromatic bromine on the oxygen index of low-density polyethylene $(FR/Sb_2O_3 = 3 : 1)$, where MI = melt index.

Figure 20 compares the flammability of two low-density resins of different melt index. The flame-retardant additive contains aromatic bromine. As can be seen, higher oxygen index values can be obtained using the higher melt flow resin. Evaluation in the UL-94 test shows that about 4.5% Saytex BT-93 is required to give a V-2 rating for the 11 melt index resin, whereas 6% additive is required for the 0.5 melt index resin. The oxygen index values for the two resins at these V-2 points differ by about one unit. From a practical application, however, one may wish to use the same concentration of flame retardant for each resin. The UL-94 V-2 points fall on the steep slope of the curves. In production, a slight error in weighing or poor dispersion could result in a burning composition. Operation at the peak or plateau of the curve with both resins is consistent with good practice, and the use of approximately 7% flame-retardant additive is recommended.

The effect of antimony oxide concentration on the flame retardancy of LDPE is shown in Table 24. These data show that antimony is much more efficient in polyethylene than we had seen earlier in polypropylene. A 4 : 1 ratio of flame retardant (FR) to antimony oxide is quite sufficient to give a UL-94 rating of V-2 with an oxygen index of 27.2. Using the 25 melt index resin, the oxygen index values were 27.8 and 27.5 for 2 : 1 and 4 : 1 FR/antimony oxide, respectively. The average burning time in the UL-94 test for both of these samples was only

Table 24 Effect of Antimony Oxide Concentration on Flame-Retardant Properties of Low-Density Polyethylene

Melt index	Concentration (%)		Average burn time, UL-94 (s)	UL-94 rating	Oxygen index
	BT-93	Sb_2O_3			
4.5	9	4.5	3	V-2	26.2
	9	3.0	1	V-2	26.9
	9	2.25	1	V-2	27.2
25	8	4	1	V-2	27.8
	8	3	1	V-2	27.8
	8	2	1	V-2	27.5
	8	1	17	Burns	27.2

1 s. Laboratory data indicate that LDPE having a 0.5 melt index is difficult to flame retard.

The products of Fig. 20 contain an aromatic bromine compound and are rated UL-94 V-2. The addition of low-cost inert fillers such as talc can be used to produce UL-94 V-0 low-density polyethylene. However, additional flame retardant and antimony oxide are needed. A possible explanation is that the filled polymer is not cooled by hot polymer dripping from the flame zone, as discussed in Section II. A composition containing 30% inert talc, 12% Saytex BT-93, and 6% antimony oxide gives a UL-94 V-0 compound with an oxygen index value of 28.

Polyethylene can be cross-linked with radiation or with peroxides such as dicumyl peroxide. One of the major applications is wire and cable insulation and jacketing. The preferred flame retardant has been the alicyclic chlorine-containing compound Dechlorane Plus. Aromatic bromine-containing compounds are more effective flame retardants as measured by the oxygen index and UL-94. However, when coated wire is tested by the UL-44 VW-1 flammability test, the Dechlorane Plus system passes. Large levels (50 phr) of medium thermal (MT) carbon black frequently are added, with little effect on flame retardancy. Dechlorane Plus, decabromodiphenyl oxide, and Saytex BT-93 are compared for wire and cable application in Table 25.

Many flame retardants bloom extensively from polyethylene. For example, in unfilled cross-linked polyethylene at 65°C, Dechlorane Plus showed bloom in 115 h, and decabromodiphenyl oxide in less than 75 h. Saytex BT-93 (aromatic bromine) does not bloom from polyethylene (Table 25).

Table 26 shows the properties of cross-linked wire and cable compounds containing Dechlorane Plus as the flame retardant. The two grades of Dechlorane Plus differ in particle size (see Section IV).

Table 25 Flame-Retardant Cross-Linked Polyethylene

	Formulations (parts)		
	1	2	3
Ingredients			
Polyethylene	100	100	100
Dechlorane Plus	30	—	—
Decabromodiphenyl oxide	—	30	—
Saytex BT-93	—	—	30
Antimony oxide	15	15	15
Vol-Cup 40 KE	5	5	5
Flame Retardancy			
Oxygen index	24.6	26.1	26.1
UL-94	Burns[a]	V-0	V-0
Bloom	Slight	Very heavy	None

[a] Passes UL-44 VW-1 vertical flammability test on wire.

Table 26 Cross-Linked Low-Density Polyethylene

	Formulations (parts)		
	1	2	3
Ingredients			
Polyethylene	100	100	100
Dechlorane Plus 25	—	60	—
Dechlorane Plus 2520	—	—	60
Antimony trioxide	—	20	20
Thermax MT carbon	20	20	20
Diphos	2	2	2
Dicup-R	2.7	2.7	2.7
Flame Retardancy			
Oxygen index (%)	18.9	29.2	28.8
UL-94 at 1/8 in.	Burns	V-0	V-0
Tensile Properties			
100% modulus (psi)	1300	1065	1055
300% modulus (psi)	1875	1260	1430
Tensile strength (psi)	2490	1470	1480
Elongation	400	310	315

Source: Data sheet, Occidental Chemical, Niagara Falls, NY.

Table 27 Flame-Retarding Polyethylene with Hydrated Fillers

| | UL-94 | Oxygen index | Smoke density (D_m) | |
			Flaming	Smoldering
Alumina trihydrate				
40%	V-1	31	—	—
60%	V-0	31	147	214
Magnesium hydroxide				
40%	V-1	28	—	—
50%	—	—	79	231
60%	V-0	37	—	—

Source: Ref. 30.

Tables 27 and 28 shows that it takes 40–65% alumina trihydrate or magnesium hydroxide to flame retard polyethylene.

Low-density polyethylene film is conventionally flame retarded with halocarbons and antimony oxide. Obviously, these films are opaque. A transparent film has been reported using a dithiopyrophosphate (Sandoflam 5060) flame retardant. The flame retardant is used commercially in viscose rayon. Concentrations of 5–15% will give self-extinguishing products that are light transparent, resistant to migration, free of halogen and antimony oxide, and compatible with antistatic agents and photostabilizers [26]. Table 29 shows the oxygen index as a function of concentration. The properties of Sandoflam 5060 are shown in Table 30.

Flame-retarding, high-density polyethylene is similar to the process for LDPE, and the same flame retardants and principles apply.

Flame retardants release a low concentration of halogen acid during high-temperature processing. As a result, certain maintenance procedures are suggested for downtime to inhibit corrosion of molds and dies. For short-duration downtime, the barrel should be purged with a nonretarded polymer containing 1% sodium

Table 28 High-Density Polyethylene[a] (35%) Flame Retarded with Alumina Trihydrate (65%)

Property	Value
ASTM radiant panel flame spread index	20
NBS Smoke density, D_s (4 min)	10
Flammability, UL-94	V-0

[a] Including stabilizer package.

Table 29 Flame-Retardant Low-
Density Polyethylene Blown Film with
Dithiopyrophosphate (Sandoflam 5060)

FR concentration (%)	Oxygen index
0	17.5
5	22.8
10	25.0
15	28.3

Source: Ref. 26.

stearate, followed by purging of the die and mold. For long-term downtime, the dies and molds should be cleaned and then washed with a saturated solution of sodium bicarbonate. After rinsing and thorough drying, a light coating of oil or grease should be applied. Another procedure is to polish mold cavities and runners with Noxon or other metal polish containing ammonia.

C. General-Purpose Polystyrene

General-purpose polystyrene is used for a myriad of applications. The requirement for flame retardancy is generally limited to polystyrene foam insulation. This material is of two types: extruded foam (e.g., Styrofoam) and expandable polysty-

Table 30 Properties of Sandoflan 5060, a Dithiopyrophosphate

Structure	
	CH_3, CH_2O, OCH_2, CH_3 — C, P—O—P, C — CH_3, CH_2O, S, S, OCH_2, CH_3
Appearance	White crystalline solid
Analysis	C: 34.7%; H: 5.8%; P: 17.9%, S: 18.5%
Melting point (°C)	228–229
Density at 20°C (g/cm^3)	1.38
Thermogravimetric analysis (N_2, 6°C/min) weight loss	5%/220°C
Solubility (g/L)	
Methylene chloride	9
Water	<0.5

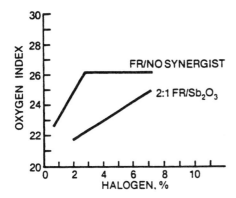

Figure 21   Effect of antimony oxide on flame-retardant general-purpose polystyrene.

rene (EPS) bead board. These construction materials need to be qualified by the ASTM-E-84 25-ft tunnel test, and they compete with polyurethane rigid foam insulation.

Three alicyclic bromine-containing flame retardants are used commercially: pentabromochlorocyclohexane, hexabromocyclododecane, and dibromoethyldibromocyclohexane. Pentabromochlorocyclohexane has been used to flame retard both extruded foam and expandable bead board, whereas the other two are used mainly in the latter application. Very low concentrations of flame retardant are used in expandable bead board, generally less than 1%. The use of these flame retardants in extruded or molded products is limited by their poor thermal stability. Hexabromocyclododecane is the major commercial product.

No antimony oxide synergist is used in general-purpose polystyrene. In fact, halogen and antimony oxide can appear to be antagonistic, as shown in Fig. 21 [38]. The reduction in oxygen index may be due simply to inhibition of dripping.

In a UL-94 test, these products would pass V-2; they are very drippy, and flaming plastic drops away from the sample, aiding in extinguishing the burning material. In the E-84 tunnel test and in any horizontal test, the foam both drips and shrinks away from the burner. Properties of flame-retardant additives which have been used commercially in polystyrene foam insulation are shown in Table 31.

D.   Styrene–Acrylonitrile Copolymer

Alicyclic bromine-containing compounds are highly efficient flame retardant additives for styrene–acrylonitrile (SAN) resins. Compositions containing about

Table 31 Flame Retardants Used Commercially for Polystyrene Foam Insulation

Property	FR-651-P (Dow): pentabromochloro- cyclohexane	CD-75P (Great Lakes chemical): hexabromo- cyclododecane[a]	Saytex BCL-462 (Ethyl): dibromoethyldibromo- cyclohexane
Molecular weight	513.5	642	428
Bromine content	77.9	75	74
Chlorine content (%)	6.9	—	—
Appearance	White powder	White powder	White crystalline powder
Melting range (°C)	>180	185–195	65–80
Decomposition point (°C), TGA[b]	230	—	135

[a] Also available from Ethyl and Ameribrom.
[b] TGA: thermogravimetric analysis.

4% alicyclic bromine exhibit oxygen index values of about 28 and give ratings of V-2 by the UL vertical test. The products are transparent. As with general-purpose polystyrene, the addition of antimony oxide results in decreased oxygen index values. The poor thermal stability of the additives needs to be considered.

E. Impact Polystyrene

High-impact polystyrene (HIPS) differs from the general-purpose grades discussed earlier in cost, properties, and applications. Impact polystyrene, which contains rubber, is also more difficult to flame retard and requires antimony oxide as a synergist [38]. Applications that require flame retardancy include television receiver cabinets and business machine housings. Structural foam grades for these applications also need to be flame retardant.

The flame retardant used commercially almost to the exlusion of all others is decabromodiphenyl oxide. This product contains 83% aromatic bromine and must be used with antimony oxide as a synergist for maximum flame-retardant efficiency. The flame-retardant specification is generally UL-94 V-0, which is attainable with about 12% decabromodiphenyl oxide and 4% antimony oxide. For small housings, a V-2 product may be acceptable.

Laboratory data comparing decabromodiphenyl oxide with two other aromatic bromine compounds is shown in Table 32. These would be satisfactory replacements and would not be expected to yield dioxins or furans during combustion. But they are about 2.5 times more expensive.

Table 32 Comparison of Flame Retardants in HIPS[a]

	DBDPO	S-8010	BT-93
%Bromine	10	10	8
Oxygen index	26.6	28	25.7
UL-94 @ 1.6 mm	V-0	V-0	V-0
Izod Impact (ft-lb/in.)	1.8	0.8	0.5
Izod impact[b] (ft-lb/in.)	2.6	2	1.5
Gardner impact (ft-lb/in.)	3.7	2.8	2.2
DTUL (°F)	165	167	169
UV stability, xenon arc 300 hs ΔE	54.9	26.2	6.3

Note: DBDPO = dcabromodiphenyl oxide; S-8010 = decabromodiphenyl ethane; BT-93 = ethylene bis-tetrabromonorbornane dicarboximide.
[a] 12% flame retardant/4% antimony oxide.
[b] With 3% impact modifier.
Source: FA Pettigrew, SD Landry, JS Reed, FR Conf. 92, New York: Elsevier Applied Sciences, 1992, pp 156–167.

The addition of flame-retardant additives to HIPS has a deleterious effect on the impact properties of the polystyrene. An additional impact modifier can be added to help maintain the impact resistance. When a plasticizing flame retardant is used, the impact may be retained and the resin may process better, but the heat-distortion temperature drops, perhaps below specification. Properties of commercial grades of flame-retardant impact polystyrene used in television receiver cabinets are shown in Table 33. The key properties are Izod impact strength and heat-deflection temperature.

Table 33 Properties of Commercial Grades of Flame-Retarded Impact Polystyrene

	Property/resin				
	Styron 779	PS702	PS707	PS779	PS8080
Manufacturer	Dow	Huntsman	Huntsman	Huntsman	Mobil
Density (g/cm³)	1.16	1.17	1.17	1.18	1.16
Melt flow (g/10 min)	7.5	4.5	4.5	2.5	9.5
Izod impact, notched (ft-lb/in.)	2.8	1.9	3	1.7	2
Vicat softening point (°F)	208	207	203	215	207
Heat-deflection temp. @ 264 psi (°F)	185	183	181	185	180

Aromatic bromine-containing compounds such as decabromodiphenyl oxide absorb in the sunlight wavelength of light. As a result, flame-retardant polystyrene shows very poor UV resistance. The usual UV stabilizers are not very effective in these systems. The use of Saytex BT-93 may give products with better UV resistance. Television receiver cabinets and office machine housings are coated, for example, by painting. Television receiver cabinets are acceptable in black and dark brown.

The European market has been searching for nonhalogen replacements for decabromodiphenyl oxide because of its potential to form the toxic brominated dioxins and furans during combustion. One response has been to use about 5% of a triaryl phosphate ester to promote dripping for an UL-94 V-2 rated product. Another suggestion is to eliminate all flame retardant and make the part thicker just so it passes the V-2 rating. Another approach is to isolate and protect select areas such as the high-voltage area, obviating the need to flame retard the housing.

F. Acrylonitrile–Butadiene–Styrene Resin

Acrylonitrile–butadiene–styrene (ABS) is an impact resin with greater impact strength and slightly higher heat-distortion temperature than impact polystyrene. It is for these reasons and perhaps because of its better surface properties that ABS commands a higher price. ABS and impact polystyrene compete in many applications, and flame retardancy is a specification for many of the applications.

The ABS resins have greater impact resistance than impact grades of polystyrene but are very sensitive to pigment fillers. For example, the conventional flame-retardant system used for impact polystyrene, decabromodiphenyl oxide/antimony oxide, reduces the Izod impact strength from about 2.8 ft-lb/in. for the virgin impact polystyrene to about 2.0. In ABS, this flame-retardant system reduces the Izod values to about a third of the virgin resin. The effect of antimony oxide on Izod impact in ABS is shown graphically in Fig. 22. As a result, the flame retardants of choice for ABS are melt-blendable materials, which presumably plasticize the resin.

The three additives used commercially in ABS have been bis(tribromophenoxy)ethane (FF-680), octabromodiphenyl oxide, and tetrabromo-bis-phenol A. Brominated epoxy oligomers have replaced the octabromodiphenyl oxide. Also available is an ABS/PVC alloy in which the poly(vinyl chloride) is the flame retardant.

Table 34 compares the properties of the resins containing these bromine flame retardants. About 22% bis (tribromophenoxy) ethane plus 3% of antimony oxide is required for a UL-94 V-0 rating. A major problem with this flame retardant is that it tends to plateout during molding. This necessitates frequent mold cleanout, resulting in relatively long molding cycles. This plateout can be avoided by using octabromodiphenyl oxide, which, however, suffers from severe UV light discoloration and presents toxicity problems. Octabromodiphenyl oxide has been

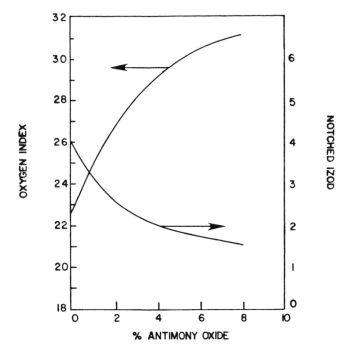

Figure 22 Effect of antimony oxide concentration on the properties of acrylonitrile–butadiene–styrene containing 13 wt% flame retardant.

replaced with a brominated epoxy oligomer which does not bloom, offers a high level of UV resistance, and has good thermal stability, good heat-deflection temperature, and excellent flow. A potential problem is black specks and the oligomer may need to be chain capped to improve it's thermal stability. Tetrabromo-bisphenol A gives ABS compounds of lower heat-distortion temperature, even below that attainable by flame-retardant HIPS. The impact resistance is also reduced significantly. The compound is unsuitable for television cabinet application. The ABS compound also shows poor heat and light stability. Choosing a flame retardant for ABS is therefore a compromise.

An ABS/PVC alloy is used mainly in sheet molding applications. The active flame retardant is chlorine from the poly(vinyl chloride). These alloy have very high impact resistance, with notched Izod impact strength values of 10.5 ft-lb/in. and Gardner impacts of 26+ in.-lb. UL-94 ratings of V-1 at 0.058 in. thickness are attainable. The oxygen index is 30.

Decomposition of the PVC may be caused by simple overheating, excessive frictional heat, or excessive dwell time at elevated temperatures. The recom-

Table 34 Typical Formulations and Properties for Flame-Retardant
Acrylonitrile–Butadiene–Styrene Resins

		Formulations			
Flame retardant	None	Bis(tribromo-phenoxy) ethane	Octabromo-diphenyl oxide	Tetrabromo-bisphenol A	Brominated epoxy oligomer
Ingredients (%)					
FR	—	21.5	18.8	18	21
Antimony oxide	—	3.0	3.1	3.6	6.8
Bromine (theoretical)	—	15	15	10.6	11
Flame Retardancy					
UL-94 at 1/16 in.	HB[a]	V-0	V-0	—	V-0
Average burn time (s)	—	1	0	—	—
Oxygen index	18.5	—	—	29.5	—
Physical Properties					
Izod impact strength, notched (ft-lb/in.)	6.7	3.3	3.7	1.6	2.2
Heat-distortion temperature (°F)	197	186	193	161	189
UV stability, Gardner ΔE	1.1	3.1	—	9.7	—

[a] Horizontal burn.

mended processing temperature is 370–400°F, not to be exceed 400°F. This product is lower in cost than ABS that has been flame retarded with bromine-containing compounds.

Flame-retardant ABS gives injection-molded parts of high gloss as compared with a dull surface for flame-retardant impact polystyrene. Although the
former has better UV stability, most flame-retarded ABS and HIPS housings are
painted and dark in color; painting adds substantially to the cost.

G. Modified Polyphenylene Oxide

Alloys of polyphenylene oxide (PPO) and HIPS are produced by General Electric
under the Noryl trade name. Commercial grades contain 35–45% PPO and
65–55% HIPS. PPO itself is very difficult to process and the addition of HIPS

gives a good processing resin. The composition is determined by the properties desired. PPO chars during the pyrolysis.

Modified PPO is compared with the other two commodity impact resins, HIPS and ABS. The flame retardants in HIPS and ABS are solid bromine additives with antimony oxide as synergist. Modified PPO uses liquid phosphate esters as the flame retardant and antimony oxide is not needed. Commercial grades of modified PPO contain about 13% of a phosphate ester to give products with a UL-94 rating of V-0 at 1.6 mm. The oxygen index of these compounds are 30–31, the highest of the group.

The modified PPO resins have notched Izod impacts of about 7 ft-lb/in., twice that of ABS (3.5–4.0), which is considerably higher that that for HIPS. All three resins have similar heat-distortion values with the HIPS being 5–10°F lower than ABS. Higher heat-distortion resins are available from the modified PPO resin depending on the concentration of the HIPS in the blend. The density of the HIPS and ABS resins is significantly higher because of the high specific gravity of bromine compounds and antimony oxide. A comparison of these three resins in given in Table 35.

The alkylated phosphate esters are thermally stable at processing tempera-ture but are volatile and can "juice" during injection molding. If the phosphate ester condenses on the mold and the molded part at a high stress area, the part will stress crack. These monophosphates, therefore, have been replaced, almost in total, with the less volatile resorcinol diphosphate (RDP). Stress cracking is

Table 35 Comparison of Flame Retardant Impact Resins

	Resin type		
Property	Impact polystyrene[a]	ABS[b]	Modified PPO[c]
Density (g/cm^3)	1.18	1.22	1.08
Izod impact strength (notched ft-lb/in.)	1.7	4.0	7.0
Heat-deflection temperature at 264 psi (°F)	185	190	190
UL-94 rating/thickness (in.)	V-0/0.058	V-0/0.060	V-0/0.060
Flame retardant	Decabromodiphenyl oxide and antimony oxide	Bis(tribromophenoxy) ethane (FF-680) and antimony oxide	Phosphate ester

[a] PS 351, Huntsman.
[b] KJT General Electric.
[c] Noryl N-190, General Electric

Table 36 Flame-Retardant Noryl Resin (Modified PPO)

Description	
Flame retardant	Reofos 50[a]
FR concentration (%)	13
Flame Retardancy	
UL-94 rating at $\frac{1}{16}$ in.	V-0,
	5.0 s
Oxygen index	30.9
Physical Properties	
Heat-deflection temperature at 264 psi (°F)	165
Impact strength	
Izod, notched (ft-lb/in.)	4.1
Gardner (in.-lb)	160
Flexural strength (psi)	9490
Flexural modulus ($\times 10^{-5}$ psi)	3.39

[a] Isopropylated triphenyl phosphate.

considerably reduced. bis-Phenol A could replace the expensive and single source resorcinol. This new compound would be more hydrolytically stable. Tables 36 and 37 show properties of commercial grades of Noryl resin produced by GE Plastics. The resins with the higher heat distortion presumably have a higher concentration of PPO.

Phosphine oxides have been reported to be effective flame retardants for modified PPO. The resultant properties are consistent with those obtained with phosphate esters. These phosphine oxides contain about 14% phosphorus, compared to about 8% for phosphate esters and therefore less additive is required to obtain the same level of flame retardancy [24,25].

Table 37 Properties of Modified PPO Resins[a]

Property/Grade	731	N190X	SE100	N225	SE1	N300
UL-94 rating (in.)	HB/0.058	V-0/0.058	V-1/0.060	V-0/0.060	V-1/0.058	V-0/0.068
Specific gravity	1.06	1.12	1.08	1.09	1.08	1.09
Heat-distortion temp., 264 psi (°F)	260	190	210	215	255	295
Izod impact, notched	4	5.5	5.5	4.5	3.5	5

[a] GE Plastics data sheet.

H. Polycarbonate Resins

Polycarbonates are high-impact resins with Izod impacts of about 15. They are inherently flame resistant, having an oxygen index of about 25 and a UL-94 self-extinguishing rating of V-2. Products with UL-94 ratings of V-0 are attainable by the addition of a low concentration of different types of additives.

The commercial flame retardants include a polycarbonate oligomer of tetra-bromobisphenol A chain capped with phenol for a bromine content of 51.3% (BC-52/Great Lakes Chemical), a similar oligomer chain capped with tribromo-phenol for a bromine content of 58% (BC-58/Great Lakes Chemical), phosphate esters such as triphenyl phosphate and a brominated phosphate containing 60% bromine and 4% phosphorus (tris-dibromophenyl phosphate [24]). As shown for modified PPO resin, phosphorus is a very effective flame retardant in polymers containing oxygen. To obtain a V-0-rated clear product, about 7% flame retardant is needed. The addition of about 0.2% of a fibrillated Teflon will give an opaque V-0-rated product with only 3% flame retardant [25]. Antimony oxide is not used in polycarbonate resins because it can function as a depolymerization catalyst.

Polycarbonates can be difficult to process especially if sheet or thin mold-ings are needed and if the polymer is filled with conductive carbon or with glass. The brominated phosphate is melt blendable (m.p. = 110°C) and serves as a processing aid.

Another method of flame retarding polycarbonate resin is to add an alkali metal salt of an halogenated benzene sulfonic acid. A concentration of less than 1% yields a product with an oxygen index of 33–35. The product is V-0 at 3.2 mm but only V-2 at 1.6 mm.

The thermal degradation of polycarbonates can proceed by two mecha-nisms. Isomerization leads to linear oligomeric ethers with phenol end groups. Subsequent loss of carbon dioxide and water leads to crosslinking:

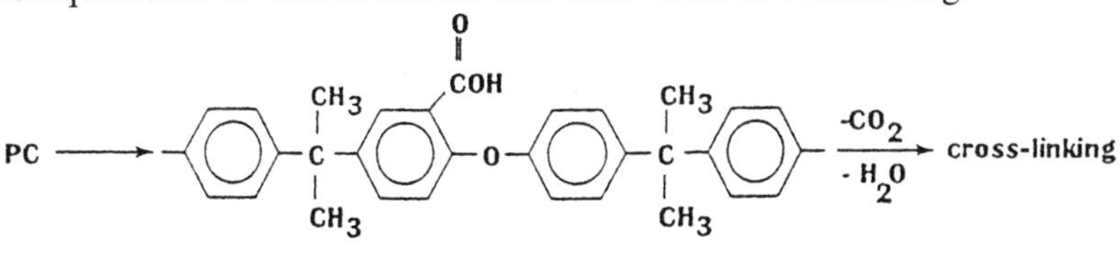

Intramolecular exchange leads to formation of cyclic oligomers which react with water vapor and liberate carbon dioxide to form aromatic phenols:

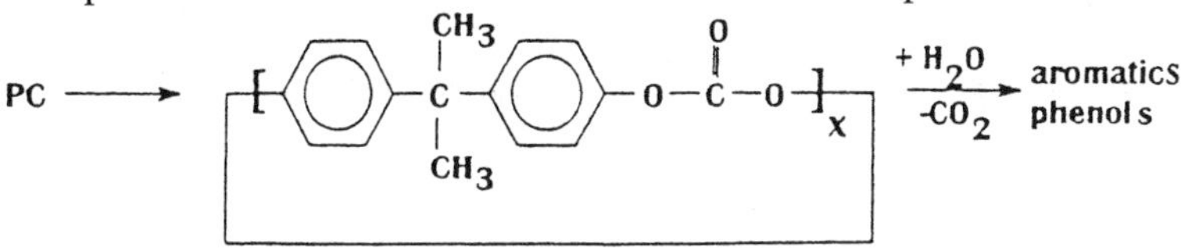

The addition of catalytic levels (0.1%) of aromatic sulfonate salts changes the mechanism of thermal degradation; oxygen transfers to the ortho position, form-

Table 38 Properties of Polycarbonate Resins

Property	Lexan 141[a]	Lexan 500[a]	Lexan 940[a]	Macrolon 6410[b]	Macrolon 6560[b]
UL-94 rating at 1/16 in.	V-2	V-0	V-0	V-2	V-0
Oxygen index	25	32.5	35	28	37
Density (g/cm^3)	1.200	1.250	1.210	1.250	1.200
Izod impact strength, notched (ft-lb/in.)	16	2	12	1.8	14[c]
Heat-distortion temp. at 264 psi (°F)	270	288	270	277	268

[a] General Electric.
[b] Bayer.
[c] Sensitive to notch dimensions.
Source: Company data sheets.

ing ketones (Fries rearrangement). This takes place at a lower temperature and char formation by cross-linking takes place earlier than without the catalyst:

The TGA data show that the sulfonate salt accelerates the decomposition and lowers the temperature at which carbon dioxide evolves. Aromatic sulfonates are not effective in PC/ABS blends; the ABS lowers the decomposition temperature of the polycarbonate.

Lexan 500, a bromine-containing resin, has an oxygen index of 32.5 compared with 25 for the virgin resin (Table 38). When flame retardant is added to the polycarbonate resin, the Izod impact is apparently degraded from about 15 to 2. Lexan 940 is apparently flame retarded with the sulfonate salt. Table 39 shows data for glass-reinforced polycarbonate resins.

I. Polybutylene Terephthalate

Polybutylene terephthalate has a low glass transition temperature (104°F) and, therefore, a low heat-distortion temperature under load (125°F at 264 psi). Reinforcing PBT with 30% glass increases the heat-distortion temperature to over 400°F. PBT is a very brittle resin exhibiting a notched Izod impact strength of 0.7 ft-lb/in. Glass-reinforced resins have Izod values up to 2 ft-lb/in.

Polybutylene terephthalate (PBT) is generally used with reinforcing fillers such as glass or mineral. Glass reinforcement improves the strength, stiffness, and toughness properties, whereas reinforcement by mineral and mineral/glass provides increased strength and stiffness. Mineral filler also aids in reducing warp significantly over products reinforced with glass alone.

Table 39 Properties of Reinforced Polycarbonate Resins

Property	Lexan 3412[a]	Makrolon 8325[b]	Makrolon 9415[b]
Glass content (%)	20	20	10
UL-94 rating at $\frac{1}{16}$ in.	V-1	V-0 ($\frac{1}{8}$ in.)	V-0
Oxygen index	30	25	30
Density (g/cm^3)	1.350	1.340	1.279
Izod impact strength, notched (ft-lb/in.)	2.0	3.2	2.0
Heat-distortion temp. at 264 psi (°F)	295	290	284

[a] General Electric.
[b] Bayer.

Flame-retardant additives for PBT include decabromodiphenyl oxide, Pyro-Chek 68PB (brominated polystyrene), and BC-58 (brominated polycarbonate oligomer). Usually 8–10% bromine from an aromatic bromine compound is required to obtain a V-0 rating for a part thickness down to 1/16 in. Less bromine is required for glass-reinforced and mineral-filled grades. For thinner sections, substantially more flame retardant is needed, along with additives that inhibit dripping.

Decabromodiphenyl oxide blooms to the surface of PBT resin. In accelerated aging, this flame retardant will show moderate bloom after only 5 h at 138°C in an air-circulating oven. Pyro-Chek 68PB shows only slight bloom after 500 h at 121°C and no bloom after 650 h at 65°C. This excellent resistance to migration or bloom for Pyro-Check 68PB is presumably due to the polymeric nature of the flame retardant (molecular weight of 200, 000–300,000).

Both BC-52 and BC-58 can be used in reinforced and nonreinforced PBT resin systems. The former oligomer is chain capped with phenol and the latter with tribromophenol. BC-58 offers advantages in higher flow and flame-retardant efficiency, whereas BC-52 has an advantage in physical impact and toughness. Table 40 compares the properties of these flame-retardant resins with the base resin. These oligomers do not bloom from PBT.

Polybutylene terephthalate containing 14% Pyro-Chek 68PB and 4.7% antimony oxide shows a V-0 rating at 1/32 in. thickness in the UL-94 flammability test. The flame-retardant additives further reduce the low Izod impact value for PBT. Glass-reinforced flame-retardant grades show Izod impact values of 1–2 ft-lb/in.

The PBT resins containing tetrabromobisphenol A epoxy oligomers maintain excellent color impact, flow, and thermal stability properties of the virgin resin (Table 41).

A brominated phosphate containing 60% bromine and 4% phosphorus was reported to be a very effective because of the combination of bromine and phos-

Table 40 Flame-Retarding Polybutylene Terephthalate

Property	Base resin	PBT BC-52[a]	PBT BC-58[b]
Flame retardant/Sb$_2$O$_3$ ratio	—	18/3	16/4
UL-94 rating at $\frac{1}{32}$ in.	Burns	V-0	V-0
Izod impact strength, unnotched (ft-lb/in.)	>80	7.9	7.0
Gardner impact strength (in.-lb)	210	8	6
Heat-distortion temperature (°F)	124	146	152
Spiral flow (in.)	16.3	12.6	16.2
Flexural modulus ($\times 10^5$ psi)	3.27	3.75	3.82

[a] 52% bromine.
[b] 58% bromine.
Source: R. C. Nametz. Fire Retardant Chemicals Association Meeting, 1984.

Table 41 Flame-Retarding Polybutylene Terephthalate with Brominated Epoxy Oligomer

	Formulations 1	Formulations 2
Ingredients		
PBT	57.6	50
Impact modifier	18.4	—
Glass fiber, 0.25 in.	—	30
Thermoguard 240	17.8	13.5
Antimony oxide	6.2	6.5
Flame Retardancy		
Oxygen index	30.0	31.5
UL-94 rating at $\frac{1}{8}$ in.	V-0	V-0
Flame-out (s)	0.5	1.0
Physical Properties		
Izod impact strength, notched (ft/lb-in.)	3.2	1.6
Tensile strength (psi)	4840	1700
Tensile modulus ($\times 10^{-3}$ psi)	190	430
Elongation (%)	162	—
Flexural strength (psi)	8150	1650
Flexural modulus ($\times 10^{-3}$ psi)	70	180

Source: Company Data sheet.

Table 42 Flame-Retarding Mineral-Filled PBT

	Compositions			
PBT mineral filled	84	84	84	84
Brominated polycarbonate	12	—	16	—
Brominated phosphate	—	12	—	16
Antimony oxide	4	4	—	—
Oxygen index	31.8	29.7	29.1	31.2
UL-94				
Rating ($\frac{1}{16}$ in.)	V-0	V-0	B	V-0
Time (s)	0	3.7	—	3.1

phorus [25]. With mineral-filled PBT no antimony oxide is needed with the brominated phosphate for a V-0 rated product (Table 42). In glass-filled PBT, the brominated phosphate improves processability and impact resistance, distinct advantages over the brominated flame retardants. Approximately 10% flame retardant plus 3% antimony oxide is used [59] (Table 43). Properties of commercial grades are shown in Table 44.

Table 43 Flame-Retardant Polybutylene Terephthalate

	Control	1	2	3
PBT–30% Glass	100	87.5	87.5	86.5
Teflon 6C	—	0.5	0.5	0.5
Antimony oxide	—	2.5	2.5	3.5
Brominated polystyrene	—	10	—	—
Brominated polycarbonate	—	—	10	—
Brominated phosphate	—	—	—	10
Bromine/phosphorus of FR (%)	—	68/0	58/0	60/4
Oxygen index	20.7	26.7	28.5	28.5
UL-94				
Rating ($\frac{1}{16}$ in.)	B	V-0	V-0	V-0
Time (s)	—	2.6	0	0
Izod impact	1.32	0.80	1.07	1.26
Deflection temp. (°C)	207	205	201	201
Melt index (g/10 min)	13.4	16.0	11.6	19.1
Spiral flow (in.)	37	36	26	35
Tensile strength (psi)	17,100	11,600	15,600	15,400
Elongation at break (%)	3.9	2.3	3.3	4.5
Flexural strength (psi)	29,420	20,900	28,350	28,790
Flexural modulus ($\times 10^6$ psi)	1.20	1.20	1.20	1.06

Table 44 Properties of Reinforced Polybutylene Terephthalate Resins

	PBT resins					
Property	Valox 420a	Valox 420-SEO[a]	Celanex 3210[b]	Celanex 3310[b]	Valox 760[a]	Valox 752[c]
Filler	Glass fiber	Glass fiber	18% glass	30% glass	30% mineral	Glass/ mineral
Density (g/cm^3)	1.520	1.580	1.60	1.690	1.470	1.800
UL-94 rating at 0.030 in.	Burns	V-0	V-0	V-0	V-0[c]	V-0
Izod impact strength, notched (ft-lb/in.)	2.2	1.8	1.0	1.3	1.2	0.8
Heat-distortion temp. at 264 psi (°F)	415	415	395	406	170	380

[a] General Electric.
[b] Hoechst Celanese.
[c] At 0.60 in.

J. Polyethylene Terephthalate

Glass-reinforced polyethylene terephthalate (PET) is flame retarded with brominated polystyrene (Pyro-Chek 68PB) and sodium antimonate. Antimony oxide cannot be used because it degrades the polymer. The use of a combination of bromine and phosphorus obviates the need for an antimony synergist, and the sodium antimonate can be replaced with are additional flame retardant such as a brominated phosphate [24] (Table 45). The properties of commercial grades of filled PET resin is given in Table 46.

Table 45 Flame-Retardant PET/30% Glass Resin

	Compositions		
PET/30% glass	80	82	80
Brominated phosphate	15	18	20
Sodium antimonate	5		
Teflon 6C	0.5	0.5	0.5
Oxygen index	29.4	30.6	36
UL-94			
Rating (0.062 in.)	V-2	V-2	V-0
Time (s)	5.1	1.2	0.1

Table 46 Flame-Retarded Polyethylene Terephthalate

Property/grade	Rynite 430	Rynite FR530	Rynite FR945
Flame retarded	No	Yes	Yes
Filler	30% glass fiber	30% glass fiber	45% glass/ mineral
Density (g/cm^3)	1.49	1.67	—
Izod, notched (ft-lb/in.)	2.6	1.5	—
Heat-deflection temp. @264 psi (°F)	410	435	392

K. Polycarbonate/ABS Blend

Polycarbonate (PC) resin can be difficult to process and polycarbonate/ABS blends are used to mold computer housing. Resin with high polycarbonate content are used because they are easier to flame retard. Bromine, phosphorus, and brominated phosphates can be used without the need for antimony synergist. The commercial flame retardants suitable for this application include the bromine-containing BC-52 and BC-58, the phosphorus-containing triphenyl phosphate (TPP) and resorcinol diphosphate (RDP), and the synergistic tris-dibromophenyl phosphate.

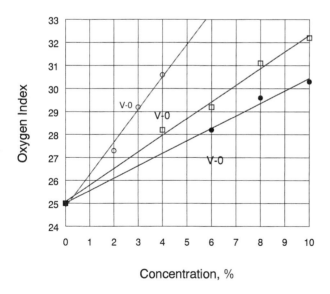

Concentration, %

Figure 23 Flame-retarded PC/ABS blends; impact modified 8/1 PC/ABS ratio. ● Resorcinol diphosphate, □ BC-58, ○ tris-dibromophenylphosphate.

RDP is the major flame retardant used in PC/ABS computer housing. The major resin blend used for computer housing is about an 8/1 blend of PC/ABS.

All-phosphorus, all-bromine, and brominated phosphate compounds were compared as flame retardants in 8/1, 5/1 and 3/1 blends [39]. To attain an oxygen index of 30 with the 8/1 blend 4.1% of the brominated polycarbonate oligomer, 7.5% of the resorcinol diphosphate, and 3.1% of the brominated phosphate were required (Fig. 23). A higher concentration of flame retardant was required with the other blends, but the comparative results were similar. Comparison on an elemental basis demonstrated bromine–phosphorus flame retardant synergy (Figs. 24 and 25) [40].

L. Polycarbonate/Polybutylene Terephthalate Blend

Figure 26 compares the flame-retardant efficiency of two all-bromine flame retardants with a brominated phosphate at varying PC/PBT ratios. The brominated compounds are a polycarbonate oligomer containing 58% bromine (BC-58) and a brominated polystyrene containing 68% bromine (68PB). As can be seen, the brominated phosphate containing 60% bromine and 4% phosphorus (tris-dibromophenyl phosphate) is a much more efficient flame retardant. For example, at 1 : 1 blend of PC/PBT, the brominated polycarbonate-containing resin burns, the brominated polystyrene-containing resin has a V-2 rating and the resin containing the brominated phosphate is V-0. The flame-retardant concentration is 12%. The oxygen index values are consistent with these results [24].

M. Polycarbonate/Polyethylene Terephthalate Blend

Polycarbonate/polyethylene terephthalate blends can be flame retarded with all-bromine, all-phosphorus, a mechanical blend of the two, or a brominated phosphate where both the bromine and the phosphorus are in the same molecule. For example, one can use a brominated polycarbonate oligomer (BC-52 or BC-58), triphenyl phosphate or resorcinol diphosphate (RDP), or brominated phosphates such as Reoflam PB-370 (70% bromine and 3% phosphorus). Additionally, 0.2–0.5% of a fibrillated Teflon is added as a drip inhibitor and about 5% of an impact modifier. Table 47 shows the relative efficiencies of these different flame retardants in a 2/1 PC/PET blend [9]. The phosphate esters would be expected to plasticize the resin, reducing the HDUL (heat distortion under load). The product containing only 6% additive would be expected to have the best mechanical properties.

The high flame-retardant efficiency of the brominated phosphate has been shown to be due to phosphorus–bromine synergy. Figure 27 compares the oxygen indices of resins containing the all-bromine, all-phosphorus, and brominated phosphate additives. On a weight basis, the brominated polycarbonate oligomer and the triphenyl phosphate coincidentally give equivalent results. The former con-

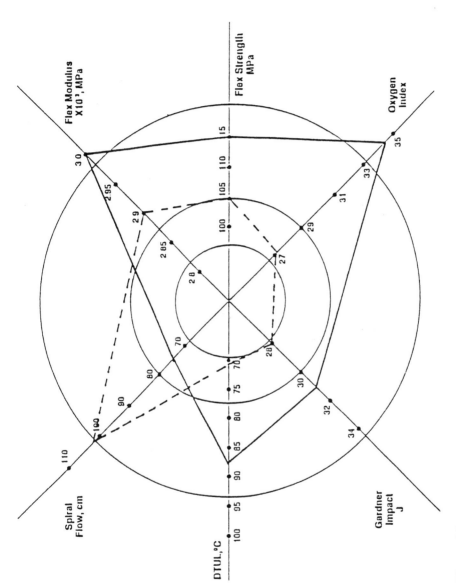

Figure 24 Comparison of triphenyl phosphate and brominated phosphate flame retardants in 3/1 polycarbonate/ABS blend at 14% FR: (---) TPP; (—) BrP.

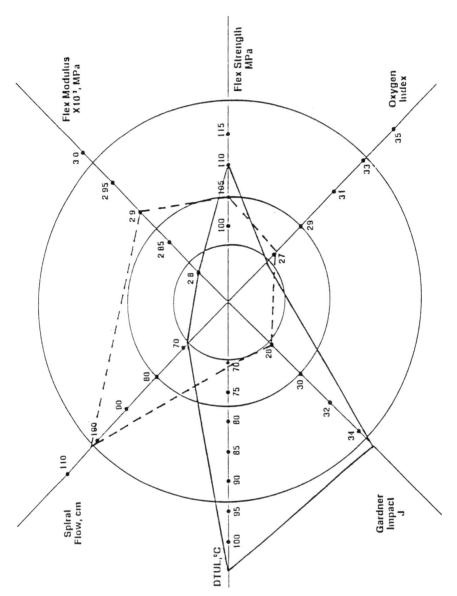

Figure 25 Comparison of a 3/1 polycarbonate/ABS blend containing 6% BrP with 14% TPP: (---) TPP; (—) BrP.

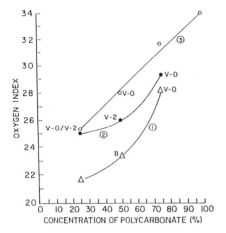

Figure 26 Flame-retarding PC/PBT polyester blends (12% flame retardant). 1: Brominated polycarbonate oligomer; 2: brominated polystyrene; 3: brominated phosphate ester.

Table 47 Flame-Retarding Impact Modified 2/1 Polycarbonate/PET Blend

Flame retardant[a]	BrPC	TPP[b]	BrP 60/40[c]	BrP 70/3[d]
Oxygen index	24.9	28.5	29.5	28.9
UL-94 @ 1.6 mm				
Rate	B	B	V-0	V-2
Time(s)			2	6.2

Note: Blend consists of 59.3% polycarbonate, 29.9% PET, 5% impact modifier, 0.5% Teflon 6C, and 6% flame retardant.
[a] BrPC, BC-58. BrPC = brominated polycarbonate oligomer.
[b] TPP = triphenyl phosphate
[c] BrP 60/4 = tris-dibromophenyl phosphate.
[d] BrP 70/3 = tris-tribromoneopentyl phosphate.

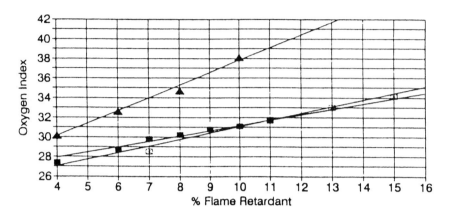

Figure 27   Flame-retarded 2/1 PC/PET Blend; oxygen index versus FR concentration.
■ brominated polycarbonate oligomer, □ triphenyl phosphate, ▲ tris-dibromophenyl phosphate.

tains 51.3% bromine and the latter contains 9.5% phosphorus, suggesting that phosphorus is a much more efficient flame retardant in this polymer blend. The brominated phosphate is much more efficient; for example, 11% of the brominated flame retardant is required for an oxygen index of 32, which is attainable with only 5.5% of the brominated phosphate [9].

Figure 28 shows the data in a different way. The bromine increases from left to right and the phosphorus increases from right to left. They cross at about 5% bromine and 0.5% phosphorus. The graph also shows that equivalent oxygen indices are attained at about 10% bromine and 1% phosphorus, indicating, again, that phosphorus is about 10 times more effective than bromine. This is not inconsistent with known technology.

Because the bromine/oxygen index and phosphorus/oxygen index curves are linear, blends of bromine and phosphorus compounds would be expected to fall on the theoretical additive line connecting the two extremes. However, a mixture of the two compounds with a composition of 6% bromine and 0.4% phosphorus gives an oxygen index significantly higher than predicted by the curve. When bromine and phosphorus are in the same molecule as in a brominated phosphate, an even higher oxygen index is obtained. These data conclusively demonstrate that bromine and phosphorus are synergistic in this polymer blend and that the synergy is further advanced when the bromine and phosphorus are in the same molecule [9].

N. Polyamides

Polyamide 6 and 6,6 are flame retarded commercially by different types of flame retardants: specifically, the chlorine-containing Dechlorane Plus, red phosphorus,

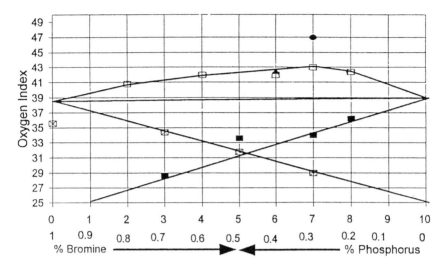

Figure 28 Bromine–phosphorus synergy in a 2/1 polycarbonate/PET blend. ■ brominated polycarbonate oligomer (BrPC), ⊠ triphenyl phosphate (TPP), □ BrPC/TPP blend, ▲ tris-dibromophenyl phosphate, ● tris-tribromoneopentyl phosphate.

melamine cyanurate, and magnesium hydroxide. This is demonstrated in Table 48.

Red phosphorus can be pyrophoric in air and also can react with moisture to yield toxic phosphine gas. As a result, red phosphorus is heavily encapsulated.

Melamine cyanurate alone is not effective in glass-filled polyamide. Formulations showing melamine cyanurate flame retardant in glass systems is demonstrated in Table 49.

Table 48 Flame-Retarding Unfilled Polyamides

Property/flame retardant	Dechlorane plus (Cl)	Red P[a]	Melamine cyanurate[b]	Mg(OH)$_2$
FR concentration (%)	19	7	9.5	32
Antimony oxide (%)	5			
Glass fiber (%)	28	35	None	13
Oxygen index	32	32	29	25
UL-94 V-0 thickness (in.)	0.03	0.03	0.03	0.06
Impact strength				

[a] Encapsulated and stabilized.
[b] Inefficient with glass fiber.

Table 49 Flame-Retardant Additives in
UL-94 V-0 Polyamides

Polyamide 6	41		
Polyamide 6,6		38	47
Glass fiber	27	27	25
Melamine cyanurate	10	15	
Melamine			18
Zinc borate	10	8	4
Dechlorane Plus	10	12	6
Antimony oxide	2		

In unfilled polyamides 6 and 6,6, melamine cyanurate will give products
with a UL-94 rating of V-0 down to 0.4 mm.

Dechlorane Plus, an alicyclic chlorine-containing flame retardant, has been
used for many years to flame retard Nylon 6/6. These nylon compounds have
good electric properties, improved heat-deflection temperatures, improved thermal aging, essentially unchanged hardness, and only a moderate lowering of
impact strength. Tables 50 and 51 show formulations and performance properties
for flame-retardant Nylon 6/6 and Nylon 6 resins, respectively. The suggested
levels of flame-retardant additives in these formulations, including both Dechlorane Plus 25 and a metal oxide synergist, are from 15% to 22% for Nylon 6/6
and 25% for Nylon 6. The ratio of flame retardant to synergist used is 4 : 1.

These formulations incorporate either antimony trioxide or ferric oxide as
the synergistic additive; the latter yields red compounds. Ferric oxide as a synergist shows a greater flame-retardant effectiveness as measured by the oxygen
index.

Several brominated polymeric additives are used as flame retardants for
nylon. These include Pyro-Check 68PB, PO-64P, and brominaded epoxy oligomer. These polymeric additives are nonmigrating and nonblooming [41].

Pyro-Chek 68PB is a brominated polystyrene containing 68% aromatic
bromine (specific gravity = 2.8). It is effective in flame retarding nylon at an
add-on level of 10 to 15% plus 3–5% antimony oxide.

PO-64P, polydibromophenylene oxide, is very stable and can be compounded into Nylon 6/6 at 270–280°C. This bromine-containing compound is so
stable that a flame-retardant compound containing 50% regrind can be processed
without serious property impairment. The recommended loading in unfilled Nylon
6 and 6/6 to achieve a V-0 rating at 1/16 in. thickness is about 18%, with 4.5%
antimony oxide. With glass-reinforced resin, less flame retardant is needed. No

Table 50 Flame-Retardant Nylon 6/6 Resin

		Unfilled		Glass filled	
Property	Control	Using antimony trioxide	Using ferric oxide	Using antimony trioxide	Using ferric oxide
Ingredients (wt%)					
Nylon 6/6[a]	100	80	85	48	55
Dechlorane Plus 25	—	16	12	17.5	12
Metal oxide	—	4	3	4.5	3
Glass fiber	—	—	—	30	30
Flame Retardancy	27	31	33	31	36
Oxygen index	27	31	33	31	36
UL-94 rating at 1/16 in.	V-2	V-0	V-0	V-0	V-0
Physical Properties					
Heat-deflection temp. at 264 psi (°C)	64	76	76	211	207
Tensile strength (psi)	7,500	6,370	7,460	13,200	11,600
Elongation (%)	2.8	2.2	2.7	2.3	2.3
Flexural strength (psi)	13,400	12,500	9,910	18,300	19,700
Flexural modulus ($\times 10^3$ psi)	338	454	430	960	1,050
Izod impact strength, notched (ft-lb/in.)	0.76	0.46	0.34	1.2	0.70
Hardness, Rockwell	95L	96L	97L	85M	84M
Electrical Properties					
Volume resistivity ($\times 10^{14}$ Ω-cm)	4.0	3.1	1.0	1.25	1.28
Dielectric strength (V/mil)	407	422	444	385	452
Dielectric constant at 10^4 Hz	3.72	3.21	3.23	2.97	3.79
Dissipation factor at 10^4 Hz	0.025	0.021	0.023	0.0124	0.0057
Arc resistance (s)	169	156	170	128	116

[a] Material is Nylon 6/6 and Nylon 6/12 in a 4:1 ratio.

Table 51   Flame-Retardant Nylon 6 Resin

Property	Control	FR resins Using antimony trioxide	FR resins Using ferric oxide
Ingredients (wt%)			
Nylon 6	100	75	75
Dechlorane Plus 25	—	20	20
Metal oxide	—	5	5
Flame Retardancy			
Oxygen index	23.4	29.4	31.3
UL-94 rating at 1/16 in.	V-2	V-0	V-0
Physical Properties			
Heat-deflection temp. at 264 psi (°C)	54	63	61
Tensile strength (psi)	8,820	7,440	7,620
Elongation (%)	4.0	3.4	3.5
Flexural strength (psi)	14,040	14,600	13,770
Izod impact strength, notched (ft-lb/in.)	0.71	0.50	0.42
Hardness			
Rockwell L	93	92	93
Rockwell R	113	112	113

migration of this polymeric additive is observed during aging at 150°C. The heat-distortion temperature at 264 psi is 165–175°F. The notched Izod values are about 0.6 ft-lb/in.

PO-64P has a softening range below or near the melting temperature of Nylon 6 and 6/6. It therefore compounds readily into these resins. Actually, it improves the flow of resin during molding.

The results with a brominated epoxy resin in flame-retardant nylon are given in Table 52.

O.   Poly(methyl methacrylate)

Transparent poly(methyl methacrylate) (PMMA) molding compounds with flame-retardant properties have been reported. PMMA containing 15% Sandoflam 5085, a chlorine containing phosphorinane, gives an injection-molded product with a UL-94 rating of V-2 and an oxygen index of 22 (Table 53). The transpar-

Table 52 Flame-Retarding Nylon with a Brominated Epoxy Oligomer

Property	Nylon 6	Nylon 6/6
Ingredients (%)		
Flame retardant	17.3	15.4
Antimony oxide	5.8	7.7
Flame Retardancy		
Oxygen index	26.3	35.3
UL-94		
Rating	V-0	V-0
Time(s)	1.0	1.0
Physical Properties		
Izod impact strength, notched (ft-lb/in.)	—	3.3
Tensile strength (psi)	—	9750
Tensile modulus ($\times$ 10^3 psi)	—	325
Elongation (%)	—	5.4

Source: Company data sheet.

ency of the product remains relatively unaffected, but the Vicat temperature is significantly decreased. The water absorption and light exposure, important for outdoor application, remain quite good. The use of a bromine containing phosphorinane does not appear to be any more effective (Table 9) [26].

P. Poly(vinyl chloride)

Poly(vinyl chloride) contains 57% chlorine and is inherently flame retardant. Rigid PVC products require stabilizers and processing aids but do not need addi-

Table 53 Poly(methyl methacrylate) Molding Compound Containing Sandoflam 5085 Flame Retardant

Flame retardant concentration (%)	Oxygen index	UL-94 rating	Vicat temperature (°C)	Water absorption 24 h/70°C (wt%)	48 days/20°C (wt%)
0	17.2	—	110	1.1	1.6
5	19.2	—	—	—	—
10	21.6	—	98	1.3	1.7
12.5	21.8	V-2	95	1.4	1.8
15	22	V-2	93	1.5	2.0

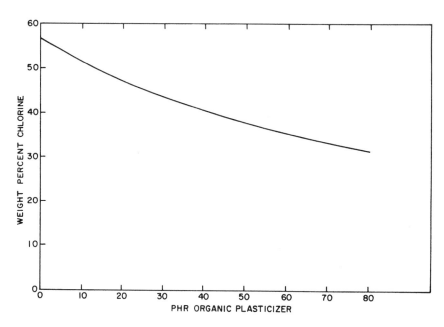

Figure 29 Chlorine dilution of poly(vinyl chloride) by addition of plasticizer.

tional flame retardants to meet flammability specifications. Flexible PVC contains high concentrations of organic plasticizers; hence, it is flammable. For example, rigid PVC has a chlorine content of 57% and an oxygen index of about 37. This product is considered to be flame retardant by the usual laboratory tests. When plasticized with 60 parts of an organic plasticizer such as dioctyl phthalate, the chlorine content drops to 36% and the oxygen index to about 22 (Figs. 29 and 30). Such a film burns readily.

Many additives have been used to flame retard plasticized PVC, including alumina trihydrate, antimony oxide, zinc and barium borates, chlorinated paraffins, and phosphate esters. As measured by the oxygen index, the most effective flame retardants are antimony oxide, which functions as a synergist for the chlorine in PVC, and phosphate esters, which replace in whole or in part the organic plasticizer. This is demonstrated in Table 54.

The flame-retardant efficiency of antimony oxide in PVC does not increase linearly with increasing concentration. The oxygen index value curve plateaus at about 27 to 28 (Fig. 31). These high levels of antimony oxide may contribute to "afterglow," a glowing combustion that is maintained after flaming has ceased. Replacement of some of the antimony oxide with zinc borate may inhibit the "afterglow."

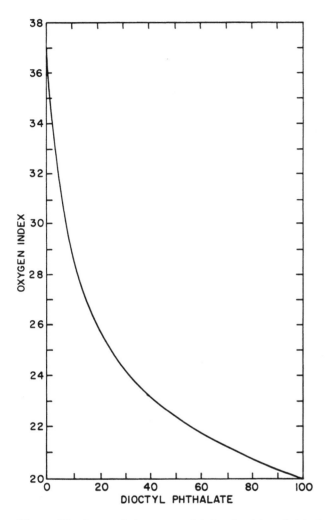

Figure 30 Oxygen index versus phthalate level in poly(vinyl chloride) sheet 10 mils thick.

High-tinting antimony oxide cannot be used to produce clear PVC products. In other compounds, the white pigmenting effect of the oxide may necessitate a higher concentration of expensive pigments. Low tinting grades partially offset this problem.

Transparent film can be made by using phosphate esters as flame-retardant plasticizers. These are classified into three types: mainly triaryl, alkyl diaryl,

Table 54 Effect of Various Flame Retardants in Plasticized Poly(vinyl chloride)

	Concentration (phr)	Oxygen index
Control	—	20.8
Antimony oxide	6	25.0
Alumina trihydrate	25	22.6
Zinc borate	25	21.2
Isopropylphenyl diphenyl phosphate	30	24.2
Isopropylphenyl diphenyl phosphate	45	25.4

Note: For each 100 parts of PVC, the following were added: stearic acid, 0.4 phr; Ba/Cd stabilizer, 2.0 phr; $CaCO_3$, 25 phr minus solid FR above; and dioctyl phthalate, 75 phr minus phosphate.

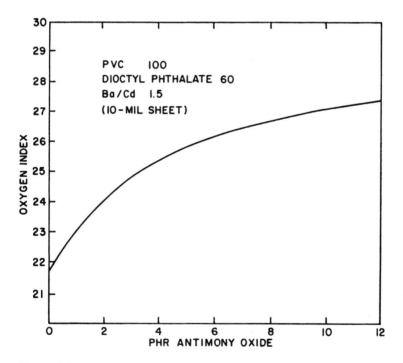

Figure 31 Oxygen index versus antimony oxide level.

and trialkyl phosphates (Table 6). These plasticizers replace the organic ester plasticizer in whole or in part, depending on the required level of flame retardance.

The flame-retarding efficiency of the phosphate esters depends on the aromatic content; the alkyl content is detrimental to flame retardancy (Fig. 32). However, there is a trade-off in that the alkyl groups, although detracting from

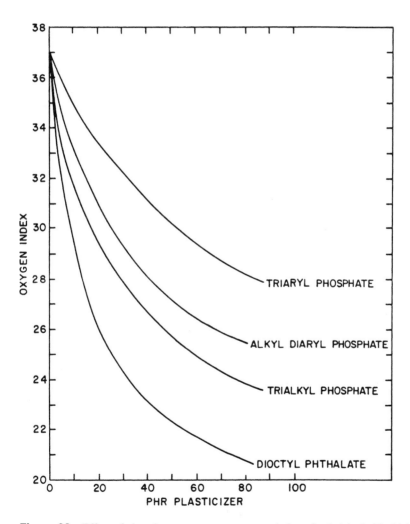

Figure 32 Effect of phosphate ester type on oxygen index of poly(vinyl chloride) sheet 10 mils thick.

flame retardancy, give improved plasticizer efficiency and better low-temperature properties, both highly desirable characteristics [42].

Although the phosphate esters were used as the sole plasticizer in this comparison, rarely is this done in commercial compounds. In actual practice, the phosphates are combined with the more economical phthalate plasticizers, replacing them only in amounts necessary for the desired flame-retardant requirement. Figure 33 shows how flammability of a DOP-plasticized sheet is reduced by partial substitution with phosphate esters. In this situation, oxygen index increases almost linearly with increasing phosphate level (in contrast to the plateau effect of antimony oxide).

The relative flame-retardant efficiencies of the various types of phosphates also are evident in Fig. 33; they fall in precisely the same order as shown in Fig. 32, with the triaryl esters being the most effective.

Table 55 lists different phosphate esters, in order of decreasing phosphorus content, along with their respective oxygen indices. It is apparent that flammabil-

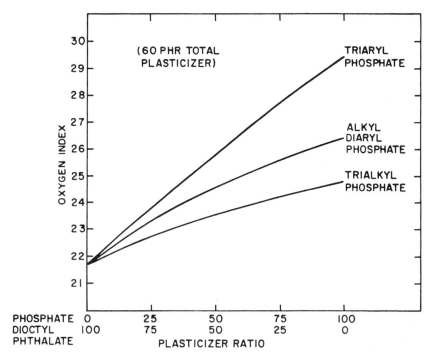

Figure 33 Effect of phosphate/phthalate ratio on flammability of poly(vinyl chloride) sheet 10 mils thick.

Table 55 Effect of Phosphorus Content of Ester on Flammability in Polyvinyl Chloride

Phosphate ester	Phosphorus content (%)	Oxygen index
Isodecyl diphenyl	7.95	26.4
Diisopropylphenyl phenyl	7.6	30.0
Trioctyl	7.14	25.0

Note: For each 100 parts of PVC, the following were added: stabilizer, 1.5 phr; calcium carbonate, 30 phr; phosphate ester, 61 phr.

ity is not a direct function of phosphorus content. For example, the isopropylphenyl ester, which contains 7.6% phosphorus, gives a higher oxygen index than isodecyl diphenyl phosphate of higher phosphorus content [28].

Table 56 presents a further comparison of some of the same esters. Here, the esters were substituted for DOP in the amount necessary to provide 1% phosphorus in the total compound. If flame retardancy were strictly a function of phosphorus content, each of these compounds should have the same oxygen index. This is not the case [28].

Consideration of the various types of ester shows that those containing alkyl ester groups are less flame retardant than those that have only aryl ester groups. Presumably the alkyl groups cleave and burn readily in the flame zone. Consequently, the formulator should not choose a phosphate ester solely on the basis of its phosphorus content.

Typical properties for a triaryl phosphate (isopropylphenyl diphenyl phosphate—Reofos 100) are shown in Table 57. Typical properties for a triaryl phosphate plasticized poly(vinyl chloride) resin are shown in Table 58.

Table 56 Flammability of PVC Sheeting Compounds Containing 1.0% Phosphorus

Phosphate ester	phr for 1% P	Oxygen index
Trioctyl	27.0	24.6
Isodecyl diphenyl	24.2	24.8
Tricresyl	22.9	25.9
Isopropylphenyl diphenyl	24.8	26.0

Note: For each 100 parts of PVC, the following were added: Ba/Cd stabilizer, 1.5; $CaCO_3$, 30; dioctyl phthalate, 61 minus phosphate ester.
Source: Ref. 28.

Table 57 Properties of Reofos 65, a Typical Triaryl Phosphate

Molecular weight (approx.)	400
Specifications	
Specific gravity at 20°C/20°C	1.150–1.165 g/cm^3
Moisture	0.1 wt% max
Color, Pt–Co (APHA)	75 max
Odor	Very slight
Acidity[a]	0.1 wt% max
Average Properties	
Boiling range at 4 mm	220–270 °C
Mid-boiling point at 4 mm	250°C
Pour point	−25°C
Flash point	470°F
Viscosity at 20°C	90 cPs
Refractive index N_0	1.552
Weight per gallon at 20°C	9.6 lb

[a] As acetic acid.

Table 58 Flame-Retardant Poly(vinyl chloride) Plasticized with Phosphate Esters

	Formulations		
Property	1	2	3
Composition			
PVC (parts)	100	100	100
Stabilizers (phr)	6	6	6
Santicizer 148 (phr)	50	—	—
Reofos 65 (phr)	—	50	—
Reofos 3600 (phr)	—	—	50
Flammability			
Oxygen index	28.5	31.5	28.5
Vertical burn[a] (CCCT 191B, 5903)			
After flame (s)	0.8	1.1	0.6
Char length (in.)	1.8	1.8	2.6
Physical Properties			
Hardness, Shore A	82	89	82
Modulus (psi)	1510	2090	1603
Low-temperature flexibility (°C)	−26	−11	−27
Carbon volatility, 24 h/90°C	7.3	5.7	5.6

[a] See Section IIIC.

Table 59 Typical Properties of a PVC Plenum Cable Jacket
Compliance with UL 910

Producer	Teknor Apex	AlphaGary
Compound No.	Fireguard 910	Smokeguard II
		6920
Specific gravity	1.61	1.57
Durometer hardness		
Shore C	85	
Shore D		62
Tensile (psi)	2515	2600
Elongation (%)	225	220
Brittle Point, D-746 (°C)	0	-13 to -19
Oxygen index	48	47
Smoke, D-4100 (%)	4.59	6
Dielectric constant, 1 MHZ	3.27	3.6
Dissipation factor, 1MHZ	0.0145	0.0383
Halogen content (%)	24.9	

A PVC jacket for plenum application must pass the U.L. 910 Steiner tunnel test with low smoke and low flame spread. The oxygen index of these compounds is over 45. They are heavily loaded with flame retardants and smoke suppressants. The flame-retardant function of the chlorine is boosted with a brominated additive (tetrabromophthalate ester), and aluminum trihydrate; zinc borate and ammonium octamolybdate (AOM) are added as smoke suppressants. The concentration of the expensive AOM ranges from 12 to 40 phr. Santicizer 2148, a low-volatile alkyl diaryl phosphate, is generally used as a plasticizer, at least in part, for low-temperature properties. The properties of two commercial compounds are shown in Table 59.

Q. Red Phosphorus as Flame Retardant for Various Thermoplastics

Red phosphorus has been suggested as a flame-retardant additive for a wide variety of thermoplastics (Table 60). Low concentrations of red phosphorus will produce polymer products that meet the UL-94 V-0 standard (Table 61). Red phosphorus appears to be particularly effective in polyethylene terephthalate [43], polycarbonate, and phenolic resins.

The flame retardant is a purple-red powder, with a red phosphorus content of 85%, a yellow phosphorus content of less than 0.005%, a density of 2.0 g/cm^3, and a mean particle size of 20–30 nm.

Table 60   Amgard CRP (Encapsulated Red Phosphorus) Concentration
Required for UL-94 V-0 Rating

Resin	Amgard CRP concentration (%)
Polyamide	7
Polycarbonate	1.2
Polyethylene	10
Polyethylene terephthalate	3
Polystyrene	15
Filled phenolic resin	3

Source: Data sheet, Albright and Wilson, Ltd., West Midlands, England, B69 4LN;
encapsulated red phosphorus.

On hydrolysis, red phosphorus can evolve highly toxic phosphine gas and
may burst into flame on blending with polymers. Amgard CRP, a product of
Albright and Wilson (Richmond, VA), is a microencapsulated form of red phosphorus claimed by the producer to be specially stabilized for easy handling.

Albright and Wilson state that with appropriate safety precautions, Amgard
CRP may be added directly to a polymer melt, or solution, or by mixing in a
powder blend. Master batches can be made and added at the extrusion or molding
stages.

The use of halogen-containing compounds with Amgard CRP increases the
flame-retarding efficiency. The use of a 1 : 1 halogen/phosphorus mole ratio is
suggested for polymers such as methacrylates, polyesters, polyolefins, poly(vinyl
chloride), and styrenic polymers.

Table 61   Oxygen Index of Polymers Containing 0% and 10% Amgard CRP

	Oxygen index (%)	
Polymer	0% Amgard CRP	10% Amgard CRP
ABS	18.5	22.5
Epoxy resin	21.0	26.0
Polyamide	20.8	24.0
Polyethylene terephthalate	20.4	31.0
Polypropylene	17.0	19.5
Polystyrene	19.0	22.0

Source: Data sheet, Albright and Wilson, Ltd., West Midlands, England B69 4LN; encapsulated red phosphorus.

R. Epoxy Resin

Epoxy resins are condensation polymers that use 54 parts of tetrabromobisphenol A as a monomer to attain flame retardancy. This flame retardant is the largest volume organic flame retardant in the industry. The major application is epoxy glass laminate circuit boards. Dimethyl methyl phosphonate (DMMP), which contains 25% phosphorus, is also used. Phosphine oxides are highly efficient in this resin although at this time it is not used commercially (Ref. 16).

S. Unsaturated Thermoset Polyesters

Unsaturated polyesters are condensation polymers and flame retardancy can be obtained by replacing the monomers with halogenated diols or dicarboxylic acids. Thermoset polymers also have the ability to accept high concentrations of fillers and both inert and active fillers are used for flame retardancy.

Halogenated monomers which are used include tetrabromophthalic anhydride (to replace in part or in whole the phthalic anhydride), tetrachlorophthalic anhydride, chlorendic anhydride, dibromoneopentyl glycol (to replace the ethylene or propylene glycol), and postbromination of resin containing tetrahydrophthalic anhydride. Alicyclic bromine is the most effective of these, and aromatic chlorine the least effective.

Triethyl phosphate and DMMP (dimethyl methyl phosphonate) are also used as additives; they also function as diluents to improve processability. High concentrations of inert fillers such as calcium carbonate will reduce the polymer or fuel level. Alumina trihydrate at very high levels are used. It functions both to reduce the fuel level and to generate high volumes of water during decomposition to cool the system. Alumina trihydrate has 30% water of hydration.

T. Polyurethane Foam

Polyurethanes fall into three types: rigid foam, flexible foam, and thermoplastic elastomer. Rigid foam has used tetrabromophthalate diol as a reactive flame retardant and chloroalkyl phosphates as an additive and plasticizer. The chlorinated blowing agent which remains in the closed cells also presumably imparts flame retardancy to the foam. The flame-retardant requirements for flame-retardant rigid foam is now met with polyisocyanurates, which are inherently flame resistant. A small quantity of DMMP or a chlorophosphate is added as a nonflammable plasticizer to the friable polyisocyanurate.

Flexible polyurethane foam uses chlorinated phosphates at a concentration of about 12 parts per hundred of polyol. Examples of these include Fyrol FR-2 tris(1,3-dichoro-2-propyl) phosphate and Antiblaze 100 (Albright and Wilson). The aliphatic chlorine compounds are not very thermally stable and the high exotherms (175°C) generated during foaming may scorch the foam, especially when thick sections are produced. DE-60F Special (Great Lakes Chemical), a blend of pentabromodiphenyl oxide and a phosphate ester, does not scorch but is less effective and about 16–18 parts is required. The major application for these foams is construction insulation. As such, the flame-retardant requirement is a flame spread rating in the ASTM E84 25 foot Steiner tunnel.

U. Inorganic Hydrates

Inorganic hydrates such as alumina trihydrate and magnesium hydroxide contain 30–35% water of hydration. The largest volume of ATH goes into carpet backing and unsaturated polyesters. A typical unsaturated polyester may contain 20% glass and 40% ATH. Polyethylene and polypropylene require 60–65% of the hydrates. Ethylene–propylene rubber may contain 50% ATH. Surface-treated grades are available commercially; they provide for reduced viscosity and dusting, as well as improved dispersion and properties. A commercial low-voltage-power cable insulation based on polyethylene contains 58–60% of surface-treated magnesium hydroxide (Union Carbide). Surface treatments include 0.25% oleic acid and 2% of organic phosphate ester salts.

VI. SMOKE SUPPRESSANTS

Polymers differ in the quantity of smoke they produce during combustion. Table 62 shows maximum specific optical density (D_m) data or smoke obtained for a large variety of polymers [44]. Polymers that degrade to monomer such as the acetals, Nylon 6, and poly(methyl methacrylate) burn cleanly and give very little smoke. By contrast, the aromatic polymers, or those that form aromatics on burning, such as PVC, give very high levels of smoke. The polyolefins give relatively little smoke. It is interesting to note the difference between poly(vinyl chloride) and poly(vinylidene chloride); it is assumed that these polymers decompose by different mechanisms, with the former forming an aromatic structure upon dehydrochlorination.

Table 63 lists approaches to smoke suppression. Additives that can function as catalysts to modify the pyrolysis mechanism appear to be effective with PVC, cross-linked unsaturated polyesters, and polyurethanes. One can modify the polymer structure; for example, compare poly(vinyl chloride) with poly(vinylidene chloride), or polyurethanes with polyisocyanurates. Fillers can be added, either inert or active (ATH), reducing the amount of combustible polymer in the blend. Finally, one can coat the part with an intumescent coating that forms a thick porous char during burning, insulating the substrate from oxygen, heat, and flame.

Table 62 Smoke Chamber Results for Various
Materials Using Rohm & Haas XP-2 Chamber

Material	Smoke density (D_m)
Acetal	0
Nylon 6	1
Poly(methyl methacrylate)	2
Polyethylene	
Low density	13
High density	39
Polypropylene	41
Poly(vinylidene chloride)	98
Polyethylene terephthalate	390
Polycarbonate	427
Polystyrene	494
Acrylonitrile–butadiene–styrene	720
Poly(vinyl chloride)	720

Source: Adapted from Ref. 44.

Table 63 Examples of Polymer Smoke Suppressants According to Type of Approach

Type of approach	Example	Application	Probable mechanism	Ref.
Additive	Ferrocene	PVC	Char promotion, soot combustion	46
Additive	Fumaric acid	Polyurethane foam	Pyrolysis modifier	47
Filler (inert)	Silica	Various thermoplastics and elastomers	Noncombustible diluent	
Filler (active)	Alumina trihydrate	Various thermoplastics and elastomers	Char promoter, reduction of mass burning rate, diluent	
Surface treatment	Intumescent coatings	Various plastics	Surface insulation, reduction of mass burned	48
Structural modification	Polyurethanes/ isocyanurate	—	—	49
Structural modification	Chlorinated PVC	—	—	50,51

Source: Ref. 45.

The following discussion concentrates on the use of active additives or catalysts, specifically zinc borate, molybdenum compounds, and tin compounds.

Considerable effort has been expended to develop smoke suppressants, but without much success. Poly(vinyl chloride) and cross-linked unsaturated polyesters are among the few polymers that respond to a significant extent. Only the former are discussed here. The smoke-suppressant additives with some commercial significance include alumina trihydrate, magnesium hydroxide, zinc borate, zinc phosphate, molybdenum compounds, and tin compounds [52].

Flame retardants that act by the radical trap mechanism in the gas phase almost always contribute to an increase in smoke, because they suppress oxidation. These flame retardants (e.g., halogen and antimony oxides) are the ones used in thermoplastics, and increased smoke is observed in their presence.

Smoke suppressants that promote the oxidation reaction also interfere with the flame-retardant mechanism. They, therefore, work in opposition to flame retardants. For a major reduction in smoke that would also function simultaneously to inhibit burning, one may consider the use of intumescent coatings.

The inorganic hydrates are used in concentrations of 50–60%. Presumably they function in large part by reducing the fuel component of the plastic blend. It has also been suggested that the aromatic compounds that form as precursors of soot particles are adsorbed by the high-surface-area filler.

The first step in the degradation of poly(vinyl chloride) is dehydrochlorination leading to the formation of polyene structure. Further degradation gives benzene and other aromatic compounds responsible for the formation of significant smoke, PVC compounds containing zinc, molybdenum, and tin compounds give significantly less smoke. It is believed that the metal compounds react with the liberating hydrochloric acid to form Lewis acids which alter the mode of PVC decomposition. Trans polyenes form [53] and subsequently cross-linked polymer and, finally, char (see Fig. 34).

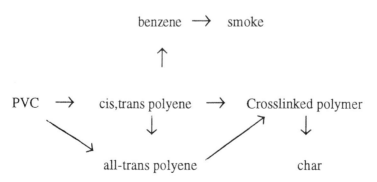

Figure 34 Decomposition of poly(vinyl chloride).

Table 64 Zinc Borate as Smoke Suppressant in Dioctyl Phthalate
Plasticized Polyvinyl Chloride

Additive (phr)	$D_m/5$ g[a,b]	Oxygen index
None	337	24.6
3 Antimony oxide	438	27.8
3 Firebrake zinc borate	257 (41)	24.7
1.5 Sb_2O_3 + 1.5 zinc borate	343 (22)	27.1
6 Antimony oxide	465	28.2
3 Sb_2O_3 + 3 zinc borate	438 (6)	28.4

Note: For each part of PVC, the following were included: DOP, 50 phr; stabilizer; additives.
[a] Flaming mode.
[b] Numbers in parentheses give smoke reduction (%).
Source: Data sheet, U.S. Borax, Los Angeles, CA.

A 50% replacement of antimony oxide with zinc borate gives a significant
reduction in smoke as measured in the NBS smoke chamber [54]. Although zinc
borate alone gives a significant reduction in smoke, it does not function as a
flame retardant. This is true for both dioctyl phthalate and a blend of dioctyl
phthalate and phosphate ester-plasticized PVC (Tables 64 and 65).

Table 66 gives data for an all-phosphate-ester plasticized PVC compound.
Increasing the zinc borate additive from 0.5 to 5 parts reduces the smoke signifi-

Table 65 Zinc Borate as Smoke Suppressant in a Dioctyl Phthalate/Phosphate Ester
Plasticized Poly(vinyl chloride)

Additive (phr)	NBS smoke density[a,b]		Oxygen index
	D_m	D_s (4 min)	
None	—	—	25.7
6.2 Antimony oxide	295	291	31.8
6.2 Firebrake ZB	—	—	26.5
2.1 Sb_2O_3 + 4.1 zinc borate	213 (28)	214 (26)	29.6

Note: For each 100 parts of PVC, the following were included: isodecyl diphenylphosphate, 20 phr;
DOP, 26 phr; $CaCO_3$, 41 phr; stearic acid, 0.3 phr; TiO_2, 16 phr; Ba/Cd/Zn stabilizers, 3 phr; additives.
[a] Flaming mode.
[b] Numbers in parentheses indicate smoke reduction (%).
Source: Data sheet, U.S. Borax, Los Angeles, CA.

Table 66 Effect of Zinc Borate in Reducing Smoke of Phosphate Ester Plasticized
Poly(vinyl chloride)

	Formulations			
	1	2	3	4
Composition (phr)				
Polyvinyl chloride	100	100	100	100
Stabilizer	2.5	2.5	2.5	2.5
Reofos 65	50	50	50	50
Firebrake ZB	0.5	1	2	5
Flame Retardancy				
Smoke, smoldering (D_m/g)	92	79	71	60
Oxygen index	30.6	30.9	31.2	32.1

Source: FMC, Princeton, NJ.

cantly. Here, the oxygen index values increase with increasing zinc borate concentration, indicating both increased flame retardancy and reduced smoke (Fig. 35).

Zinc borate contains 14–26% water of hydration, which is retained at temperatures up to 500°F, allowing this compound to be incorporated safely into many polymers.

Molybdenum compounds behave as good smoke suppressants and moderate flame retardants in halogenated polymers. Smoke test data show that low levels of molybdenum compounds reduce smoke formation by more than 50% for typical flexible and rigid PVC compounds [55,56]. The magnitude of the effect depends on the type and level of plasticizer. Molybdenum compounds act in the solid phase [56] as char formers, most likely by a Lewis acid mechanism. Compared to control samples, the amount of char formed may be more than doubled. Little or none of the molybdenum is volatilized [57] contrary to what may be expected by analogy with antimony oxide.

Tables 67 and 68 show that molybdenum will reduce smoke of rigid PVC by 30–60%. Table 68 shows that extended molybdenum compounds are highly effective in combination with zinc. Table 69 shows excellent smoke reduction results using combinations with ATH. Molybdenum is also a highly effective smoke suppressant for phthalate-plasticized PVC; similar results are obtained with phosphate ester-plasticized PVC.

Molybdenum also has been claimed as a flame retardant for PVC. Although this claim is valid, oxygen index values are not as high as are obtainable with antimony oxide (Table 70).

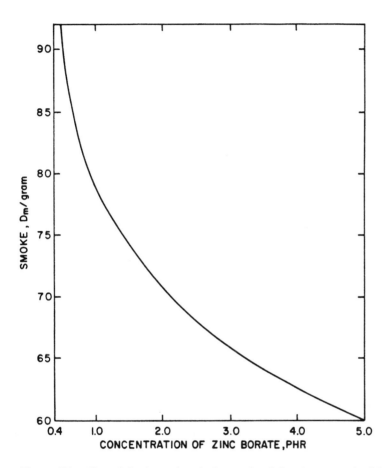

Figure 35   Effect of zinc borate in reducing smoke of phosphate ester plasticized poly(vinyl chloride).

Compounds based on tin have been proposed as flame retardants and smoke suppressants. Compounds studied include anhydrous and hydrous tin (IV) oxide, and various metal stannates and hydroxystannates. The best results were demonstrated with zinc hydroxstannate, $ZnSn(OH)_6$, and zinc stannate, $ZnSnO_3$. The tin compounds appear to act primarily in the condensed phase by a char-promoting mechanism leading to a significant decrease of pyrolysis products. They also release water during combustion:

$$ZnSn(OH)_6 \rightarrow ZnSnO_3 + 3H_2O$$

Table 67 Smoke Suppression by Ammonium Octamolybdate (AOM) and Molybdenum Trioxide (Rigid PVC Window Profile)

Formulation	Oxygen index	Test data on samples 125 mils thick		
		Arapahoe[a] (% smoke)	OSSR total smoke release[b] (smoke/m^2)	
Control	41	7.44	632	
0.7 phr AOM	45	4.90 (34)	—	
1.3 phr AOM	45	4.93 (34)	341 (46)	
2.5 phr AOM	46	4.84 (35)	292 (54)	
1.3 phr MoO$_3$	45	5.38 (28)	310 (51)	
2.5 phr MoO$_3$	46	4.25 (43)	271 (57)	

[a] Numbers in parentheses indicate smoke reduction (%).
[b] See Section III.F. For each 100 parts of PVC, the following were included: processing aids, 8.5; fillers, 12; additives, 2.
Source: Amax Speciality Chemicals Division, Ann Arbor, MI.

Table 68 Smoke Reduction of Rigid Poly(vinyl chloride) with Molybdenum and Zinc Compounds

Additive (phr)	Test data[a]	
	Arapahoe (% smoke)	NBS chamber D_m (flaming)
3 Calcium carbonate	8.0	310
3 KemGard 911C[b]	4.7 (41)	207 (33)
3 KemGard 911A[c]	4.3 (46)	217 (30)

[a] Numbers in parentheses indicate smoke reduction (%).
[b] Zn$_3$Mo$_2$O$_9$ on talc core.
[c] CaMoO$_4$ + ZnO on CaCO$_3$ core. For each 100 parts of PVC, the following were included: acrylic processing acid, 1.0, additives, 2.7.
Source: Sherwin-Williams, Coffeyville, KS.

Table 69 Molybdenum Plus Alumina Trihydrate as Smoke Suppressants (Rigid PVC Conduit)

	Arapahoe test data	
Formulation (phr)	% Smoke	% Reduction
$25CaCO_3$	7.8	—
25ATH	5.5	29
$10CaCO_3$ + 15ATH	6.9	12
$25CaCO_3$ + $2.5MoO_3$	5.3	32
25ATH + $2.5MoO_3$	3.5	55
$10CaCO_3$ + 15ATH + $2.5MoO_3$	3.3	58

Source: Ref. 55.

Table 70 Flame-Retarding Plasticized Poly(vinyl chloride) with Molybdenum Compounds

		Arapahoe test data	
Flame retardant (phr)	Oxygen index	% Smoke	% Reduction
None	27.5	23.2	—
$2Sb_2O_3$	32.5	26.7	$(15)^a$
$2MoO_3$	30.5	4.8	79
2 FR-21	29.0	8.3	64
$1Sb_2O_3$ + 1 FR-21	31.5	6.8	71

Note: For each 100 parts of PVC the following were added: diisodecyl phthalate, 30 phr; other, 7.8 phr.
[a] Increase.
Source: Climax Molybdenum, Ann Arbor, MI.

VII. MARKETS

The consumption of flame retardants in plastics in 1996 in the United States is forecast at 300 million pounds, not including alumina trihydrate (Fig. 36). World consumption of flame retardants is 788 million pounds (1992), not including ATH (Fig. 37).

The major market for ATH at 290 million pounds is unsaturated polyesters and polyacrylates. Polyolefins and PVC, especially in wire and cable, represents other sizable uses.

Decabromodiphenyl oxide is the largest organobromine additive even though it has been under attack, especially in Europe, because of toxicity concerns. Octabromodiphenyl oxide used in ABS has been replaced perhaps with the brominated epoxy oligomer. The Polycarbonate/ABS blend used for business machine housing now uses a nonhalogen flame retardant—specifically, resorcinol diphosphate. Tetrabromobisphenol A is the largest-volume reactive organobromine compound. It is used primarily in epoxy laminates for printed circuit boards. Another major reactive organobromine compound is tetrabromophthalic anhydride used in unsaturated polyesters.

The major types of organophosphates are triaryl and alkyl diaryl phosphates. Major applications include PVC, in which they function as nonflammable plasticizers, and modified PPO resin, in which the triaryl phosphate functions as vapor-phase flame retardants. A higher-molecular-weight alkyl diaryl phosphate has found special use in PVC plenum wire jacketing, where low smoke is required.

Organomomonophosphates traditionally used in modified PPO resin has been replaced with resorcinol diphosphate because it is less volatile and eliminates, in large part, the stress cracking problem. It is also the preferred flame

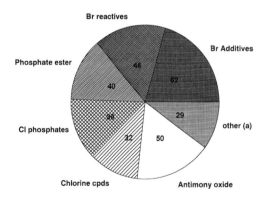

Figure 36   U.S. consumption of FRs for plastics (millions of pounds) (1996 forecast).

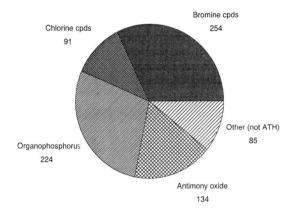

Figure 37 World consumption of flame retardants in 1992 (millions of pounds).

retardant for polycarbonate/ABS resin used for business machine housings. RDP may be replaced with a diphosphate based on bisphenol A, which is much lower in cost than the single-source resorcinol. The product is also more hydrolytically stable.

Halogenated phosphate esters are used primarily in rigid (wallboard insulation) and flexible (cushioning) polyurethane foam. A blend of pentabromodiphenyl oxide and a triaryl phosphate has replaced a large part of the chloroalkyl phosphate market because it is more thermally stable and has better scorch resistance.

The major chlorinated hydrocarbon flame retardants are chlorinated paraffins and Dechlorane Plus. The former is used in applications were their low cost and plasticizer properties are an advantage. Dechlorane Plus is a solid additive with much better thermal stability; it is used in nylon, polyethylene, and a small amount in polypropylene. Reactive chlorinated compounds include chlorendic acid and anhydride for application in unsaturated polyesters.

The markets for flame retardants by polymer application are as follows:

Product	%
Plastics	66
Textiles	5
Coatings/adhesives	3
Rubber	24
Wood/paper	2

The large market shown for rubber is due to the large volume of alumina trihydrate used in carpet backing.

REFERENCES

1. G Avondo, C Vorelle, R Delbourgo. Combust Flame 31:7, 1978.
2. FA Williams. In: Encyclopedia of Physical Science and Technology. San Diego, CA: Academic Press, 1987, Vol 3, p 211.
3. JJ Pitts, PH Scott, DG Powell. J Cell Plast 6:35, January/February 1970.
4. SK Brauman, AS Brolly. J Fire Retard Chem 3:66, 1976.
5. J Green. Plast Compound 10(3):57, 1987.
6. J Carnahan, W Haaf, G Nelson, G Lee, V Abolins, P Shank. Fourth International Conference of Flammability and Safety, San Francisco, 1979.
7. JW Hastie. J Res Natl Bur Stand 77A(6):733, 1973.
8. J Green, J Fire Sci 12:257–267, 1994.
9. WJ Miller. Combust Flame 13:210, 1969.
10. ME Morrison, K Scheller. Combust Flame 18:3, 1972.
11. JW Lyons. J Fire Flammabil 1:302, 1970; The Chemistry and Uses of Fire Retardants. New York: Wiley–Interscience, 1970, pp 20–24.
12. RM Aseeva, GE Zaikov. Combustion of Polymer Materials. New York: Hanser, 1985, pp 232, 239.
13. CP Fenimore. Conference on Flammability Characteristics of Polymeric Materials, 1970.
14. British Patent 792, 997 (1958) (to Associated Lead Ltd.).
15. KR Barton, WC Wouten, CC Donnelly. Text Chem Color 4(1):22, 55, 1972.
16. JW Vanderhoff, AK John. US Patent 3,058, 929 (Oct. 16, 1962) (to Dow Chemical).
17. J DiPietro, H Stephneczka. Soc Plast Eng J 27:23, February 1971.
18. ASTM D 2863-77; Annual Book of ASTM Standards, Vol. 08. 02. Philadelphia: American Society for Testing and Materials, 1977, p 635.
19. DW van Krevelin. Polymer 16:615, 1975.
20. S Wada. Gosei Juski Kogyo (Synth Resins Ind), 132–138, November 1985.
21. Standards for Tests for Flammability of Plastic Materials for Parts in Devices and Appliances, Subject 94. Underwriters Northbrook, IL: Laboratories, 1973.
22. EE Smith. Fire Flammabil 8:309, 1977.
23. JJ Brenden. J Fire Flammabil 6:50, 1975.
24. J Green. In: GL Nelson, ed. Fire and Polymers, ACS Symposium Series 425. Washington, DC: American Chemical Society, 1990, pp 253–266.
25. J Green. J Fire Sci 12:388–408, 1994.
26. R Wolf. Kunstst German Plast 10:76, 1986.
27. Thermoguard Data Sheet, M & T Chemicals, Rahway, NJ.
28. JP Hamilton. Plast Compound 1(3), 54–59, 1978.
29. J Green. In: M Lewin, SM Atlas, EM Pearce, eds. Flame Retardant Polymeric Materials. Vol 3, New York: Plenum Press, pp 1–34.
30. L Keating, S Petric, G Beekman. Plast Compound 9(4):40, 1986.

31. PR Hornsly, CC Watson. Plast. Rubber Process Applic 6(2):169, 1986.
32. Y Halpern. US Patent 4,154,930 (May 15, 1979) (to Borg–Warner).
33. CT Fleener Jr. US Patent 4,253,972 (March 3, 1981) (to Borg–Warner).
34. G Bertelli, P Roma, R Locatelli. US Ptaent 4,312,805 (Jan. 26, 1982) (to Montedison).
35. HW Bost. US Patent 4,216,138 (Aug. 5, 1980) (to Phillips Petroleum).
36. RB Bush. Plast Eng 42(4), 29–32, 1986.
37. RJ Schwartz. In: WC Kuryla, AJ Papa, eds. Flame Retardancy of Polymeric Materials. New York: Marcel Dekker, 1973, Vol. 2, p 88.
38. J Green. J Fire Flammabil Fire Retard Chem 1:194, 1974.
39. J Green. Plast Compound, 16(1), 32-4, 1993.
40. J Green. Polym Degrad Stabil 54:189–193, 1996.
41. I Touval. Proceedings of SPE 43rd Annual Technical Conference, 1985, p 968.
42. J Green. Plast Compound 7(7), 30–40, 1984.
43. A Granzow, JF Cannelongo. J Appl Polym Sci. 20:689, 1976.
44. LG Imhof, KC Stueben. Polym Eng Sci 13(2):146, 1973.
45. DF Lawson, EL Kay. J Fire Flammabil Fire Retard Chem 2:132, 1975.
46. JJ Kracklauer, CJ Sparkes. Plast Eng 11(6):57, 1974.
47. HP Doerge, M Wismer. J Cell Plast 8:311, 1972.
48. HL Vandersall. J Fire Flammabil 2:97, 1971.
49. AJ Papa. Ind Eng Chem Prod Res Dev 11:379, 1972.
50. EJ Quinn, DH Ahlstrom, SA Liebman. Polym Prepr 14(2):1022, 1973.
51. MM O'Mara. Polym Prepr 14(2):1028, 1973.
52. S Karpel. Fire Prevent 143, 1981.
53. WH Starns, D Edelson. Macromolecules 12:797, 1979.
54. KK Shen. Plast Compound 8(5):66, 1985.
55. GA Skinner, PJ Haines. Fire Mater 10:63, 1986.
56. FW Moore, RA Ference. Twelfth International Conference on Fire Safety, Millbrae, CA, 1987.
57. FW Moore, DA Church, Third International Symposium on Flammability and Flame Retardants, Toronto, 1976.
58. Chemical Economic Handbook. SRI International, Menlo Park, CA, 1993.
59. J Green. J Fire Sci 8:254–265, 1990.

7
Heat Stabilizers for Halogenated Polymers

Richard F. Grossman
Halstab Division, The Hammond Group, Hammond, Indiana

Dale Krausnick
NSF Enterprises, Toledo, Ohio

I. INTRODUCTION

In very few cases can poly(vinyl chloride) (PVC) be fabricated into products without prior addition of a heat stabilizer. In this regard, PVC is unlike most thermoplastics. Milling or other processing of unstabilized PVC at temperatures needed for plastic flow results in rapid degradation, with symptoms including discoloration, odor, elimination of hydrogen chloride, and irreversible adhesion to equipment surfaces [1]. Other halogenated polymers such as chlorinated polyethylene (CPE), chlorosulfonated polyethylene (CSM), PVDC, and chlorinated PVC are more or less similar. The object of stabilization is to control this degradation both during processing and end use.

Stabilizer efficiency is the speed and extent to which an additive can repair or limit damage before competing degradative reactions occur. The additive must, to be efficient, have high compatibility with, and high mobility within, the polymer matrix. It should have a similar solubility parameter, about 19 SI units (reciprocal megapascals). Efficiency, however, depends also on stabilizer demand. This factor includes the extent of energy input, the rate at which it is applied, and the time it is maintained. A process with relatively low stabilizer demand (e.g., coating and fusing of plastisols) can make do with a stabilizer system that would be inadequate for high-speed calendering of semirigid film. Saying that a stabilizer is efficient denotes only that it is able to control or repair damage in a given situation more rapidly than the development of competing degradative reactions.

The claim that one stabilizer is more efficient than another means only that a lesser quantity provides control or repair, or provides it longer, or more cost-effectively in a specific situation. In a different process, the conclusion may no longer be correct.

Stabilization during service life also involves competing reactions. It differs from process stability not only in the type of energy input (heat or light versus heat and shear) but also in the locus of damage, often the surface rather than the bulk of the composition. Mobility and compatibility requirements are not identical. As a result, stabilizers used in situations where properties and appearance during service life are important (e.g., flooring, wall coverings) are often different from those used where stabilizer demand relates mostly to processing.

If a polymer can eliminate small molecules, it is usually the case that elimination occurs before other modes of degradation. In the case of PVC, this reaction is critical for two reasons: the small molecule that is eliminated, HCl, is a strong acid that catalyzes further elimination, and the resultant dehydrochlorinated product, which now contains an additional allylic chlorine, is more prone to further degradation. Central to practical compounding is that this self-catalyzed elimination, referred to as "unzipping," begins at specific sites: carbon–chlorine bonds more reactive than average, designated "labile" chloride [2].

A. Degradation of Halogenated Polymers

The elimination of HCl from organic molecules has been studied for many years. A comprehensive summary is given in the well-known textbook by March [3]. The thought that HCl elimination from PVC is likely to be a chain reaction was advanced some years ago by Michel [4]. General acceptance of this view, however, awaited the publication of a series of articles by Fisch and Bacaloglu that provided a theoretical justification for chain-reaction elimination, leading to a statistical distribution of products with varying extents of unsaturation:

$$—(CH_2CHCl)x \rightarrow —(CH{=}CH)x + xHCl \tag{1}$$

Their calculations and review of the literature strongly supports this description [5]. At first glance, the concept of chain-reaction elimination of HCl is hardly reassuring. Almost no processing of PVC can be carried out without some loss of HCl. The thought that all PVC processing must involve the creation of conjugated unsaturation, with its implications regarding color formation and ease of oxidation, is unsettling. On second thought, it is clear that a chain reaction favors the effective operation of additives, simply because chain reactions are usually easily interfered with. Such interference is the *modus operandi* of all antioxidants and many flame retardants.

The elimination of HCl is most likely initiated at an allylic defect site in the polymer, the allylic double bond activating the loss of chloride. Recently, it has been suggested that stabilizers complex with allylic chlorides, reducing the extent of activation [6]. In this scenario, the initial function of the stabilizer may be to increase the energy of activation of HCl elimination, increasing the probability of the transition state returning to reactants, completely analogous to the proposed initial function of antioxidants (see Chapter 1). In processes having low stabilizer demand, it is possible that increasing the activation energy associated with elimination may be the most important stabilizer function.

Many, perhaps most, actual situations require additional stabilizer activity. This is the case because even the mildest processing steps can run out of control at times, particularly in long runs. An important stabilizer function is to protect against catastrophic degradation during such excursions. The catalysis of elimination by HCl, as described by Bacaloglu and Fisch [5],

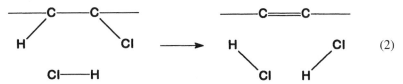

(2)

provides opportunity for the next function of the stabilizer, scavenging HCl, in competion with the above reaction:

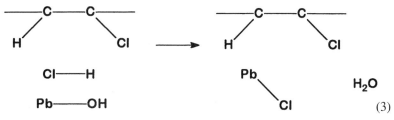

(3)

In reaction (3), a lead stabilizer is shown, but analogous reactions can be drawn with other systems. The key to efficiency in reaction (3) is the ability of the stabilizer to interfere with HCl (or other acid catalysis) *at the elimination reaction site*. The mere presence of an HCl-reactive species in the composition (i.e., removal of HCl by its diffusion to an active acid absorber) is insufficient. Thus, active acid absorbers such as $Mg(OH)_2$ and PbO are relative inefficient heat stabilizers for PVC [7]. Both are effective, however, in CPE and CSPE, indicating that these polymers, with greater freedom of internal movement (lower T_g), are able to coordinate to insoluble fillerlike additives more efficiently. PVC stabilizers that are effective in preventing catalysis of HCl elimination seem invariably to possess special structures that enable HCl capture efficiently at the reaction site.

With some applications, HCl scavenging may be sufficient for effective stabilization. This is often the case with dark colors and anaerobic processes, such as extrusion of black insulated wire or cove base. Most applications, however, require that the stabilization system also maintain good color, by terminating (rather than just interfering with) development of unsaturation. To do this, the stabilizer functions by replacing labile chloride. It does this by displacement with a ligand less easily removed by the effects of heat, light, or shear. (Obviously, if the new ligand were less stable to elimination, one would not expect the displacement reaction to proceed readily.) The stabilizer must also convert the displaced chloride to a PVC-unreactive state while retaining its power to stabilize further (i.e., displace again with another "good" ligand) [8]. From this simple picture, it is clear that the effective stabilizer must be able to coordinate rapidly with the polar carbon–chlorine bond and yet have high mobility. This implies a molecule with polar and nonpolar domains. Most stabilizers fit this description, which is also typical of lubricants. Additives used as heat stabilizers are either in themselves lubricants or readily form complexes with lubricants; these complexes are important in providing stabilizer efficiency by transporting the active stabilizer to the site of potential degradation.

Stabilization systems also provide an antioxidant function, which becomes increasingly important as stabilizer demand increases. The stabilizers may be antioxidants themselves, such as tin mercaptides or dibasic lead phosphite, or they may contain added antioxidants. Industrial usage of particular systems will be discussed application by application with reference to the various types of stabilizers.

Poly(vinyl chloride) stabilization may be summed up as comprising the following steps:

1. Increasing the energy of activation of initiation of degradation by complexing with an allylic site, countering the activating effect of the allylic double bond. All classes of PVC stabilizers appear to do this. The result is the preservation of essentially unchanged appearance and properties for a length of time appropriate to many service and processing conditions.

2. Removing traces of HCl at the degrading site, interfering with catalysis of further elimination. All classes of PVC stabilizers contain functional groups which do this. A distinction should be drawn between systems where the stabilizer itself appears to capture HCl, and others in which predominant action seems to be transfer of trace HCl to a second species. The first type of behavior leads to good early color, and the second to long processing life. Actual stabilizers are usually systems that compromise by providing both effects.

3. Terminating the development of unsaturation by replacing labile chloride with other functional groups. All classes of PVC stabilizers can be formulated for such behavior. The results include longer maintenance of initial appearance and increased processing life.

4. Providing antioxidant protection during processing and service life. Here, it is necessary to distinguish between stabilizers that are in themselves good antioxidants and those where antioxidants are added. Those of the first type (dibasic lead phosphite, organotins) are very good at early color retention. Those where antioxidants are added, such as mixed-metal stabilizers containing phosphites, generally provide a balance of long process stability and good color retention.

Although a great deal of speculation has appeared in the literature as to the exact chemical reactions involved in PVC degradation, aside from the initiation step, very little is known with certainty. The PVC formulator and user is, therefore, better off in a practical sense to regard degradation and stabilization in the general terms given above.

II. LEAD STABILIZERS

Lead stabilizers are typically described in the literature as double salts of PbO (e.g., tribasic lead sulfate as $3PbO \cdot PbSO_4 \cdot H_2O$) (see Ref. 9 for example), despite it being known for many years that their infrared spectra indicate the presence of Pb—OH groups [10]. They are extremely efficient HCl scavengers:

$$—Pb—OH + HCl \rightarrow —Pb—Cl + H_2O \tag{4}$$

In addition, lead stabilizers complex allylic chlorides, lowering the extent of allylic activation, as indicated by the infrared spectra of such complexes [6]. The shift in C=C stretching frequencies to lower energy through conjugation with —Cl, observed generally with allylic chlorides, is largely reversed in the complex with stabilizer.

The very long processing stability provided by lead stabilizers stems not only from their HCl capture efficiency but also from the resistance of the reaction products to thermal decomposition, which would regenerate chloride. Such regeneration has long been considered important in the generally shorter processing times associated with mixed-metal stabilizers, particularly those based on zinc compounds [11]. The reaction products are, initially, not lead dichloride, but the stabilizer molecule itself, in which one or more Pb—OH groups has been replaced with Pb—Cl. These products are far less water soluble than lead dichloride [12]. This insolubility and unreactivity underlies the use of lead stabilizers in products that must resist swelling and hydrolysis by water, such as roofing, pond liners, and insulated wire. A further useful characteristic is that ultimate failure is almost invariably gradual; the batch may darken and become unprocessable, but rarely chars or burns, unlike with most other stabilizer systems.

The oldest of these stabilizers is white lead, basic lead carbonate, found in textbooks as $2PbCO_3 \cdot Pb(OH)_2$. The infrared spectrum of this molecule confirms the presence of Pb—OH; its ^{1}H-NMR (nuclear magnetic resonance) spectrum indicates a single type of proton [13a]. Its structure is most likely

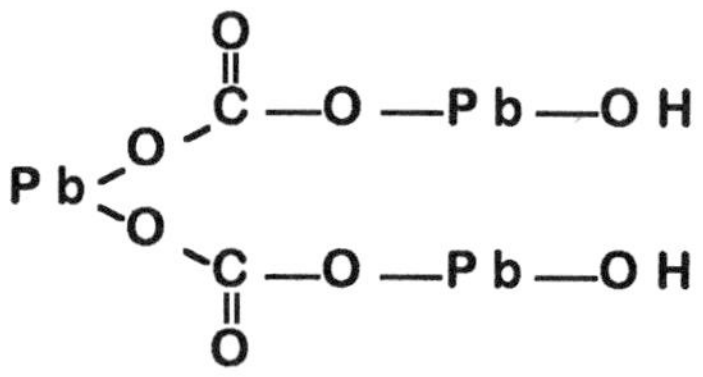

The structure of PbO is that of a trigonal pyramid, with the lead atom at one apex, two oxygens linking one lead another, and an unshared electron pair, forming the other corners. In basic lead carbonate, as in all the lead stabilizers, there are pyramids of —O—Pb—OH groups. Some, such as the above, also contain —O—Pb—O— pyramidal groups. In basic lead carbonate, two planar carbonate anions link a —O—Pb—O— to two —O—Pb—OH pyramids. The relationship to litharge is clear, but the stabilizer is by no means a simple complex of PbO.

On reaction with HCl, replacement of —OH with —Cl generates water. The combination of water and HCl leads to hydrolysis of basic lead carbonate, evolving CO_2 and leaving a mixture of $PbCl_2$ and $Pb(OH)_2$. The latter is thermally unstable, yielding PbO and water, continuing the process. Although thermal analysis differential thermal analysis (DTA) indicates that basic lead carbonate is itself stable to above 200°C, the presence of HCl no doubt lowers this [14]. At one time, basic lead carbonate was widely used to stabilize compounds for automotive low-voltage wiring because of its low cost and strong value as a white pigment. With higher-temperature ratings now prevalent and higher temperatures used in more rapid extrusion of wire, the tendency toward gas evolution has made it necessary to replace basic lead carbonate. Its commercial value is now limited to its use as an artist's white pigment, where the depth and subtlety of the shadings that can be obtained exceed that provided by competitive materials.

In contrast, normal lead carbonate, without Pb—OH groups, has no value as a heat stabilizer in PVC, despite high lead content:

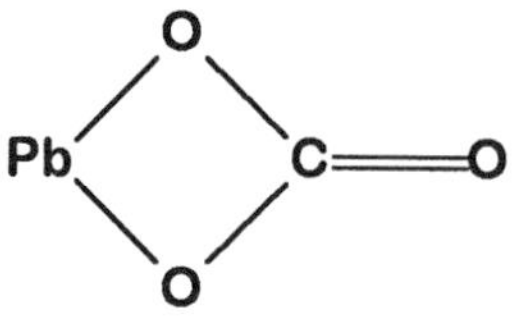

In the above representation, bonds between lead and oxygen are drawn as a convenience, not to suggest a completely covalent character. The actual crystal may be better represented as [—O—Pb—O—CO—].

The lead sulfate family is more diverse. Normal lead sulfate is analogous to the above carbonate and has no stabilizer value:

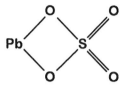

Again, in the crystal, [—O—Pb—SO$_3$—] may be a better description. When PbO and sulfuric acid are reacted in the proper stoichiometry, monobasic lead sulfate results. This has been written in textbooks as PbO·PbSO$_4$. Its infrared (IR) and NMR spectra [13], however, indicate Pb—OH groups present. It is most likely

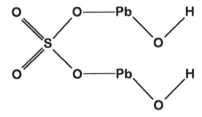

Monobasic lead sulfate is not a particularly efficient heat stabilizer and has found only limited use, primarily as a blowing agent activator.

As the PbO/sulfuric acid stoichiometry is adjusted to higher lead levels, the primary product becomes tribasic lead sulfate, appearing in the literature as 3PbO·PbSO$_4$·H$_2$O. The IR and NMR spectra [13] nevertheless indicate not water, but Pb—OH groups. Sulfur may be at the center of a square pyramid, oxygen at the apex, and four —O—Pb—OH pyramids linked to the bottom corners:

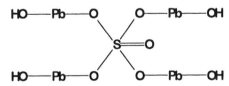

There is some condensation between —Pb—OH groups, giving rise to dimers, trimers, and so forth linked at corners of —O—Pb—O pyramids. Recent x-ray and neutron diffraction studies of tribasic lead sulfate crystals are consistent with the following structure within the crystal [15]:

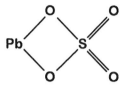

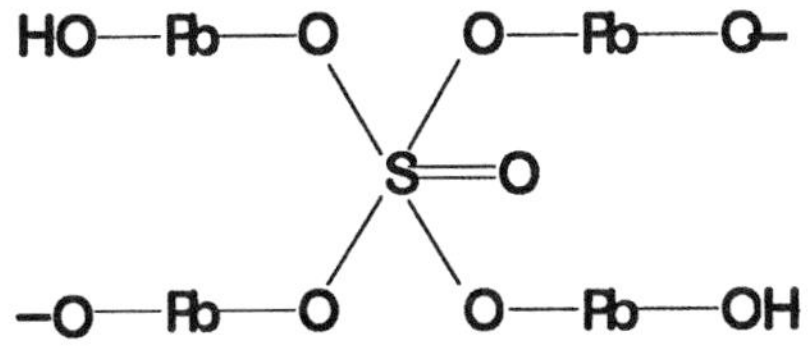

In actual use, there will always be sufficient water available to convert the above to the monomeric form.

Tribasic lead sulfate is the most widely used PVC stabilizer. It is used almost universally in wire insulations rated for service below 90°C and, throughout most of Europe, South America, and Asia, as the workhorse stabilizer for rigid PVC. This usage is based on the following thermal stability, decomposition temperature above 230°C [14]; high efficiency, leading to extremely long process safety; low cost; and insensitivity to water. It is not a coincidence that tribasic lead sulfate is also the most economical general-purpose stabilizer.

The drawbacks to the wider use of tribasic lead sulfate include the following: its opacity, preventing use in clear products; the investment needed to ensure that workers neither ingest nor inhale lead-containing compounds; and strong reactivity toward the hydrolysis of ester plasticizers, particularly phthalates, during high-temperature heat aging or processing. The last of these is corrected by the use of stabilizers based on weaker acids than sulfuric.

At a higher ratio of PbO to sulfuric acid, tetrabasic lead sulfate results. This has been written as $4PbO \cdot PbSO_4 \cdot H_2O$. Again, the IR and NMR spectra [13] indicate not water, but Pb—OH groups. In addition, the NMR spectrum discloses a proton peak not found in monobasic or tribasic lead sulfate, possibly corresponding to S—OH. The data could be accounted for by

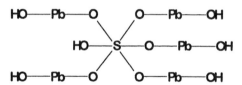

Tetrabasic lead sulfate is off-white to yellow and is photosensitive, quickly darkening in the presence of ultraviolet light. Nevertheless, it is widely used in Europe to stabilize dark or strongly colored wire compounds, particularly power cable jackets.

Two salts of phosphorous acid are know. Normal lead phosphite shows strong P—H absorption (2340 cm^{-1}) in the infrared, as does the parent compound, phosphorous acid. This proton also appears in the NMR spectrum [13]. The structure is reasonably represented as

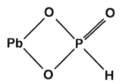

Again, bonds drawn between lead and oxygen are for convenience, rather than

to suggest a pure covalent character. Although of some activity in ultraviolet light stabilization [16], normal lead phosphite has little or no value as a heat stabilizer. The correct ratio of PbO to phosphorous acid will yield, instead, dibasic lead phosphite, previously written $2PbO \cdot PbHPO_3 \cdot \frac{1}{2}H_2O$. As may be surmised by analogy with the above, the IR spectrum indicates not water, but Pb—OH and P—H [13]. The NMR spectrum shows these two types of protons almost exactly in the ratio of 2 —OH to 1 P—H. This suggests the structure

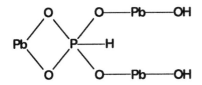

By analogy to tribasic lead sulfate, the crystal structure may be better represented as [—O—Pb—O—PH—(O—Pb—OH)$_2$—]. In addition to being a heat stabilizer of about the same effectiveness as tribasic lead sulfate, dibasic lead phosphite is a strong antioxidant, comparable in PVC to organic phosphites (see Chapter 1). The P—H bond abstracts peroxy oxygen from hydroperoxides, yielding dibasic lead phosphate, (HO—Pb—O—)$_3$P=O, a relatively efficient heat stabilizer. Dibasic lead phosphite is used in flexible PVC (e.g., bright colored wire jackets and electrical tape, roofing, and pond liners) and, other than in North America, in rigid PVC applications (siding, window profile) where ultraviolet (UV)-light resistance is significant. Many of these applications use combinations of tribasic lead sulfate and dibasic lead phosphite for greater economy, because phosphorous acid is considerably more expensive than sulfuric acid. Dibasic lead phosphite is thermally stable to above 300°C [14]. It is the only basic lead stabilizer that provides excellent color hold at high temperatures or in the presence of UV light (i.e., under oxidizing conditions). With other basic lead salts, oxidation leads to Pb(II)Pb(IV) mixtures, generally with a characteristic beige–pink coloration.

A number of basic salts are also formed with organic acids. At a 1:1 equivalence, dibasic phthalic acid and PbO yield monobasic lead phthalate. The NMR spectrum of this species [17] shows two types of proton (aromatic and —OH) in the ratio of 2:1. Because there are four aromatic protons on the ring, there must be two —OH groups. Thus, monobasic lead phthalate is best described, not as $PbO \cdot Pb(C_6H_4CO_2)$, but as $C_6H_4(CO—O—Pb—OH)_2$. As with monobasic lead sulfate, it is not a particularly efficient stabilizer and has found only limited use. Its extreme thermal stability, decomposition temperature about 400°C [14], suggests potential as an acid scavenger in fluoropolymers.

With higher levels of PbO, phthalic acid (or anhydride plus water) yields dibasic lead phthalate. The NMR spectrum of this molecule [17] shows equal numbers of aromatic and —OH protons; therefore, four —OH groups. A possible structural formula is

In the above, it is visualized that one carboxyl group is coplanar and conjugated with the aromatic ring, and linked to one —O—Pb—OH group. The second is tetrahedral and linked to two such groups; ^{13}C-NMR supports this assignment [17]. A characteristic of all of the high-efficiency basic lead stabilizers is that they seem, at least in part, to be salts of ortho acids [i.e., —C(OH)$_3$, rather than —CO$_2$H]. Analogous ortho structures seem to be the case with dibasic lead phosphite and the polybasic lead sulfates. In other words, these stabilizers contain more —OH groups than one would initially guess. (The last is true, even if alternatives to ortho acid structures are drawn.) Back coordination of electron density from oxygen to lead may account for this extra abundance of —OH groups, which appears to be an essential element of lead stabilization. Related structural dispositions are found with other types of stabilizer.

Dibasic lead phthalate is used throughout the world for stabilization of wire insulation where service temperature ratings of 90°C and higher are required and for a number of other applications where moisture resistance is important. Dibasic lead phthalate itself is thermally stable to about 315°C [14]. Its rate of saponification of diisodecyl phthalate (DIDP) plasticizer is three orders of magnitude lower than that of tribasic lead sulfate [7]. This combination of thermal stability and unreactivity to plasticizers underlies the suitability of dibasic lead phthalate for high-temperature service as well as processing.

Dibasic lead phthalate and tribasic lead sulfate can be coprecipitated in approximately a 1 : 1 molar ratio to yield a product with a NMR spectrum different from a physical blend of the ingredients [18]. This product, sold under the trade name Halstab 60, has a reaction rate with DIDP at 165°C only slightly greater than that of dibasic lead phthalate [7]. In addition, thermal analysis of Halstab 60 indicates a single decomposition, starting at about 300°C [18]. The coprecipitate seems to consist of a core of tribasic lead sulfate and a shell of dibasic lead phthalate. In many cases, dibasic lead phthalate may be replaced with Halstab 60, at lower cost, with no loss in stabilizer efficiency.

Where color retention is important, lead-stabilized PVC compounds, both rigid and flexible, typically contain a metallic stearate to terminate growing conjugated unsaturation by replacement of labile chloride (see Section I). The stearate content can be provided by coating the lead stabilizer with barium or calcium stearate, or by the addition of one of the two common forms of lead stearate. Dibasic lead stearate is more commonly used. Although commonly written as $2PbO \cdot PbSt_2$ (St for the complex stearic acid mixture), the NMR spectrum of dibasic lead stearate indicates the presence of aliphatic to —OH protons in a ratio of about 10 to 1 [17]. Its structure is probably

Dibasic lead stearate serves as a lubricant and additional stabilizer in addition to being a source of stearate. It is almost invariably used in concert with another lead stabilizer.

Normal lead stearate, $PbSt_2$, is devoid of Pb—OH groups and is, therefore, of only modest efficiency as a stabilizer. At low levels, it is soluble in PVC, although functioning as an external lubricant. Combinations of lead and barium stearates appear to be more effective than either individually as PVC stabilizers. Such combinations are commonly used in phonograph records, where the solubility of the stabilizer leads to low surface noise, and in thin-wall wire coverings, particularly bright colors. Barium/lead stearate blends have found some use in clear applications such as unpigmented lampcord. Degradation, however, leads to insoluble lead chloride, causing haziness to develop. It should be noted that this can also be the case with mixed-metal stabilizers in clear PVC, through development of insoluble barium or calcium chloride.

In summary, basic lead salts are effective as PVC stabilizers because they provide the following:

1. Form complexes with allylic chloride segments in the polymer, countering the activating effect of the allylic double bond toward HCl elimination

2. Scavenge HCl at the degradation site very effectively, forming thermally stable products, leading to long processing safety and very gradual failure

3. Terminate developing conjugated unsaturation in the polymer by trans-
    ferring poor leaving groups, usually stearate, to activated carbon atoms
    bearing labile chloride.
4. Provide antioxidant protection at the degradation site, in the case of
    dibasic lead phosphite, by means of the antioxidant phosphite group;
    otherwise, by gradual oxidation of Pb(II) to Pb(IV)

A. Using Lead Stabilizers Safely

Lead is found throughout the crust of the Earth at concentrations, in soil-forming
rocks, of 2–25 mg/kg [19]. In soil, typical background concentrations average
10–30 mg/kg [20]. Similar low levels are found in food, water, airborne dust,
and the human body. At overall body levels of about 0.1–0.2 mg/kg, lead appears
to be an essential element in mammals [21]. The level of lead in human blood
considered normal is similar, 0.1–0.3 mg/kg, usually expressed, however, as
10–30 μg/100 g whole blood. At these concentrations, the body is able to maintain
a balance, eliminating lead (among other heavy metals) at about the same rate it
is acquired. There are upper limits of tolerance for all essential trace elements.
At levels of lead above 80 μg/100 g blood, symptoms of lead poisoning may
develop. These can include abdominal pain, metallic taste, loss of weight, muscu-
lar pain and weakness, constipation, and nausea. If recognized, treatment with
chelating agents such as EDTA will reverse heavy metal buildup. With reasonable
precautions in handling lead stabilizers, blood lead levels of workers are readily
maintained in the normal range, well below the OSHA Biological Limit Value
of 50 μg/100 g blood. Similarly, with properly formulated compounds, the loss
of lead stabilizers to the environment is very low, lower than with any other type
of stabilizer [22].

To prevent inhalation, the best method of use is to meter lead stabilizers
from semibulk containers by action of gravity through dispensing equipment
directly to the mixing chamber, usually a high-intensity mixer or ribbon blender.
Such systems have been installed by many processors and enable operation below
the American Conference of Governmental and Industrial Hygienists (ACGIH)
recommended Time Weighted Average Threshhold Limit Value (TLV–TWA)
for lead compounds in air, currently 0.15 mg/m^2 (as Pb), and the Permissible
Exposure Level (PEL), currently 0.05 mg/m^2 (as Pb), specified by U.S. Code of
Federal Regulations, Vol. 29, Par. 1910.94 and Par. 1910.1025. Industrial experience is that maintenance of such lead-in-air standards, combined with appropriate
industrial hygiene, will result in normal worker blood lead levels. The safe handling of lead stabilizers has been the subject of a detailed publication [23].

Semibulk handling and metering equipment, of course, involves capital
investment. Temporarily, it may be convenient instead to use lead stabilizers heat
sealed in low-melt ethylene vinyl acetate (EVA) bags, adjusting the batch to

accommodate an integral number of 25- or 50-lb bags. Special bags have been devised that enable heat sealing of EVA (or other films) without the stabilizer contacting the sealing line; otherwise, lead compounds strongly interfere with this process [24]. Using this approach, the sealed bags are usually added to the mixer at the beginning of the cycle with the PVC resin, as is generally also the case with semibulk addition. Lead stabilizers have a strong affinity to PVC grains and are readily adsorbed early in the cycle in a closed mixer. Plasticizer, filler, and small ingredient additions are then carried out without permitting the escape of the stabilizer prior to the batch being fluxed. In large-scale operations, the use of low-melt bags cannot compete with semibulk handling, because of the higher cost involved in filling bags.

In small-scale plasticized PVC operations, it may be convenient to use lead stabilizers in nondusting forms, such as prills, extruded strands, pressed tablets, or paste dispersions in plasticizer. All except the last are, in general, less easy to disperse in PVC than the pure stabilizers. In flexible compounds, prills or strands are best added after part or all of the plasticizer has been incorporated. The addition of the plasticizer broadens the solubility parameter of the composition, favoring more rapid incorporation of less polar additives (e.g., waxy binders used to encapsulate lead stabilizers in prills or strands). On the other hand, if sufficient shear is developed early, as with many high-intensity mixers, encapsulated lead stabilizers can be added successfully with the polymer. It is a question of knowing the capabilities of a given piece of mixing equipment.

Rigid PVC compounds that are lead stabilized routinely use prills, strands, or tablets. These often incorporate lubricants and pigments with stabilizers. Such products (referred to as "one packs") when used with reasonable care provide suitably low levels of lead in air. Lead stabilizers are generally sufficiently cost-effective in rigid PVC to justify the added cost of dispersion preparation. In flexible PVC, European practice is to use lead stabilizers as pumpable paste dispersions, added with plasticizer. Only a few manufacturers in the United States have adopted this practice, mainly because of the competitive nature of the most significant market, wire and cable, operating against the increased raw material cost of the dispersion.

All users of lead stabilizers should avail themselves of the practical precautions presented in Safety in Handling Lead Chemicals [25] and in Controlling Lead Exposures in the Workplace [26]. Adherence to these day-to-day guidelines has enabled routine usage of lead stabilizers without danger to workers or to the environment.

III. ORGANOMETALLIC STABILIZERS

Organometallics (i.e., compounds containing carbon–metal bonds) tend to be compatible with polar polymers such as PVC. Most organometallics have disadvantages: extreme toxicity (alkyl lead, zinc, and mercury compounds), violent reactivity to air or water (organoaluminum and organotitanium compounds), or high reactivity with polar substrates (silanes). An exception is the case of monoalkyltin and dialkyltin derivatives. In the typical stabilizer, tin, in the (IV) oxidation state, is covalently bound to either one or two alkyl groups, and either to two or three reactive ligands. The latter are capable of displacing labile chlorine and scavenging HCl. The resultant monoorganotin and diorganotin compounds form a group of valuable stabilizers, principally for rigid PVC, but with some flexible PVC uses. The generalized stabilizer formula is R_2SnX_2 or R—SnX_3 [27].

Typical organotin stabilizers have solubility parameters very close to that of PVC. This matched polarity underlies their high efficiency in many applications. Tin stabilizers are not, generally, effective lubricants. They may coordinate with lubricants also functioning as secondary stabilizers (e.g., calcium stearate). This coordination (complex formation) enables facile replacement of a ligand on tin with stearate after the original ligand has functioned in stabilizing PVC, generating a slightly altered but still effective stabilizer. In actual practice, calcium and, occasionally, other stearates are found in most tin-stabilized formulations.

In the first generation of tin stabilizers to appear (in the United States in the late 1930s), the R— group used in the above general formula, was n-butyl, because of the availability and relative low cost of n-butyl chloride and its suitability in the Grignard reaction with magnesium [reaction (5)]:

$$R—Cl + Mg \rightarrow R—Mg—Cl \tag{5}$$

The resultant Grignard reagent reacts rapidly with tin halides, replacing chloride with the R— group. In practice, tin tetrachloride is converted to tetrabutyltin, originally with butylmagnesium chloride and more recently with the corresponding aluminum trialkyl [28]:

$$SnCl_4 + 4R—MgCl \rightarrow R_4Sn + 4 MgCl_2 \tag{6}$$

$$3SnCl_4 + 4AlR_3 \rightarrow 3R_4Sn + 4AlCl_3 \tag{7}$$

The resultant tetraalkyltin is then reacted further with tin tetrachloride, causing disproportionation into a range of alkyltin chlorides, which are separated by fractional distillation:

$$R_4Sn + SnCl_4 \rightarrow RSnCl_3 + R_2SnCl_2, \tag{8}$$

and so forth. The disproportionation process and fractionation can be controlled to yield mixtures of almost entirely mono-organotin and diorganotin chlorides, ranging from an equilibrium 65/35 di/mono ratio to pure mono or di, depending on reaction conditions. These intermediates are then reacted with carboxylic acids or with ligands containing mercaptan groups to yield actual stabilizers.

At the present time, tin stabilizers are available with a choice of R— groups from among n-butyl, n-octyl, methyl, and acrylic esters. All have high compatibility with PVC. In the absence of secondary stabilizers to scavenge chlorine, the reaction products are substituted organotin chlorides, which are relatively weak Lewis acids that, as with lead-stabilizer reaction products, do not catalyze further dehydrochlorination of PVC.

Dimethyltin dichloride intermediate is now produced directly from tin and methyl chloride [28]:

$$Sn + 2CH_3Cl \rightarrow (CH_3)_2SnCl_2 \tag{9}$$

Similarly, monomethyltin trichloride is synthesized from stannous chloride:

$$SnCl_2 + CH_3Cl \rightarrow CH_3SnCl_3 \tag{10}$$

These methods are cost-effective and, in addition, directly yield intermediates without the need for fractional distillation of a mixture. Yields are good only with methyl chloride. The cost advantage is magnified by the lower molecular weight of the methyl group, providing higher molar activity per given weight.

Estertin intermediates are also produced by direct one-step reactions at a cost advantage. In the case of monoestertins, this is done by alkylation of stannous chloride by an acrylate ester in the presence of anhydrous HCl:

$$RO\!-\!CO\!-\!CH\!=\!CH_2 + SnCl_2$$
$$+ HCl \rightarrow RO\!-\!CO\!-\!CH_2CH_2\!-\!SnCl_3 \tag{11}$$

In the case of diestertins, two equivalents of acrylate ester are reacted with metallic tin, again in the presence of dry HCl:

$$2RO\!-\!CO\!-\!CH\!=\!CH_2 + Sn$$
$$+ 2HCl \rightarrow (RO\!-\!CO\!-\!CH_2CH_2)_2SnCl_2 \tag{12}$$

Unfortunately, this ingenious reaction scheme has been discontinued in the United States, possibly as a result of the estertin products having unique chemistry, and thus not dropping in exactly for butyltin or methyltin analogs.

Varying the nature of the alkyl group produces other effects as well. It was thought that the long chain length of the n-octyl group limited water extraction from PVC articles and was thus important in obtaining FDA sanction. Recently, per Code of Federal Regulations 21, Par. 178.2650, FDA sanction has been extended to certain methyltin stabilizers and to estertins as well. Chain length bears also on volatility. If the other ligands are kept constant, decreasing the chain length of the alkyl group increases volatility. In practice, this is compensated for by the choice of the remaining ligands.

The migration of tin compounds to the environment from PVC is almost as low as with lead stabilizers [22]. The most recent study concludes that ''there will be no adverse environmental or human effects resulting from the use of methyltin stabilizers'' [29]. Nevertheless, in some parts of the world, organotin stabilizers are viewed as emotionally as lead compounds. This has resulted from confusion of nontoxic mono-organotin and diorganotin compounds with biotoxic triorganotins and tetraorganotins. The latter have biocidal activity because they are strong alkylating agents, destroying microbes as they are converted to mono-organotin and diorganotin reaction products. It is very much like considering sodium sulfate to be hazardous because it can be made from sulfuric acid and caustic soda; in other words, not considering the irreversibility of that reaction. Ignorance of entropy is typically at the heart of the bad approximation.

The original tin stabilizers were di-n-butyltin dicarboxylates, most often the dilaurate, prepared from dibutyltin oxide:

$$R_2SnO + 2C_{11}H_{23}COOH \rightarrow R_2Sn(OCOC_{11}H_{23})_2 + H_2O \qquad (13)$$

Tin dilaurate stabilizers have excellent PVC compatibility, provide some lubrication, and are useful stabilizers, limited only by their relatively high cost. Dibutyltin dilaurate is now used mainly as an esterification and hydrolysis catalyst, for example, in polyurethane cross-linking and in hydrolytic cross-linking of silanes grafted to olefinic polymers. Usage in PVC has been almost completely superseded by more advanced tin stabilizers.

Replacement of laurate with unsaturated maleate ligands led to greatly increased stabilizer efficiency and, through reduced usage levels, lower cost. Because maleic acid is difunctional, polymeric products are generated:

$$R_2SnO + \underset{\underset{CHCOOH}{\|}}{CHCOOH} \rightarrow [-R_2Sn-O-CO-CH=CH-CO-]_n \qquad (14)$$

These did not provide the (modest) lubrication value of tin laurates; therefore, combination maleate–laurate stabilizers were synthesized:

$$R_2SnO + \underset{\underset{CHCOOH}{\|}}{CHCOOC_{12}} \rightarrow R_2Sn-O-CO-CH=CH-CO-OC_{12} \qquad (15)$$

Today, the tin maleates remain of great value in cases where highly efficient nonsulfur-containing stabilizers are desired. Both maleates and maleate–laurates are commercially important in flexible PVC applications where maximum clarity and lowest possible odor are needed. In addition, the polymeric maleates are used to stabilize halogenated flame retardants commonly used in polyolefins.

Various explanations have been proposed for the improvement in efficiency gained in going from the original laurate to maleate ligands. It is clear that whatever mechanism dominates must involve improved coordination to labile chlorine sites, probably through complexing with existing or developing allylic unsaturation.

A. Organotin and Antimony Mercaptides

Further improved efficiency of organotin stabilizers is gained by the replacement of carboxylate with substituted mercaptide ligands. Not only is the mercapto group a better nucleophile for displacement of labile chlorine, it is also well known for antioxidant activity. The initial stabilizer of this type was dibutyltin bis(isooctyl mercaptoacetate) [30]. Its high efficiency led to a dominating position in demanding applications such as rigid PVC extrusion and injection molding. Like tin carboxylates, the mercaptoacetates (also called thioglycolates) are highly compatible and are not particularly lubricating. The most economical synthesis is from the dichloride (or mixture of monochloride and dichloride), but dialkyltin oxides will also react:

$$R_2SnCl_2 + 2HS\text{---}CH_2CO_2C_8H_{17} \rightarrow R_2Sn(SCH_2CO_2C_8H_{17})_2 \qquad (16)$$

Tin carboxylates were, however, not completely driven from the market by mercaptides because of two characteristics typical of organic sulfur-bearing compounds: a generally unappealing odor and somewhat inferior UV resistance. These have been addressed by suppliers and compounders: odor, by reduction of traces of starting materials, and UV resistance, by compounding with suitable fillers and UV absorbers. A further problem, cost, has been mitigated by introduction of methyltin mercaptoacetates. Although there are areas in which these attributes remain troublesome, this class of stabilizers forms an excellent tool for the PVC technologist.

Higher-speed processing of rigid PVC extrusions led to the development of tin stabilizers of increased lubrication value. The primary representatives are "reverse ester" mercaptoethyl oleates, made from fatty acid esters of β-mercaptoethanol. The primary reason for this approach to inclusion of long side chains is that the synthesis is much less expensive than that, for instance, of oleyl mercaptoacetate:

$$R_2SnCl_2 + 2 HSC_2H_4OCOC_{17}H_{33} \rightarrow R_2Sn(SC_2H_4OCOC_{17}H_{33})_2 \qquad (17)$$

The reason is the lower cost of long-chain acids as compared to the corresponding alcohols. When combined with the cost savings of, for example, using methyltin, this class of stabilizers has developed secure domination of the U.S. PVC pipe market.

In actual practice, the rigid PVC fabricator describes the application and receives, for evaluation, a proprietary organotin stabilizer selected or developed by the supplier. The supplier will normally state a characteristic of the stabilizer such as whether or not the composition contains sulfur, the nature of the alkyl group bonded to tin, or, typically, the percent tin in the product. In addition, a technical data sheet will recommend usage levels and provide such regulatory agency sanctions that have been obtained and a specification based on observable properties such as specific gravity, refractive index, color, and so forth. In some cases, the stabilizer supplier will provide information on compounding and processing for typical applications which may extend to consulting services by the supplier's technical staff.

An important innovation in such proprietary products is the structural modification of certain tin stabilizers to include groups which perform the functions of antioxidants and secondary stabilizers. This has been done, for example, by introduction of sulfur bridging, as shown in

$$CH_3\text{—}Sn\text{—}(S\text{—}CH_2CH_2\text{—}OOC\text{—}C_{17}H_{33})_2$$
$$|$$
$$S$$
$$|$$
$$CH_3\text{—}Sn\text{—}(S\text{—}CH_2CH_2\text{—}OOC\text{—}C_{17}H_{33})_2$$

Thiobis[monomethyltin bis(2-mercaptoethyloleate)]

The sulfur bridge readily reacts with HCl eliminated from the polymer to form volatile H_2S, a very weak acid. Thus, a maximum of the mercapto- ligands are available for displacement. This "internal buffering" makes possible the use of alkyltin mercaptides at reduced and highly cost-effective levels.

A further innovation is the development of combination of ingredient or "one-pack" tin stabilizers. These consist of a tin stabilizer plus, usually, calcium stearate, and other lubricant(s). Calcium stearate is included as a secondary stabilizer as well as an internal lubricant. At early stages in degradation, the secondary stabilizer displaces chloride from tin, replacing it with a "good" stearate ligand, as occurs with more typical "mixed-metal" systems. Additional lubricants may be a balance of external (paraffin) and internal or multipurpose lubricant (oxidized polyethylene). In such cases, a one-pack recipe containing 20% tin stabilizer, 20% calcium stearate, and the remainder a blend of lubricants would be typical; or, the lubricant may consist of a proprietary multipurpose species. Such products, when they can be used successfully, offer factory convenience and potential for reduction of errors in formulation. They are best used in situations where simultaneous internal and external lubrication is required [31]. The alternative is compounding by the user toward the same ends. This is preferable in two situations:

1.   Optimum processing is obtained with sequential internal and external lubrication rather than through both simultaneously. The effects of using lubricant blends versus sequential use of individual lubricants may, in some cases, be very marked [32].

2. Variation is encountered, for example, from one production line to another, from lot to lot of polymer (or other ingredient), or seasonally, which is most readily compensated for by minor on-the-spot compounding changes. There are certain situations in which attempts to eliminate all variation become cost-ineffective. A case in point is the processing of compounds during periods of PVC shortage, with reliance on a number rather than a single grade of resin and on use of regrind of marginal consistency.

B. Antimony Analogs of Tin Stabilizers

Although antimony carboxylates are not sufficiently stable for practical use, it has been long known that antimony mercaptides provide many of the properties of the analogous tin compounds [33]. A typical example is antimony tris(isooctylmercaptoacetate):

$$\begin{array}{l} S-CH_2COO-i-C_8H_{17} \\ | \\ Sb-S-CH_2COO-i-C_8H_{17} \\ | \\ S-CH_2COO-i-C_8H_{17} \end{array}$$

In comparison with a typical dialkyltin stabilizer, the major difference is that the $+3$ oxidation state of antimony accommodates three active ligands. In place of an alkyl group, trivalent antimony, in a tetrahedral configuration, has an unshared electron pair. It is not, therefore, an organometallic compound. (Structures such as $RSbX_2$ are not sufficiently stable for practical use.) An important consideration is that antimony is a less expensive metal than tin and may be converted to a stabilizer from antimony oxide, rather than a synthesized intermediate economically. Further, the presence of three active ligands leads to a lower equivalent weight (as much as 30%) than the corresponding dibutyltin derivative. This effect, in some cases, offsets the observation that, in high-demand applications, antimony analogs are inferior in efficiency compared to tin [34]. A factor involved in this is that antimony trichloride is a stronger Lewis acid than any of the dialkyltin dichlorides [35]. Antimony mercaptides, like those of tin, have solubility parameters very close to that of PVC and are not strongly dipolar. They are (in the United States) used at very low levels in rigid pipe and conduit to minimize cost. If process and service requirements can be met with a low level of stabilizer, then the lower equivalent weight and basic low cost of antimony mercaptides become attractive.

Similar in action to lead stabilizers, organotin and antimony stabilizers are effective because they complex with and deactivate labile chloride sites, increasing the activation energy for the initiation of degradation. They scavenge HCl at the reaction site before elimination occurs, usually by acting as a charge transfer catalyst and conducting the acid to a secondary stabilizer (e.g., calcium stearate) but in some cases, such as with sulfur-bridged stabilizers, react with HCl themselves. Labile chloride is replaced with mercaptide or carboxylate ligands. The mercaptides are active on-the-scene antioxidants. Almost all organotin and antimony stabilizers contain antioxidants, most often butylated hydroxytoluene (BHT), to protect the stabilizers themselves.

Except for the polymeric maleates, most members of this class of stabilizers are mobile liquids. Exposure to damp air results in a mixture of hydrolysis and oxidation, leading to the formation of a solid phase, appearing as a white sludge. With heating, this can often be mixed in, with little loss of stabilizer efficiency. Nevertheless, once shipped, few customers are equipped (or motivated) to combat such problems. As a result, organotin and antimony stabilizers typically contain antioxidants and proprietary additives to inhibit or delay sludge formation.

With organotin mercaptides, a high monoalkyl level favors bright early color at the expense of long processing safety. The reverse is true with high dialkyl levels. (It should be noted that this conclusion does not apply to estertins.) Thus, the ratio of monoalkyltin to dialkyltin is adjusted in actual stabilizers to provide a specific balance for a given application.

IV. MIXED-METAL STABILIZERS

Mixed-metal stabilizers comprise pairs of compatible salts. In nonionizing media, such as the typical polymeric composition, metal salts exist undissociated, in polar domains. In the mixed-metal blend, one component must have appropriate polarity to coordinate with a potential degradative site on the PVC polymer; it should be based on a cation that has affinity for labile chloride (i.e., one whose chloride is a Lewis acid). This component, referred to as the primary stabilizer, is able to displace labile chloride. The secondary metal is chosen to form a chloride that is a very weak Lewis acid, but quite stable thermally. In both components, the anions impart sufficient mobility to permit a close approach and complex formation between the two salts. Primary metal cations include mainly cadmium and zinc and occasionally stannous tin. Secondary metals include barium and calcium and occasionally strontium or magnesium [36].

In the classical view, after the primary metal salt has displaced chloride, the secondary metal provides a ligand in exchange. The displacement mechanism was first proposed by Frye and Horst in 1964 [37]. Following their discussion, labile chloride is represented as allylic in the following, although other modes of activation are possible. The most common ligands donated by the metal stabilizer are carboxylate anions:

$$-CH=CH-\underset{\underset{Cl}{|}}{CH}- + \tfrac{1}{2}M^{+2}(RCOO-)_2 \rightarrow \tfrac{1}{2}M^{+2}Cl_2 + -CH=CH-\underset{\underset{RCOO}{|}}{CH}-$$

$$(18)$$

Reaction (18) is known to proceed strongly to the right, if the metal has Lewis-acid character (Zn, Cd, Sn, Sb) as a result of experiments with model allylic compounds [38,39]. It is, on the other hand, very slow with carboxylates, such as those of calcium or barium, where the metal in the (II) oxidation state is not particularly electrophilic (i.e., not a Lewis acid). In mixed-metal blends (Ba/Cd, Ba/Zn, Ca/Zn), it is thought that either zinc or cadmium participates in reaction (18), followed by replacement of a carboxylate ligand donated by barium or calcium. In this rationalization, the Lewis-acid strength of zinc or cadmium is involved with acceptance of electron density from the allylic double bond, weakening the metal–carboxylate bond, as the latter group displaces on the allylic carbon. In their original discussions, Frye and Horst recognized that displacement by primary stabilizers and ligand replacement by secondary stabilizers could occur stepwise or might involve a one-step, concerted process. Most texts have advanced the simpler stepwise scheme:

$$-CH=CH-\underset{\underset{Cl}{|}}{CH}- + \tfrac{1}{2}Cd(RCOO)_2 \rightarrow \tfrac{1}{2}CdCl_2$$

$$(19)$$

$$\tfrac{1}{2}CdCl_2 + \tfrac{1}{2}Ba(RCOO)_2 \rightarrow \tfrac{1}{2}BaCl_2 \qquad (20)$$

The fact that mixed chloride/carboxylate salts [e.g., CdCl(RCOO)] are never isolated in reactions with model compounds suggests that a concerted mechanism is more likely [40]. Recently, it has been proposed that divalent metal carboxylates exist in solution in an equilibrium of monomeric and bridged structures:

$$(21)$$

Further, mixed-metal blends involve rapid formation of mixed bridged structures, where the metals (M) in reaction (21) might be barium and cadmium, or other mixed-metal pairs [41]. It is then reasonable to think that it is the mixed complex that actually displaces labile chloride, forming, in the process, chloride-bridged instead of carboxylate-bridged structures. The chloride-bridged complex, being a stronger Lewis acid than the starting mixed-metal analog, reacts rapidly in a second displacement (at a different labile chloride site) to form barium chloride. An undesirable side reaction is that, to some extent, the primary metal stabilizer reacts directly, yielding a Lewis-acid chloride. With model compounds in dilute solution, barium–cadmium complexes were shown to react with allylic chloride at the same rate as pure cadmium carboxylates, but, with excess barium in the system, the product was pure barium chloride [42]. As the concentration of reactive chloride is increased, this relationship breaks down and cadmium chloride begins to form. Thus, as the extent of degradation passes a certain point, the mixed-metal system is unable to suppress the side reaction of activated chloride with primary stabilizer. The underlying controlling principle is to what extent a mixed-metal-bridged carboxylate is more stable than individual bridged or clustered components. The relative size and polarity of the metals bear on this.

A key factor is that if cadmium is replaced in blends with barium with the much smaller zinc(II) cation, it is found that the mixed-metal pair is less effective in competing with the above side reaction than is the barium–cadmium complex. The Lewis acid, zinc chloride, appears at a lower concentration of reactive chloride than is the case with cadmium [42]. The most significant factor in formulating barium–zinc stabilizers to replace barium–cadmium, with retention of comparable heat stability, is the use of ligands that help to drive reaction (21) to the right, maximizing reaction of labile chloride with the mixed-metal complex. The ligands used most commonly in practice are based on β-diketones, such as dibenzoylmethane and stearoyl benzoylmethane. Both exist largely as the keto-enol form at room temperature and are bidentate ligands capable of association with electrophilic groups, such as metal cations. The metal salt of an enolized 1,3-diketone is an analog of a carboxylate, the carbonyl and hydroxyl portions separated by, but conjugated through, an unsaturated link. The increased distance, as compared to carboxylate bridges, between the oxygen ligands in 1,3-diketones, and their spatial configuration may enhance the ability to complex dissimilar-sized metals. This appears to have been first noted in the case of calcium–zinc stabilizers in 1954 [43]. The literature regarding use of diketones and similar multidentate ligands in mixed-metal systems is expanding rapidly, and these costabilizers are now an important factor in the ability to use zinc in place of cadmium.

Mixed metals are considerably more polar than lead, organotin, or antimony stabilizers. Therefore, they tend not to have a high efficiency (high mobility, rapid penetrating power) in PVC compounds having a narrow solubility parameter. This is usually the case with rigid PVC. In flexible PVC, the presence of plasticizer broadens the solubility parameter of the compound, permitting rapid penetration by mixed-metal stabilizers, if the anions are selected to have high compatibility with the plasticizer.

Cadmium chloride is a moderately active Lewis acid. As degradation proceeds, as cadmium chloride builds in concentration, good early protection is followed by rapid discoloration. The strongest points with cadmium stabilization are good early color and UV resistance. The weakest points are long-term stability, discoloration if subject to sulfur contamination, and cadmium toxicity. The first, long-term process stability, is countered by blending with barium compounds. In themselves, barium carboxylates provide long processing safety, but at the expense of poor color retention. Barium/cadmium blends enable both retention of good color and excellent processing safety.

Displacement of labile chlorine may be carried out effectively with zinc carboxylates in place of cadmium. Zinc-based stabilizers are highly active in compositions of broad solubility parameter and readily permeate plasticized compositions, leading to excellent early protection. UV resistance is generally good and the danger of sulfide stain is eliminated, zinc sulfide being a white pigment. Zinc chloride is, however, a strong Lewis acid. After its concentration builds to a certain point, much lower than with cadmium, degradation becomes very rapid (sometimes referred to as "zinc burn"). The point involved is influenced by the starting concentration of zinc salt, concentration of the second metal, temperature, shear rate, and the zinc sensitivity of the compound. The latter is a function of the number of labile sites in the polymer and is increased by inclusion of zinc-reactive ingredients, notably CPE, phosphate ester, and chlorinated paraffin plasticizers.

In some instances, it is desirable to use a combination of cadmium and zinc as displacement agents. Usually a minor fraction of zinc is used to enhance early protection and UV stability, with most of the primary stabilizer being cadmium based. The use of both tends to promote flexibility in stabilizer formulating. Where the application permits the use of cadmium, combinations of cadmium and zinc are often preferable to zinc alone. Where zinc sensitivity is not overly great, there are many cases where the combination is preferable to the sole use of cadmium.

As mentioned, barium salts are in themselves inferior stabilizers, slow to act, giving poor early protection, but providing long-term processing stability. In the proper proportions, they combine with cadmium or zinc primary displacement agents to provide a range of excellent stabilizers for flexible PVC compounds. The most common proportion is about twice the concentration of cadmium, or several times that of zinc. This tends to ensure that all displaced labile chloride ends up as the innocuous barium salt. These ratios will not necessarily apply if other ingredients having a secondary stabilizing function are also present. As barium salts are usually the least soluble stabilizer component, problems with migration, in the form of plateout, during processing are common at high barium levels. This has been corrected to some extent in liquid stabilizers by the development of barium alkylphenate salts; these exhibit greater compatibility than the corresponding carboxylates [44].

As barium salts, other than sulfate, are sufficiently water soluble to extract and are relatively toxic, food contact and related applications use calcium as the secondary metal, in combination with zinc as the primary displacement agent. At one time, calcium/zinc stabilizers were used only where food contact was involved, or for applications not involving extreme heat and shear during processing. Now, however, blends of calcium and zinc stearates with powerful HCl scavengers, such as hydrotalcites, can be formulated to be the equal of high-efficiency barium/cadmium blends.

If the anions present are long-chain carboxylates (laurate, stearate), the resultant mixed-metal blend is a powder. As the length of the carbon chain is increased, the result is increased lubricant strength. The same factor, side-chain crystallization of alkyl groups having 10 or more adjacent methylenes, is responsible both for lubrication and a melting point above room temperature. Although there are examples of nonlubricating powder stabilizers, the most common (blends of solid aliphatic metal carboxylates) are strongly lubricating.

In powder-stabilizer manufacture, particle size distribution and bulk density are controlled to conform to the requirements of common PVC mixing systems. The powder blend can be mixed into liquids such as phosphite esters or epoxidized oils and milled into a fluid paste, either by the user or stabilizer supplier. In such form, the stabilizer package is easily blended into plasticized PVC, with rapid penetration.

If shorter-chain alkyl carboxylates (e.g., octoates) or alkylated phenates comprise the anions, the resultant blend is a liquid. Such blends are, in themselves, not particularly good lubricants. In practice, they are often formulated to include soluble lubricants to improve this. The resultant mixture is diluted to a highly fluid, stable viscosity with the solvent or plasticizer. Solvents are chosen to have a low vapor pressure at ambient temperature, but to vaporize during most fabrication processes. Equally important, they are chosen to provide phase and viscosity stability to the stabilizer blend. They often can absorb traces of atmospheric water in a harmless manner. In almost every case, however, more than a trace pickup of atmospheric moisture or accidental water contamination destroys the phase stability of the blend. Introduction of water also destroys secondary ingredients commonly found in liquid stabilizers: phosphite esters by hydrolysis, and epoxides by addition of water to the epoxy group. These effects can be avoided by the use of a layer of dry nitrogen or other inert gas above the liquid. This is commonly done with bulk systems. Barium/zinc and calcium/zinc liquids are especially subject to phase separation from contact with atmospheric moisture and should be protected to prevent separated sludge from clogging pipes and valves.

The choice of powder or liquid blend is highly application dependent. If the overall PVC recipe is high in liquid ingredients (i.e., plasticizer) and solid ingredients such as pigments and fillers are low or absent, the user will tend to prefer a liquid stabilizer, often added with the plasticizer. If the solvent in a liquid blend is intolerable (e.g., causes haze, fogging, or bloom), then a powder stabilizer is usually used. If the overall recipe is basically polymer and filler, then powdered stabilizers are generally useful in providing at least part of the needed lubrication. There are, of course, instances in which the choice of liquid or powder depends on the relative convenience of incorporation in a particular manufacturing situation. In general, mixed-metal liquids enable a great deal of formulating latitude and are easy to add, but have limited shelf lives; powders resist atmospheric hydrolysis, have long shelf lives, are usually more lubricating, and sometimes not as convenient to add to the batch.

Phenolic antioxidants are often included. Probably the most common are BHT (butylated hydroxytoluene) and bisphenol-A (diphenylol propane). These are used at 1–3% levels in liquid stabilizers, often up to 4–5% in powder blends. With systems containing cadmium, sulfur-containing synergists for phenolic antioxidants, such as thiodipropionates, are rarely used because of the likelihood of forming yellow cadmium sulfide. In dark compounds or those using zinc rather than cadmium, such synergists may be used without seriously affecting odor or UV resistance.

Pentaerythritol and similar polyols are common components of powder stabilizers. They are believed to function by coordination with emerging cadmium or zinc chloride, forming a complex which assists the interchange of chloride with barium. This permits use of a higher fraction of primary metal and a lower level overall of stabilizer. Polyols also function by scavenging HCl at the degradation site:

$$R—OH + HCl \rightarrow R—Cl + H_2O \qquad (22)$$

Solid carboxylic acids are sometimes added both to powder and liquid mixed-metal stabilizers. If increased lubrication is desired, minor amounts of stearic acid may be used. This may also serve to replace ligands spent in displacement; normally, improved stability results. If lubricant value is not desired, benzoic acid may be used instead. In this case, opposite to the effects of stearic acid, fusion will be promoted and melt viscosity increased.

Other solids are added to powder blends as HCl scavengers. Some, such as hydrous zeolites, exchange reactive hydroxide for chloride as in reaction (22). Others, such as hydrotalcite (synthetic magnesium/aluminum hydroxy carbonate) exchange carbonate for chloride:

$$R—CO_3 + 2 HCl \rightarrow RCl_2 + H_2O + CO_2 \qquad (23)$$

Both hydrous zeolites and hydrotalcites are considerably more efficient than polyols. They have formed the basis for a new generation of powder stabilizers, comparable in efficiency to liquid blends. Prior to their emergence, almost all mixed-metal stabilizers used organic epoxides to scavenge hydrogen chloride:

$$\overset{\displaystyle O}{\overset{\displaystyle /\,\backslash}{-CH-C}} + HCl \rightarrow -\underset{\underset{\displaystyle OH}{|}}{CH}-\underset{\underset{\displaystyle Cl}{\backslash}}{CH}- \tag{24}$$

Reaction (24) is catalyzed by primary metals, particularly zinc [45]. The activity of the epoxy group in scavenging emerging HCl provides early protection. Epoxy costabilizers also coordinate with metal chloride stabilization reaction products, which may catalyze displacement of remaining labile chlorine directly by the epoxide [46].

There are limits to the extent to which epoxy esters or epoxidized oils may be used in various recipes without onset of exudation or plateout. Mixed-metal stabilizers complex with the epoxy additive. First, the epoxy–stabilizer complex, and then, in due course, virtually the entire compound plates out. To avoid this, a balance must be struck between epoxide and stabilizer. Its point is governed by the application. Actual levels used in PVC range from about 50% of the overall stabilizer package (when the latter is in the range of several phr) to as much as 5 phr on a pure basis. This is also the case when using powdered stabilizers compounded into pastes with epoxy esters or epoxidized oils. With zeolite or hydrotalcite acid absorbers, epoxides are usually not necessary. On the other hand, zeolites and hydrotalcites cannot be used in glass clear compounds.

The synergism found with combinations of mixed metals with organic phosphites is complex. Phosphite esters are well known as antioxidants and antioxidant synergists in many polymers and, no doubt, function in this regard in PVC. Their initial use, however, was to complex insoluble metal chlorides, improving clarity [47]. Thus, they are still spoken of as "chelators." It is clear that they complex metal stabilizer salts as well, improving their efficiency through increased compatibility and probably assisting in the displacement reaction. This may be to the extent of actually contributing ligands via the Arbuzov reaction [48].

$$\underset{\underset{\displaystyle Cl}{|}}{CH=CH-C-} + (RO)_3P \rightarrow -\underset{\underset{\displaystyle RO_2P=O}{|}}{CH=CH-C-} + R-Cl \tag{25}$$

It is likely that unsaturated sites are also complexed and perhaps added to, leading to improved color retention. As with epoxides, one cannot replace mixed-metal salts with phosphite esters indefinitely without ill effects. In general-purpose compounding, they are used at about comparable levels. The most frequently used examples are diaryl monoalkyl, dialkyl monoaryl, and alkylated triaryl phosphites. These have somewhat different solubility parameters and are useful in compatibilizing the overall liquid-stabilizer package. Dialkyl monoaryl phosphite esters tend to be more valuable in conjunction with long- or branched-chain carboxylate salts. Diaryl monoalkyl phosphites tend to be used with phenates and blends of phenates with carboxylates. In plasticized PVC compounds, both the metal ligand and the phosphite must be matched to the solubility parameter of the plasticizer. Long-chain alkyl phthalates, for example, should be accompanied by relatively long-chain metal carboxylates and dialkyl monoaryl phosphites. More polar plasticizers call for the use of shorter-chain carboxylate or phenate ligands and diaryl monoalkyl phosphites. Tris-nonylphenyl phosphite (TNPP) is currently the only FDA-sanctioned phosphite for use in plasticized PVC. Although triaryl phosphites are the least efficient in PVC (aryl groups cannot participate in the Arbuzov reaction), TNPP remains the only choice for food-contact applications.

Many liquid stabilizers contain both epoxides and phosphite esters; the improvement with such blends is often more than additive. The interaction can pose hazard as well as benefits; for example, if epoxidized oils are used at near the levels at which they will migrate or exude, this tendency may be increased by the addition of phosphite esters. The balance of these ingredients is also affected by other additives. Organic acids are used as boosters in liquid as well as in powdered stabilizers, oleic and tall oil acids being common. These also help to control solvent balance of the overall blend. Such ingredients, being highly mobile, affect the behavior of other mobile species.

Normally, the mixed-metal stabilizer supplier will state the metals used and specify the percentage ranges; this will usually be true of epoxide and phosphite content as well. If applicable, sanctions by regulatory agencies will be mentioned. There will be some indication of potentially useful applications and suggestions as to use levels with perhaps additional compounding information. The range of mixed-metal stabilizers in use is very large. This reflects the relative ease of formulating a variety of stabilization systems with existing ingredients, without synthesis of new materials. Instead of 5–10 commonly used lead stabilizers or 10–20 organotins, there are literally hundreds of mixed-metal stabilizers.

V. RIGID PVC APPLICATIONS

In almost every instance, rigid PVC is used as an economical structural plastic. Therefore, it must be stabilized at minimum expense to its structural properties (heat-distortion temperature, modulus, resistance to abrasion, etc.). Common to all rigid PVC applications, consequently, is the requirement that the stabilizer not function as a plasticizer. The stabilizer must be chosen so as to be able to penetrate the polymeric matrix on its own, without requiring increased polymer mobility. The only assistance tolerable is that of the lubricant, which, by definition, affects polymer structure mainly in a transitory way, without a significant change in physical properties. A consequence is that the efficiency of stabilization is a function of the ability of the stabilizer and internal lubricant to coordinate synergistically [49].

In many instances, the moment of fabrication of a rigid PVC article is the only occasion when degradation is likely. In such processes, levels of shear and frictional heating can be very severe. Consequently, it is as important that external lubricants minimize shear rates in the vicinity of process surfaces such as for the stabilizer–internal lubricant complex to minimize shear and repair damage within the compound matrix. Therefore, in these applications, the internal–external lubricant balance is as vital as the presence of stabilizer.

In common with almost all other users of thermoplastics, fabricators of rigid PVC have certain general criteria regarding stabilizers. These are mentioned here to avoid repetition under individual applications. They are that the stabilizer be a material which is universally regarded as harmless, of pleasant odor, functional at very low concentrations, and of no economic impact on the compound. Few stabilizers meet these criteria.

A. PVC Pipe

The fabricator of extruded PVC pipe has several important criteria. Most significant is that extrusion lines run continuously at optimum output of acceptable product of very closely controlled dimensions. There should almost never be an interruption from equipment failure and certainly never one caused by compound degradation. "Acceptable," in this sense, denotes that the formulation and the product are acceptable to the National Sanitation Foundation (U.S.) or similar regulatory agency, in terms of ingredients and properties, either for potable or nonpotable water categories.

A typical state-of-the-art line uses twin-screw extrusion. In North America, the stabilizer package includes about 0.4 phr of a tin mercaptoethyl oleate stabilizer. (This term is used generically here. As discussed earlier, the actual cost-effective stabilizer will be a proprietary blend of, for example, monoalkyltin and dialkyltin sulfur-bridged mercaptoethyl esters.) Although methyltin mercaptoethyl oleates are more economical to produce, butyltin analogs are priced competitively. All provide somewhat better early color and lubrication than the corresponding mercaptoacetates in pipe recipes. More importantly, cost is significantly lower. In this application, the low stabilizer levels in use compensate for the less ingratiating odor of mercaptoethyl derivatives. The internal lubricant is most commonly 0.6–0.8 phr of calcium stearate, with external lubrication being provided by 1.0–1.2 phr of paraffin wax and 0.1–0.15 phr of oxidized polyethylene wax. The balance of these ingredients is essential for long, uninterrupted runs at high output. They may be added in combination form as a one-pack at about a 2.0-phr level at more or less similar material cost. Tin mercaptoethyl oleates can be replaced in this application with antimony mercaptoacetate. Equivalent processing usually requires slight revision of the internal–external lubricant balance.

Outside of North America, almost all pipe compounds use lead stabilizers, commonly in the form of one-packs with proprietary multipurpose lubricants. The heart of such systems is a blend of 0.3–0.5 phr tribasic lead sulfate, usually stearate coated, and 0.2–0.3 phr of dibasic lead stearate. This system does not require equipment in contact with the compound to be chrome plated. As a result, equipment wear is lower. This factor and the fact that if degradation occurs, failure is gradual, without sudden compound burning, are the major advantages in using lead stabilizers in this application. The drawback is the need to protect workers from exposure. Levels of lead extraction by the environment have been found to be extremely low, of the order of 0.01 mg/pkg product [50]. In areas of the world where both tin and lead stabilizers are viewed emotionally, calcium–zinc–acid acceptor stabilizers can be used, at some sacrifice in efficiency.

If PVC pipe is produced with a single-screw extruder, stabilizer–lubricant requirements are much increased. Here, with tin stabilization, 1.0–1.5 phr of mercaptoethyl oleate should be used, with 1.5–1.8 phr of calcium stearate as the internal lubricant and 0.4–0.6 phr of paraffin wax as the external lubricant. At these levels, some processors prefer mercaptoacetate types instead of reverse esters. With lead stabilization, 1 phr of tribasic lead sulfate is common, with 0.3 phr of dibasic lead stearate.

As they are not subject to internal pressure during service, drain, waste, and vent pipe compounds use considerable filler, resulting in a higher lubricant requirement. It is preferable to use surface-treated fillers for the best interaction with lubricants and least tendency to adsorb them [51]. Tin mercaptoethyl oleate or antimony mercaptoacetate should be used at about 0.4 phr. Calcium stearate as an internal lubricant should be in the 0.6–0.8-phr range, with 1.0–1.2 phr of paraffin wax and 0.15–0.3 phr of oxidized polyethylene wax as the external system. With the use of antimony, as previously noted, it is likely that the internal–external lubricant balance will require adjustment. Regrind containing tin and antimony stabilizers should not be mixed, so as to avoid the possibility of cross-staining. Sulfur-bridged components of tin stabilizers will react with the end products of antimony stabilization, producing colored antimony sulfide. Lead-stabilized filled pipe and conduit uses 1–1.5 phr of tribasic lead sulfate plus 0.3 phr of dibasic lead stearate, usually in a one-pack with lubricants. As the stabilizer requirement increases, the cost advantage associated with tribasic lead sulfate becomes progressively more significant. Similarly, as filler content increases, the lead-stabilizer advantage with regard to equipment wear becomes more pronounced. If neither organotin nor lead stabilizers can be used, products can be produced with mixed-metal stabilizers, based on blends of hydrotalcite with β-diketone-linked calcium and zinc stearates. These have very little actual use.

Stabilizer one-packs for the extrusion of foam layers for PVC pipe (foam core), whether based on tin or lead stabilizers, typically contain the equivalent of about 0.4 phr (on the final compound) of azodicarbonamide. Lead stabilizers are excellent activators for this blowing agent. With organotins, zinc-based activators are commonly added.

B. Siding and Profile

Outdoor siding and profile extrusions require excellent dimensional control while running at high output in thin cross sections. High stabilizer demand is complicated by the need, in service, for prolonged UV stability and heat stability at moderate temperatures. Another important property, maximum impact resistance, is gained not only from use of impact modifiers but also by choice of resin. Stabilizer usage prevalent in North America consists of 1.5–1.8 phr of tin mercaptoacetate, plus 1.2–1.5 phr of calcium stearate as the secondary stabilizer and lubricant. The external (actually multifunctional) lubricant is often 1.0–1.2 phr of an ester-type lubricant plus 0.4–1.0 phr of oxidized polyethylene wax. The UV resistance of such a stabilizer system is, in itself, inadequate and must be supplemented with 10–12 phr of titanium dioxide. This may be augmented by use of organic UV absorbers.

With lead stabilization, titanum dioxide levels of 5–6 phr are more common. Depending on service conditions, the stabilizer may be based on 2.5–3 phr of dibasic lead phosphite, or on 2 phr, plus 0.5–1 phr of tribasic lead sulfate. Both would be used with 0.2–0.3 phr of dibasic lead stearate. Where best color retention is required, such blends are often converted to one-packs with a blend of barium and cadmium stearates. In areas in which all of the above metals are subject to emotional restrictions, calcium–zinc systems have been promoted (but find little use to date).

Neither stabilizers containing cadmium nor lead may be used interchangeably with tin (or antimony) mercaptides without cross-staining from the corresponding metal sulfides. Therefore, regrind must be separated.

C. Rigid Film and Sheet

Calendering or extrusion of rigid film and sheet places a strong demand on the stabilizer and lubrication systems. These also are involved in prevention of crease whitening (stress orientation). The stabilizer must provide long color or clarity hold and not contribute uningratiating odors. Historically, food-contact applications generally used dioctyltin; others, such as credit card laminations used dibutyltin mercaptoacetates. Recently, methyltin stabilizers have been used in both. In these applications, lower-cost mercaptoethyl oleates (reverse esters) tend to give poorer color hold and excessive odor. Tin mercaptoacetates are used at about a 2.0-phr level. If high UV resistance or clarity (particularly if both) is essential, dialkyltin maleate/laurate stabilizers find some use; dioctyl derivatives for food-contact applications, dibutyl otherwise. These systems commonly use 1.0–1.5 phr of multipurpose ester-type lubricants.

Calendered rigid PVC film can also be run with 2–3 phr of tribasic lead sulfate (usually stearate coated). For best color hold with bright colors, partial replacement with dibasic lead phosphite is useful. High-speed extrusion of rigid PVC sheet and film calls for a combination of 2–3.5 phr of tribasic lead sulfate and 0.2–0.3 phr of dibasic lead stearate. Again, for the best color hold, 2 phr of the above could be replaced with dibasic lead phosphite. The most common usage of lead stabilizers in rigid PVC sheet is probably credit card stock, where high resistance to distortion of printed legend and data information by skin contact is provided.

D. Molded Products

Poly(vinyl chloride) bottles are injection blow molded from pelletized compound. The steps of mixing and fabrication impose high stabilizer demand. In addition to optimum physical properties, close dimensional control and high clarity, the product must have low odor and extractables. For general-purpose applications, special low-odor and extractable grades of dibutyl and dimethyltin mercaptoacetates have been developed. Stabilizer levels of about 2.0 phr are typical. This is used in conjunction with a combination lubricant system, typically 1.0–1.5 phr of ester-type lubricant plus 0.15–0.3 phr of oxidized polyethylene wax.

In areas where tin stabilizers are viewed with disfavor, fabricators use calcium–zinc powdered stabilizers. These are low cost and provide excellent taste and odor, but are generally marginal in efficiency for the process. Material cost savings are offset by reduced output and more frequent incidence of process instability. This has been a factor in the increasing prevalence of polyester bottles, despite the much greater material cost of the polymer.

The injection molding of pipe fittings also imposes high stabilizer demand, whether run from dry blend or pellets. A product requiring ultimate structural properties must be processed at high output, under substantial shear loading and with extended dwell time. Process stability depends not only on the stabilizer–lubricant system but also on the choice of polymer so as to have sufficient flow characteristics. Current usage in North America centers on alkyl mercaptoacetates at levels of 1.0–1.5 phr, and higher at times. Generally, these are blends of monosubstituted and disubstituted tins and may include sulfur-bridged structures. If stabilizers with very high tin content are used, as with high-tin-containing boosters, then the level may be reduced below 1.0 phr (generally at comparable overall cost.) Calcium stearate is used at 0.7–1.0 phr as an internal lubricant. In this application, simultaneous internal and external lubrication is needed; commonly oxidized polyethylene wax (0.1 phr) is used with ester lubricants (0.9

phr). This balance provides efficient lubrication and is more satisfactory than the calcium stearate–paraffin wax system used for extrusion [52]. To date, molded applications have been better served by mercaptoacetate tin stabilizers than with the less expensive mercaptoethyl reverse esters.

In other parts of the world, pipe fittings are stabilized with 2.5–3.5 phr of tribasic lead sulfate (stearate coated) plus 0.2–0.3 phr of dibasic lead stearate. These are usually supplied as one-packs containing the equivalent of 2–3 phr of lubricants, often blends of oxidized polyethylene, paraffin wax, and ester waxes. Injection-molded fittings for use with outdoor profiles are similar, except that about 2 phr of stabilizer is replaced with dibasic lead phosphite, and titanium dioxide levels of 5–7 phr are added. The cost advantage is using lead stabilizers is emphasized by the quantities required for high-speed injection molding, as compared to pipe processed by twin-screw extruders. Use of lead-stabilized fittings with tin-stabilized pipe will, however, lead to cross-staining at the joints and, therefore, could be done only with dark colored products.

Phonograph records form an interesting special-purpose application. The compound, essentially polymer and carbon black, must be stabilized sufficiently to permit mixing and stamping. During processing, internal and external lubrication are needed. Although the product must release easily from the mold, the surface must be free of lubricant in order to provide acceptable auditory characteristics. These requirements are most often met by the use of blends of barium and lead stearates, at levels of about 0.5 phr. Barium–cadmium powders have also been used, but are generally less effective, and now more expensive.

VI.   FLEXIBLE PVC APPLICATIONS

With flexible compounds, the presence of significant quantities of plasticizer greatly affects stabilizer choice. An effective plasticizer must be able to penetrate the polymer matrix and coordinate to individual chains and segments. If the stabilizer has solubility characteristics similar to the plasticizer, it will be brought with it into intimate polymer contact. Typically, useful plasticizers have broad solubility parameters. They can, therefore, associate with stabilizers which, in themselves, would not efficiently penetrate the polymer matrix. Thus, the choice of ligands for a given metal is much broader than with rigid compounds. The presence of plasticizer reduces the extent of lubrication required of the stabilizer. This factor also broadens the range of organic ligands which may be used. In some cases, it is desired that the stabilizer have no lubrication value. In such instances, alkyl carboxylate anions are commonly partly replaced with those of aromatic acids such as benzoic.

The ingredients in the mixed-metal stabilizer blend, other than the metal salts, such as phosphite esters, epoxidized oils, polyols, and organic acids, have a number of functions. In addition to serving as secondary stabilizers and providing viscosity and phase stability to the blend, they contribute to the compatibility between plasticized PVC and the metal salts by forming coordination complexes. This bears directly on stabilizer efficiency. Therefore, a given mixed-metal stabilizer cannot be said to be efficient (or inefficient) because it contains so much of metal X and so much of metal Y. Its efficiency (and all other attributes) depends on the entire composition.

Powder stabilizers for flexible PVC are made by combining the ingredients (metal stearates, laurates, or related carboxylates; polyols; acid acceptors such as hydrotalcite or zeolite; antioxidants; phosphites; special ligands such as β-diketones; wetting agents; diluents) in a ribbon blender or other low-intensity mixer. The blend may be coated with plasticizer to reduce dusting, or even prilled (especially if containing a hazardous dust, e.g., a cadmium carboxylate). In some cases, the metal carboxylate blend can be synthesized from the corresponding oxides or hydroxides *in situ* while preparing the stabilizer. Most often this is not the case. The product may be packed in bags, drums, or semibulk containers. A typical product might comprise 50% barium stearate, 25% zinc stearate, 10% stearic acid, 4% BHT, 3% β-diketone; the remainder, a coating of epoxidized soybean oil.

Liquid stabilizers are prepared by two different paths. In Western Europe, it is common to produce, for example, a barium/zinc liquid by reacting a blend of zinc oxide and barium hydroxide with an organic acid in a solvent blend, such as a moderately high-boiling hydrocarbon and alcohol mixture. The water of reaction is removed under vacuum and then most of the solvent blend, which is then replaced by liquid phosphite, and the remaining ingredients (antioxidant, β-diketone or equivalent, etc.) added, yielding a finished stabilizer in one step. This procedure is efficient and useful, particularly if a modest number of different stabilizers are to be prepared, and the differences among them are small. In North America, the range of liquid stabilizers is so large that this approach becomes impractical. In addition, the most useful organic acid for such direct syntheses has traditionally been *tert*-butylbenzoic. This ingredient is viewed as harmless in many parts of the world, but considered extremely hazardous in North America because of reported testicular damage to laboratory animals.

The alternate approach is to prepare individual intermediates and blend to arrive at finished stabilizers. This procedure is typical of North American practice. Despite various partisan claims, there is no evidence that blended stabilizers are necessarily lower in efficiency than those prepared in a single step.

The zinc intermediate most often used is zinc octoate, prepared from zinc oxide and octoic acid at concentrations equivalent to 10–18% zinc metal. At the upper range, the zinc level is more than strictly stoichiometric. Such intermediates are referred to as "overbased." They do not contain dissolved zinc oxide, but are bridged carboxylates with Zn—OH end groups. For improved lubrication, zinc oleate may be used instead of the octoate.

Cadmium octoate is prepared similarly to the oxide. Mixtures of cadmium octoate and benzoate, or octoate and oleate, or all three are common. Because cadmium salts are weaker catalysts for esterification or hydrolysis, organophosphites can be used as part of the solvent in their preparation, increasing the possible level of phosphite content of the finished stabilizer. This is generally not feasible with zinc carboxylates without substantial hydrolysis of the phosphite.

Barium octoate is also produced, but is used mainly for applications where phenolic components are undesirable, such as in flooring, wall coverings, and similar interior products. Phenols and their metal salts are extremely reactive to atmospheric oxides of nitrogen (NO_x), yielding strong darkening. For reasons not entirely clear, this discoloration is sometimes referred to as "gas fading." Trace phenolics cannot totally be avoided in most compositions, but to counter NO_x attack, simple antioxidants (BHT, bisphenol-A) cannot be used. Organophosphites must be restricted to those yielding little or no phenol on hydrolysis. The latter category includes trialkyl phosphites; the former includes dialkyl monoaryl phosphites.

The most common barium intermediate is, nevertheless, its nonylphenate salt. This can be produced at stoichiometric ratios (10–12% barium metal) and at overbased levels containing 23–34% barium. In the latter case, after reaction with nonylphenol, the excess barium is reacted with carbon dioxide. It should not be imagined that such stabilizers contain dissolved barium carbonate. They are most likely bridged structures such as

Treatment with acid liberates carbon dioxide and forms mixed phenate carboxylate intermediates. Alternatively, an overbased mixed phenate/carboxylate may be prepared, then carbonated. These types of intermediates are produced by many stabilizer manufacturers and are also sold commercially.

A typical high-efficiency liquid might consist of 35% overbased barium nonylphenate, analyzing for 26–30% barium, 15% neutral zinc octoate (12–14% zinc), 40% diphenyl isodecyl phosphite, 2–3% diphenyl phosphite (an excellent chelator often used with high barium levels to reduce plateout), 2–3% β-diketone, 1–2% reserve acid (oleic, benzoic or both), 1% hindered phenolic antioxidant, and minor amounts of solvent, plasticizer, or other compatibilizing additive.

A. Calendered and Extruded Products

Calendered flexible filled and pigmented products often still use barium–cadmium–zinc–phosphite liquid stabilizers. Levels of 1.5–4.0 phr are typical, depending on the total metal content and stabilizer demand. Such stabilizers generally contain barium to cadmium in a 2/1 weight ratio, with a low level of zinc, and phosphorus at a level comparable to cadmium. These ratios are considered optimum for processing safety and color retention. The overall metals level may vary considerably. One must consider whether bright or dark colors are involved, whether the product must be printable, whether plateout is likely, the general type of plasticizer used, the level of epoxidized oil in the recipe, and the importance placed on color retention.

Applications requiring the best color retention in white and light colors often use powdered stabilizers as "boosters." In this case, high-efficiency powders having a Ba/Cd ratio of about 1/2 are preferred. Typical usage is 0.5–1.0 phr, depending, somewhat, on the type and level of liquid stabilizer used. Often a polyol or other secondary stabilizer is included to maximize the efficiency of the cadmium salt. Powder boosters are usually lubricants; levels of other lubricants, in particular, stearic acid, must be adjusted downward to compensate for this.

Powder boosters containing, instead, a metals ratio of Ba/Cd about 2/1 are used to improve process stability in cases where long running times and high regrind levels are common. In cases in which both improved process and service stability are required, compromise Ba/Cd levels of about 1/1 are used, often with a polyol or other secondary stabilizer. Such boosters are particularly valuable where a user mixes or buys a limited number of general-purpose compounds and then blends them or adds concentrates to make a wider variety of special-purpose compounds, often on short notice. In such cases, it may be preferable to blend in a powder booster, for example, with regrind addition, rather than to reformulate for each specific case. In other instances, the same compound may be sold to a range of customers with varied and often variable process conditions. In practice, this situation may be accommodated more efficiently with booster additions than by a proliferation of recipes.

In most cases, cadmium-free stabilizers can be used successfully. These are generally barium–zinc liquids with high phosphite content, such as described in Section V, at relatively high loadings (2.5–4.0 phr). The Ba/Zn ratio is usually 2–3/1 or greater, often with high phosphite content. Where there is severe process demand, the stabilizer may be augmented with a Ba/Zn powder, such as described in Section V.

In many cases, it is possible to use the same liquid stabilizers for extrusion as for calendering. In others, process stability over long runs is a concern. In these cases, the use of low levels of zinc with stabilizers having a relatively high barium content is common. Overall levels of 1.5–3.0 phr are typical. This can be augmented with 0.5–1.0 phr of a powder booster. Stabilizers for extrusion applications tend to be more lubricating than for calendering. As lubrication values are increased, the hazards are exudation and deposition of lubricant. The latter tends to be followed by deposition of the remainder of the compound. The revision of process conditions and modification of equipment to minimize this is normally much easier with extrusion than calendering. This is also the case with a related process, injection molding of flexible compounds. In addition to being able to vary shear rate and dwell time in the extruder section, there is latitude in tooling design that can minimize deposition.

Stabilizers for flexible vinyl applications involving extended outdoor use have been based on barium–cadmium–zinc liquids. Applications having high process demand use typical 2/1 Ba/Cd ratios and contain a phosphorus level comparable to the cadmium content. Others, where UV-light resistance outweighs process demand, for example, clear applications for outdoor use, use high cadmium levels. In cases where both process and service demand are high, the effective compromise is a 1/1 Ba/Cd ratio. In all of the above, overall levels of 1.0–3.0 phr are typical. In formulations employing blends of stabilizer and epoxidized oil, levels of 2.5–4.0 phr are more common. High-efficiency barium–zinc liquids, such as previously described, are also used.

In flexible vinyl outdoor applications where water resistance is important, such as in roofing membrane and pond liners, lead stabilizers are primarily used. Dark colors use 4–5 phr of tribasic lead sulfate plus 2 phr of dibasic lead phosphite. White or bright colors require 6–8 phr of the latter.

Zinc stabilizers are used in applications requiring very low odor (e.g., refrigerator gaskets) and in those which are subject to radiation sterilization. The latter category consists mainly of various bags for medical use. These generally use liquid zinc stabilizers which are principally "FDA-grade" phosphites [e.g., tris(-nonylphenyl) phosphite] with very minor quantities of zinc salts. Such usage is most often in combination with substantial amounts of epoxidized oil. Gaskets and seals for FDA-sanctioned applications use stabilizers with high zinc levels. This is most easily accomplished by use of a zinc powder, often with a food-grade polyol as synergist. Suitable antioxidants may also be included. Where powder addition is inconvenient, such stabilizers may be readily converted to paste form by premixing with epoxidized oil. Medical tubing commonly uses stabilizers which are primarily FDA-sanctioned phosphites with minor amounts of calcium and zinc in the ratio of 1/3 or 1/4. Again, these stabilizers are used in conjunction with epoxidized oil at overall levels of 1.5–3.0 phr.

Calcium/zinc ratios in the range of 1/2 to 1/3 are typical of stabilizers used in flexible film for food contact. Powder grades provide the greatest clarity, color stability, and resistance to deposition. They are used at a 0.5–1.0-phr level or somewhat higher if diluted with epoxidized oil. A further increase in process safety can be gained, at some expense in color stability, by use of calcium/zinc ratios approaching 1/1. In powder form, such stabilizers are also used at a 0.5–1.0-phr level. A typical food packaging film stabilizer might comprise 40% zinc stearate, 25% calcium stearate, 10% stearic acid, 10% BHT, and 15% glycerol monostearate (lubricant and antistat).

The starting point for plastisol stabilization would be a reduced level of a general-purpose barium–cadmium or barium–zinc stabilizer. As the dispersion resins used in plastisols typically give good response to stabilization with zinc salts, many stabilizers use increased zinc levels. Special-purpose stabilizers are formulated with a balance of polar and nonpolar anions to assist in the control of plastisol viscosity.

For improved color stability, the barium/zinc ratio may be reduced from the typical 3/1 to 1.5–2/1 and the phosphite level increased. For very thin coatings, high levels of phosphite may be combined with a Ba/Zn ratio as low as 1/1. If no heavy metals can be tolerated, then the usual combination of high phosphite plus low zinc or low calcium–zinc may be substituted. All are typically used at 2.5–4.0 phr.

High-zinc stabilizers are also used in foamed plastisols as combination stabilizers and activators for an azodicarbonamide blowing agent. Zinc salts have little or no activating effect on sulfonyl hydrazide blowing agents. On the other hand, calcium salts are ineffective with azodicarbonamide, but they are moderate activators for sulfonyl hydrazides. Calcium–zinc stabilizers are therefore useful with blends of the two blowing agents. Where the application permits use of lead stabilizers, strong blowing agent activation is characteristic of tribasic lead sulfate and related stabilizers.

B. Wire and Cable

Its generally excellent balance of properties and modest cost make vinyl invaluable as a wire covering. When properly compounded, the resultant electrical properties are suitable for many low-voltage and some medium-voltage insulation applications. PVC compounds are readily permeated by water and water vapor, particularly at elevated temperatures often typical of service conditions. If water-soluble ionizable materials are present, the useful electrical properties of the compound rapidly deteriorate, leading to excessive power loss and ultimate failure. Such water-soluble ionizable materials include the chlorides of zinc, barium, cadmium, calcium, and tin, typical end products of stabilization. Lead stabilizers, on the other hand, form reaction products that are insoluble and inert:

HO————Pb————O   O————Pb————OH

S══O

HO————Pb————O   O————Pb————Cl

Initial reaction product of HCl and Tribasic lead sulfate

As a result, the vast majority of PVC wire and cable compounds use lead stabilizers. Insulations and jackets rated for 60–75°C service are routinely stabilized with 4–5 phr of tribasic lead sulfate. Those rated at 90–105°C use 6–7 phr of dibasic lead phthalate. If wet electrical testing of long-term insulation resistance is carried out in water at 75°C, the lower-cost coprecipitate of dibasic lead phthalate and tribasic lead sulfate may be used instead. If, instead, at 90°C, then 7 phr of dibasic lead phthalate should be used. Where color retention is important, barium stearate- or calcium stearate-coated versions are preferred. These also have lower process viscosities. With colored jackets, substitution of 2 phr of dibasic lead phosphite in the above stabilizer content leads to improved color retention.

Clear PVC insulation (e.g., clear lampcord) commonly uses blends of lead and barium stearates. Solid Ba/Cd stabilizers are also commonly, generally at a 2–3 part level. There is similar usage in flexible cords where bright colors are required. Such products are not intended for use in wet locations and their conformance to appropriate specifications does not involve long-term wet electrical testing. A related application is that of brightly colored cord jackets which must function as secondary insulators in case of failure of the primary insulation (e.g., snowblower and lawnmower cords). The 50/50 blends of barium and normal lead stearates predominate in these applications.

Nonlead replacement stabilizers based on barium–zinc and calcium–zinc plus acid acceptor systems have captured part of the wire and cable market where exposure to wet locations is not a requirement. Although these stabilizers are more expensive and less efficient than lead-based analogs, they have been helpful where it has seemed desirable to demonstrate a reduction in heavy-metal usage [12].

VII. APPLICATIONS IN OTHER POLYMERS

Lead stabilizers, in particular, are used in a number of other polymers. In chlorinated (CPE) and chlorosulfonated polyethylene (CSM), 6–10 phr of dibasic lead phthalate is common in calendering and extrusion applications. Where white or bright colored compounds are subject to outdoor exposure, dibasic lead phosphite should be used instead. In coating applications with soluble CSM grades, tribasic lead maleate has been used for viscosity control.

In ethylene propylene elastomer (EPM) and ethylene propylene diene elastomer (EPDM) wire insulation, 5–6 phr of dibasic lead phthalate is used to maximize retention of electrical properties in wet locations. Similarly, for best moisture resistance, 5–10 phr of dibasic lead phosphite finds use in epichlorohydrin (ECO) and fluoroelastomers (FKM). In all of the above, the various lead stabilizers yield better processing safety than use of litharge or red lead.

REFERENCES

1. F Chevassus, R de Broutelles. The Stabilization of Polyvinyl Chloride. New York: St. Martin's Press, 1963, p 31.
2. AH Frye, RW Horst. J Polym Sci 45:1, 1960.
3. J March. Advanced Organic Chemistry, New York: Wiley–Interscience, 1985, pp 873–941.
4. EWJ Michell. J Vinyl Tech 11:141, 1989.
5. R Bacaloglu, M Fisch. Polym Degrad Stabil 45:301, 325, 1994.
6. RF Grossman. J Vinyl Addit Tech 3:5, 1997.
7. RF Grossman. J Vinyl Addit Tech 1:228, 1995.
8. M Onozuka, M Asahina. J Macromol Sci C3:235, 1969.
9. B Engelbart In: EJ Wickson, ed. Handbook of PVC Formulating. New York: Wiley, 1993.
10. G Tranter. Infrared Spectra of Lead Stabilizers, Cookson Ltd. (UK) Bulletin AM/6/77, 1977.
11. TT Nagy et al. Polym Bull 2:77, 1980.
12. P Baker, RF Grossman. J Vinyl Addit Tech 1:230, 1995.
13. RF Grossman, D Krausnick. The Structure of Lead Stabilizers, Part 2, J Vinyl Additive Tech, 4:179, 1998.
14. RF Grossman, D Krausnick. J Vinyl Additive Tech 1:228, 1995.
15. IM Steele and JJ Pluth. J Solid State Chem 145:528, 1998.
16. J Morley and D Pearson, Cookson (UK) Bulletin PM78, 1971.
17. RF Grossman, D Krausnick. J Vinyl Addit Tech 3:7, 1997.
18. RF Grossman, D Krausnick, 4:182, 1998.
19. A Kabata-Pendias. Trace Elements in Soil and Plants. Boca Raton, FL: Lewis Press, 1992, p 365.
20. HJM Bowen. Environmental Chemistry of the Elements. London: Academic Press, 1979.
21. M Kirchgessner and AM Reichlmayr-Lais. Arch Tierernaehrung 31:731, 1981.
22. RF Grossman, GR Mitchener. Migration of Metal-Containing Additives, Hammond, IN: Halstab, 1997.

23. EJ Kincius, GR Mitchener. Plast Compound March/April Vol. 2, 1979, p 27.
24. Halstab. U.S. Patent applied for.
25. JF Cole. Lead Chemicals. New York: International Lead Zinc Research Organization (1975).
26. Available from The Lead Industries Association, New York.
27. HV Smith. Development of the Organotin Stabilizers. London: Tin Research Institute, 1953.
28. RF Bennett. Chem Bull 66:171–176, 1983.
29. MR Dominic, T Kugele. SPE Retec, Philadelphia, 1996.
30. EL Weinberg. U.S. Patent 2,648,650, 1958.
31. CH Stapfer, RD Dworkin, LB Weisfeld. SPE J 28:22, 1972.
32. EB Rabinovitch, JW Summers, JG Quisenberry. ANTEC Proceedings. Brookfield, CT: Society of Plastics Engineers, 1984.
33. EL Weinberg. U.S. Patent 2,680,726 (1954).
34. DJ Dieckmann, ANTEC Proceedings, XXII. Brookfield, CT: Society of Plastics Engineers, 1976, p 507.
35. AH Frye, RW Horst, MA Paliobagis. J Polym Sci. A 2:1765, 1964.
36. DL Clark. Mixed Metal Stabilizers. Brookfield, CT: PVC Primer, Society of Plastics Engineers, 1985, p 161.
37. AH Frye, RW Horst. J Poylm Sci A 2:1765, 1964.
38. G Ayrey. J Polym Sci. B 8: 1970.
39. DF Anderson, DA McKenzie. J Polym Sci A-1 8:2905, 1970.
40. K Prochaska, J Wypych. J Appl Polym Sci 23:2031, 1979.
41. RF Grossman. J Vinyl Tech 12:34, 1990.
42. RF Grossman. J Vinyl Tech 12:142, 1990.
43. JR Darby. U.S. Patent 2,669,548 (1954).
44. DL Clark and T Woodley. SPE Antec Proceedings, 1985.
45. RF Grossman. J Vinyl Tech 15:25, 1993.
46. D Anderson, D McKenzie. J Polym Sci A1:8, 1970.
47. WE Leistner and WE Setzler. U.S. Patent 2,564,646 (1951).
48. RF Grossman. J Vinyl Tech 14:11, 1992.
49. JE Hartitz. Polym Eng Sci 14:392, 1974.
50. Environmental Impact of Lead Stabilizers, The Nordic Plastic Pipe Association Stockholm, 1995 (available in translation from Halstab).
51. IM Danyliuk. Plast Compound 47, Nov./Dec. Vol. 4, 1981.
52. RA Lindner. Plast Compound 27, July/Aug. Vol. 6, 1983.

8
Impact Modifiers

Albin P. Berzinis
General Electric Company, Pittsfield, Massachusetts

I. INTRODUCTION

The extent to which thermoplastics have succeeded in replacing the more traditional building materials, such as wood, glass, and metal, has been due largely
to the favorable balance of properties offered by plastics. Although low density
and ease of fabrication are typical selling points for the use of thermoplastics,
the ability of many thermoplastics to resist failure when loaded at impact speeds,
commonly referred to as the "toughness" or "impact strength" of the material,
is often one of their most desirable characteristics. Indeed, technology for the
provision of toughness in thermoplastics has been practiced nearly as long as
these materials have been manufactured.

In the early days of the thermoplastics industry, the commercial polymers
such as polystyrene (PS), rigid poly(vinyl chloride) (PVC), and poly(methyl methacrylate) (PMMA) exhibited little or no impact strength in their homopolymer
form. In fact, the public associated brittle behavior with plastics as almost a
definition of the material; they were low cost, and easily formed, but limited in
their applications.

Of course, the plastics industry made great strides in overcoming this limita-
tion. In the case of PS and styrene–acrylonitrile (SAN), rubbery inclusions of
polybutadiene (PBD), which were incorporated during the polymer synthesis, act
to improve dramatically the toughness of the base polymers. This is the now
classic approach to toughening of brittle thermoplastics, which has been extensively studied and reviewed over the years since the bulk polymerization process
for toughening PS was introduced domestically by Dow in the early 1950s [1].
Although the *in situ* generation of these inclusions, which act as impact modifiers,
can be very economical when large volumes of resin must be produced, the need

313

to balance the impact-modifier synthesis against the factors controlling the base resin synthesis imposes limitations on the property balances that can be realized via such a process.

Another approach to providing toughness in plastics is to suitably construct or modify the ''backbone'' of the polymer to yield adequate toughness without the need for further modification. Thus, we have some thermoplastic polymers, such as bisphenol-A polycarbonate (PC), which exhibit good impact strength in the unmodified state. Although such an approach would seem to be the most elegant solution, the types of main-chain polymer structures leading to high impact strength do not necessarily yield attractive physical properties or ease of processing. In addition, the types of molecular structure that can be linked together to form high polymers in a commercially attractive process are not unlimited, and the development of the new monomer/polymer chemistry now represents a considerable investment in research and development.

The alternative approach to the above methods, which comprise the subject matter of this chapter, is the incorporation of impact-modifying additives that act to raise the impact strength of the base resin to the desired level. The impact modification of rigid PVC is perhaps the classic example of this approach. Rather than try to incorporate a modifier during the polymerization of PVC (which is an art in itself), rubbery impact modifiers are incorporated later into the finished resin during melt processing. This approach leaves the resin manufacturer some flexibility in terms of the polymerization process and also leaves the downstream processor some flexibility with regard to the level of impact modification desired.

In our overview of the growing number of applications for impact modifiers as thermoplastic additives, we will see that the use of impact modifiers is no longer restricted to the modification of brittle resins. Rather, impact modification is being employed in what are commonly regarded as tough polymers, such as PC, to enhance their balance of properties even further and so yield an edge in performance over the competition. The ultimate extension of the use of additive impact modifiers is in the modification of polymer blends or alloys, an application in its infancy relative to the more traditional impact modifier applications.

We begin by reviewing the theoretical aspects of impact modification. A detailed presentation on the subject is beyond the scope of this work, and the interested reader will be referred to leading references in the considerable body of literature on impact modification of plastics. We will draw on the more classic examples of impact-modified plastics to present some generalizations on choosing a likely additive impact modifier technology for a given type of thermoplastic and on its optimization for the given application.

We then review the more practical aspects of impact modification: What types of additive modifier are available, and how may they be incorporated for effective utilization? Drawing on the approaches that have been disclosed in the open literature, we will survey the application of additive impact modifier technology in each of the major classes of thermoplastic resins. Our perspective is that of the industrial researcher, seeking to choose from the variety of additives those that will best suit the matrix polymer in question and will maximize the benefits and flexibility afforded by the use of the additive-impact-modifier approach.

II. CHARACTERIZATION OF IMPACT STRENGTH

Although the impact strength, or ''toughness,'' of plastics is one of their more frequently mentioned attributes, it is not a simple materials parameter, but rather a complex deformation mode of the material that cannot be readily described by a single test procedure. This complexity is in part due to the nature of plastics themselves and in part due to the influence of part fabrication and geometry on the failure mechanism(s) involved. The speed of the test, the geometry in which the material is constrained, and the type of fracture mode that is tolerable are important variables to be considered when choosing the types of impact test to be used for materials evaluation.

For example, to a manufacturer of pickup camper shells prepared by thermoforming from an extruded sheet, the type of fracture resistance desirable is resistance to high-speed punctures, such as thrown rocks or falling tree limbs. Even in the event that an object actually pierces the material, failure should occur preferably by a ductile drawing or tearing of the material, rather than shattering into shards. In the case of the under-the-hood automotive assemblies manufactured from engineering resins, the impact resistance maintained in the face of stress-concentrating contours such as sharply defined stiffening ribs, bosses, and holes and agressive solvents is important. Here, maintenance of dimensions is critical, and failure at such stress concentrators is not permissible. In addition, the ability to maintain ductile failures at temperatures as low as $-40°F$, a severe test of impact strength, is becoming increasingly important.

To address the suitability of a thermoplastic to such applications, the most straightforward method would be to perform impact tests on the actual fabricated part. In practice, this is seldom done as the primary screening test in the laboratory because the investment in plastic material, fabrication equipment, and large-scale test apparatus is prohibitive. Consequently, polymer scientists have had to resort to using standardized tests that can be carried out on a small quantity of material and can be repeated and compared to results obtained for competing materials as they seek to develop and improve their impact-resistant materials. Probably the two most popular types of impact test method are the notched Izod impact test as described by the American Society for Testing and Materials (ASTM D 256) and the Gardner variable height impact test (GVHIT: ASTM D 3029). These tests attempt to measure toughness in the presence of stress-raising flaws and the puncture resistance, respectively. The end users of plastics must rely primarily on data generated on these relatively small samples to guide them on their materials specifications. Such an approach can easily lead to some controversy regarding the validity of a given set of laboratory test results to in-service usage. However, if an effort is made to validate the laboratory findings by comparative end-use testing, the relationships developed in small-scale testing can be used with a high degree of confidence. One must always guard against extrapolation of laboratory data into areas where such real-world validation has not been performed.

A. The Notched Izod Test

As so often happens, the simplest test to run, the notched Izod, is the most difficult to correlate with real-world application. This arises from precisely the considerations that make it simple to run: first, the deliberate introduction of a flaw into the sample via the notch; second, the constraints imposed by the sample geometry. Typical geometries are a notch radius of 0.010 in., a sample thickness of 0.125 in., and a depth of 0.50 in. along the direction the notch is propagated. The energy required to tear the sample apart from the base of the notch is measured. Often, a material that is ductile in slow tensile elongation, or even dart impact, fails in a brittle manner in the notched Izod test. The amorphous polyesters constitute an example of such polymers. Such materials are termed ''notch sensitive''; we examine the notched Izod test in more detail below.

The effect of the notch is to concentrate stress in a small volume of the sample, which, in essence, effects an increase in the strain rate experienced by this material. Havriliak [2] has calculated the approximate strain rate of typical notched Izod geometries by assuming that the radius of the notch is a crude approximation of the gauge length in a typical tensile test specimen. From the data of Table 1, it is apparent that the notched Izod test imposes a strain rate that

Table 1 Summary of Geometry Parameters for Tensil and Notched Tests[a]

Test type	Applied load rate (in./min)	Test length (in.)	Strain rate (in./in./min)
Slow tensile	0.02	0.375	0.533
Fast tensile	0.9	0.375	2.40
Tensile impact	8,420	0.375	22,500
Notched Izod			
$r = 0.250$ in.	3,890	0.250	9,500
$r = 0.100$ in. standard	3,890	0.100	38,900
$r = 0.010$ in.	3,890	0.010	389,000
$r = 0.001$ in.	3,890	0.001	3,890,000

[a] Calculated for ASTM 635 type 5 test specimens.

is at least four orders of magnitude faster than a simple tensile test. The choice of notch radius can shift the frequency of the impact event into the megahertz range. Given the viscoelastic properties of high polymers, such a high test frequency can result in a significant change in the compliance of the polymer, which can be calculated using the Williams–Landel–Ferry (WLF) time–temperature shift equations [3]. Therefore, it is not surprising to find examples of polymers that are brittle to the notched Izod test, despite evidence of room-temperature ductility by tests at slower strain. The use of additive rubber modifiers, which have a high compliance at room temperature even at the high frequencies of the notched Izod test, can be used to shift the response of the matrix polymer until ductile failure is achieved.

The classic example of utilizing additive impact modifiers to alter the ductile/brittle behavior of a polymer is found in the case of rigid PVC. A typical set of curves for the effect of increased impact modifier loading on notched Izod impact is shown in Fig. 1. The modifier used is of the Paraloid® core/shell, methacrylate–butadiene–styrene type from Rohm and Haas; these modifiers will

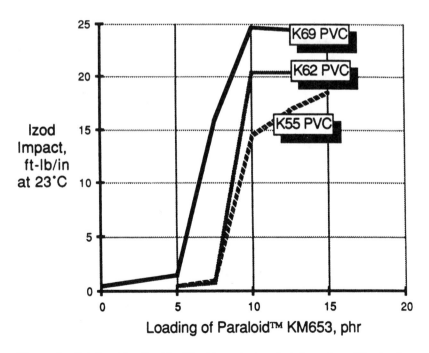

Figure 1 Relationship of notched Izod impact strength to impact-modifier loading and matrix molecular weight for rigid PVC.

be discussed in more detail with the other classes of additive modifiers in Section IV.A. All the PVC blends display brittle fracture at low loadings of impact modifier. However, over a relatively narrow range of impact-modifier loading, such as 5 parts per hundred (phr), the material transitions from brittle to ductile failure in the notched Izod test.

Notice that this brittle/ductile transition can be made to occur at lower modifier loadings if the molecular weight (MW) of the PVC resin is increased. Thus it is important when comparing the efficiency of various impact-modifier systems, to hold the molecular weight of the matrix polymer constant to obtain a fair comparison. This effect is not restricted to PVC; it appears to be operable in any impact-modified polymer blend. Examples of this effect in other resins will be shown later.

The ductile/brittle transition is also known to shift with the testing temperature: as the test temperature is reduced, the modifier loading necessary to avoid brittle failure usually increases. If the same system from Fig. 1 is examined at a constant modifier loading, the blends can be made to go back down the ductile/brittle curve by decreasing the test temperature (Fig. 2). Once again, the higher-

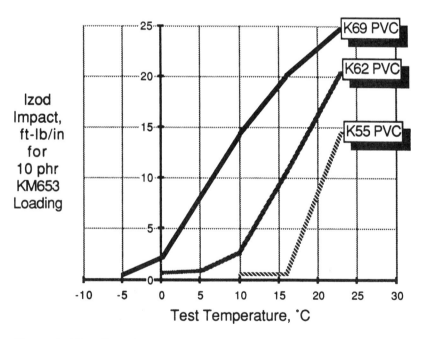

Figure 2 The effect of temperature, at constant impact-modifier loading, and PVC molecular weight on notched Izod impact strength.

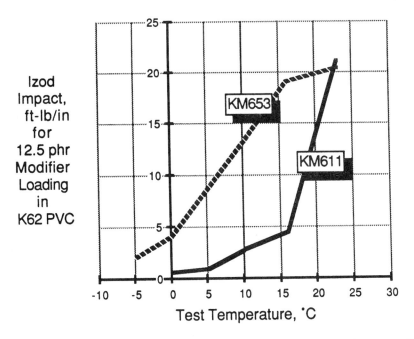

Izod
Impact,
ft-lb/in
for
12.5 phr
Modifier
Loading
in
K62 PVC

Figure 3   Comparison of two methacrylate–butadiene–styrene-type impact modifiers for impact efficiency as a function of temperature.

MW resins are more resistant to this effect. To obtain ductile impact at 0°C, the modifier loading would have to be increased enough to move the system "back up the curve" at that temperature.

However, there are limits to the extent by which impact strength at lower temperatures may be enhanced by increasing the level of impact modifier. In Fig. 3, we see the curve for the same impact modifier of the preceding two examples, Paraloid KM-653, at 12.5 phr in the medium-MW K-62 PVC resin. Compared to Fig. 2, the impact–temperature curve has indeed been shifted up by the increase in modifier level.

However, this higher level of impact modifier does not help nearly as much in the case of a related impact modifier, Paraloid KM-611, which differs from the KM-653 in that the glass transition temperature (T_g) of the rubber is higher; this change is deliberately made for the purpose of obtaining superior optical properties in PVC. The higher rubber T_g leads to a more rapid decrease in impact strength as the test temperature is lowered; the lower rubber level also means that compensation via high modifier loadings will not be as effective as in the case of KM-653.

To summarize, the notched Izod impact efficiency of an additive impact modifier can be studied from several angles. Curves may be generated by increasing the level of modifier while maintaining the test temperature and geometry constant. Alternatively, a single modified blend may be tested at progressively lower temperatures. The latter method is more economical if the modified blends are difficult to prepare; alternatively, the modifier level approach is convenient if testing at lower temperatures is either impossible or not relevant to the intended application. In any event, it should be apparent now that the system variables for the modifier, matrix, and test conditions must be carefully defined before a judgment can be made regarding the impact efficiency of a given additive impact modifier.

The geometry of the notched Izod test specimen also places restrictions on the way in which the stress applied to the notch can be relieved. The propagation of a crack from the root of the notch can be treated by applying the concepts of fracture mechanics, as reviewed by Bucknall [4]. In the simplest limiting case, the thickness of the notched specimen is small so that the sample approximates a sheet, and the matrix is linearly elastic so that the crack propagates with little local yielding at the crack tip. In this case, the applied stress is proportional to the deflection of the sample, and the crack tip is loaded in plane strain; that is, the through-thickness stress component is zero. This allows easy calculation of K_I, the applied stress at the crack tip. The theory of fracture mechanics holds that crack propagation occurs when this stress exceeds a critical value K_{IC}.

Unfortunately, in practical impact testing, both of the above fundamental assumptions are usually violated; the thickness of the notched sample is not negligible, and there is often considerable yielding at the crack tip, which complicates the calculation of K_I. Thus, the concepts of fracture mechanics can be used only in a qualitative sense to guide in the assessment of fracture toughness of plastics.

As the sample thickness is increased, the through-thickness stress component increases, which places the region in the area of the crack tip in a state of plane strain, or triaxial tension. As a result, we often observe a material that appears to go through a ductile/brittle transition in the notched Izod test at a certain sample thickness and notch radius; this results from the fact that the plane-strain contribution to the stress intensity causes the critical stress intensity to be exceeded. Polycarbonate displays a ductile/brittle transition typical of this effect, although Fernando and Williams [5] demonstrated that this transition as a function of thickness exists for other plastics such as polypropylene (PP). Not surprisingly, the thickness at which the transition occurs is influenced by the molecular weight of the PC, as an increase in molecular weight will increase K_{IC}. This effect is shown in Fig. 4.

One approach to alleviating this effect is to use an additive rubber modifier. Figure 3 shows that the addition of core/shell Paraloid modifiers can yield a

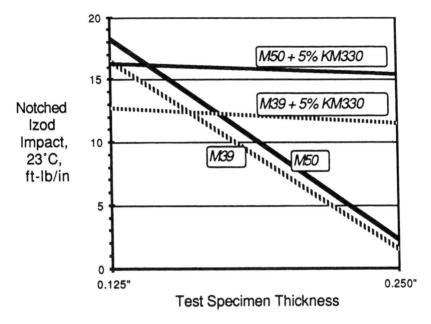

Figure 4 Increasing the critical thickness of polycarbonate via core/shell impact modification.

significant increase in the thickness range over which PC exhibits high notched impact resistance. The effect is obtained at low loadings of modifier; as a result, the physical properties of the resin are only slightly reduced. Fernando and Williams [5] showed that a similar effect could be realized in PP by addition of small amounts of polyethylene ($\leq$ 10%). They postulate that the effect is due to the reduction in yield stress accompanying the addition of polyethylene; by encouraging yielding of the matrix, K_{IC} is not likely to be exceeded, and so brittle failure is avoided.

From these considerations, it is apparent that the performance of a plastic in the notched Izod test can be manipulated by control of the sample geometry, such as the use of thin sections, and by altering the composition to encourage yielding. When using the notched Izod test as a screening tool for designing a modified plastic, care should be exercised that the polymer properties being developed will yield meaningful toughness in actual usage and that the polymer is not being designed merely to suit the method.

B. Falling Dart Penetration Tests

The other group of impact test methods, which are often portrayed as being more likely to correlate with in-use conditions, are based on resistance to dart penetration. Historically, the GVHIT, a falling dart test, has been used as a standard test method. In this test, a dart having a hemispherical tip is driven ballistically by a falling weight into a plaque of the material, which has been clamped onto a support ring. This type of test has several advantages. First, the material is placed in triaxial stress, which, as discussed earlier, is the most severe geometry for impact testing. The test conditions can be easily varied by changing the weight, drop height, dart diameter, or sample thickness. Finally, the specimen can be either a small injection-molded plaque specifically prepared for the test or a sheet sample cut from a large molding in actual service. The test is typically run to failure of the sheet, as judged by rupture of the distal surface of the sample. However, a considerable number of samples may be necessary to permit definition of the amount of energy corresponding to the mean energy of failure; this stems in part from the difficulty of judging the point of failure, particularly in ductile materials, which exhibit a considerable amount of drawing prior to rupture at the dart tip, and also in part from variations in sample preparation, thickness, and internal flaws. It commonly requires a minimum of 20–30 falling dart impacts to adequately define the mean failure height of a given sample by statistical methods, as reviewed by Moritz [6]. In addition to being time-consuming, this type of test can use up a sizable quantity of material, which may pose difficulties in exploratory programs in which exotic or difficult-to-prepare materials are being evaluated in the laboratory.

However, this type of test is now becoming increasing popular due to the advent of instrumented impact test methods, which can characterize a material's toughness to dart impact on the basis of only a few samples. The simplest systems introduce a quartz pressure transducer in the dart tip, so that the stress on the sample during dart impact and penetration can be measured continuously. The Dynatup tester manufactured by Effects Technology, Inc. is typical of this type of system. The test apparatus consists of a falling weight tower wherein the instrumented tup is guided on rails; the mass of the tup and the drop height, or velocity, can be varied. Although such a system is the simplest with regard to instrumentation, it does not take into account the deceleration of the tup that occurs upon penetration of highly ductile materials, such as PC.

More sophisticated systems incorporating accelerometers to develop a more accurate description of the force–displacement curve have been described by Winuk et al. [7]. Nonetheless, instruments such as the Dynatup represent a significant advance over the GVHIT, both in terms of material conservation and in the amount of information available to the researcher. The dart force is displayed as a function of either dart displacement or time; the result is essentially a

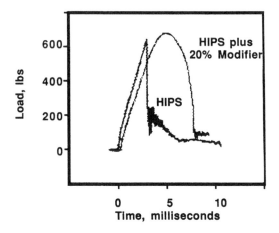

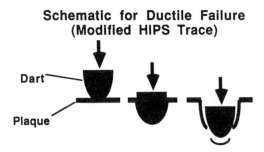

Figure 5 Dynatup impact trace comparing brittle HIPS failure to ductile, modified HIPS.

stress–strain curve at impact speed, with the "strain" being the drawing of the material by the dart. By consulting this curve, the analyst can judge the point of sample failure, although the decision as to what constitutes a failure in ductile systems is still somewhat subjective. In addition to speeding the analysis of dart impact, the initial slope of the stress–strain curve can be used as an approximate modulus measurement at impact speed.

Typical data resulting from Dynatup falling dart impact testing appear in Fig. 5. The load-versus-displacement curves for two polymer systems are compared: The first is a standard commercial "high-impact" polystyrene (HIPS), and the overlay trace is that of the same HIPS containing 20% of an additive impact modifier based on polybutadiene rubber. The trace for HIPS shows that the load builds rapidly as the dart impacts the sample, but then the sample abruptly ruptures in a brittle fashion. The elongation until break, and hence impact energy absorption, is relatively low at 10 ft-lb. The addition of the additive modifier slightly lowers the modulus of the system, as witnessed by the slightly less steep slope of the initial loading curve. However, a marked increase in elongation results, with the dart drawing through the sample in a ductile fashion, with a total energy absorption of 30 ft-lb. The load is smoothly and safely absorbed by the material, which can now truly be considered a "high-impact" material.

If desired, notched Izod samples can be similarly tested by these instruments. Figure 6 compares the notched Izod behavior of the same two samples of HIPS and modified HIPS to the Dynatup dart impact results. The commercial HIPS material behaves in a very similar fashion in this test geometry as well; that is, the load increases linearly as the hammer attempts to deflect the bar but falls abruptly as the crack propagates rapidly across the specimen in what would be characterized as a brittle failure (2 ft-lb/in.). The modified HIPS curve displays a substantial plateau in the maximum load curve, which corresponds to the energy being dissipated by crack growth across the specimen (7 ft-lb/in.). The deliberate introduction of a stress-concentrating flaw makes this propagation step much easier to see than in the falling dart test. Note also the expanded time scale for the notched Izod traces, as the time scale of the impact event in the presence of this flaw is much shortened. The bar actually remains hinged after the passage of the hammer; these "hinge breaks" are usually a hallmark of ductile impact behavior in the notched Izod test, and this drag against the hammer can be seen in the load curve not returning to zero in Fig. 6 immediately after the failure of the sample.

An approach that seeks to get around the ballistic problems of drop tower systems is the use of a hydraulically driven dart or ball to penetrate the sample. Dao [8] has shown that the use of servohydraulic crossheads to drive the tup allows precise control of the dart speed and also can be used to eliminate the effects of dart deceleration. Although commercial systems are available, such as that offered by Rheometrics, Inc., these systems cost substantially more than those of the drop tower type. One advantage of these systems is the ability to continuously vary the impact speed, which greatly facilitates the study of ductile/ brittle transitions by the dart impact test. In addition, the use of servohydraulics allows high-impact velocities to be obtained without the need for very tall ballistic drop towers. The servohydraulic systems are finding increasing favor with automotive testing laboratories, where testing at high-impact speeds is required to simulate crash situations.

Although the effects of sample geometry and loading rate present a challenge in and of themselves to the interpretation of impact test data, the situation is further complicated by the effects of sample preparation on the response of polymers to impact tests. Orientation of the polymer due to the shear stresses encountered during injection molding, relaxation of molded-in stresses, and the

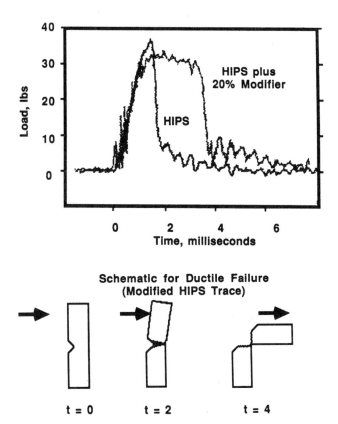

Figure 6 Notched Izod/Dynatup impact trace comparing brittle HIPS failure to ductile, modified HIPS.

size of crystallites in semicrystalline thermoplastics such as nylon and polyesters are some of the variables that can further alter the impact behavior of a thermoplastic. A recent comprehensive study by Turner [9] on the correlation between laboratory impact data and ''real-world'' impact performance yields no easy answers to the problem. The best way for the researcher to guard against conclusions based on artifacts of fabrication is to characterize the toughness of the modified polymer over the extremes of molding conditions likely to be encountered in the field. Again, the importance of validating laboratory trends with in-use evaluation and a thorough knowledge of the processing characteristics of the matrix polymer in the customer's equipment cannot be overemphasized.

III. TOUGHENING MECHANISMS IN THERMOPLASTICS

The mechanisms by which plastics may be toughened to achieve high impact strength have been extensively studied and reviewed by Bucknall [4,10] Kekkula [11], Paul and Newman [12], Manson and Sperling [13], and Deanin and Crugnola [14]. We present next an overview of the current theories regarding toughness, so that we may draw some general conclusions about choosing, or designing, an impact-modifier system for a given type of polymer. In all cases, the characteristic toughness of high-impact plastics is attained by inducing massive amounts of yielding in the vicinity of the flaw that is attempting to propagate through the material, thereby reducing the effective applied stress at the flaw and preventing catastrophic, brittle failure. Although rubber tougheners are commonly referred to as impact ''modifiers,'' they can perhaps be better described as *facilitators,* acting either to activate the matrix polymer to engage in deformation modes not attainable in the native unmodified state or to magnify greatly the extent of deformation modes already available to the matrix polymer. To study the effect of additive modifiers, we must study, first, the type of deformation modes that can be activated in thermoplastics and, second, how an additive modifier can interact with the matrix to induce the desired deformation modes.

A. Effect of Secondary Glass Transitions

Impact resistance in glassy polymers has been correlated with the presence of a secondary transition in the polymer (i.e., a transition located below the primary glass transition. Heijboer [15] and Boyer [16] show a rough correlation between the intensity and temperature of these secondary transitions and impact strength. These transitions are typically determined as local maxima in the low-frequency dynamic–mechanical loss spectrum of the material versus temperature. Empirically, it is known that such loss peaks should be located well below the testing temperature for an appreciable effect to be obtained, typically at least 50°C. The explanation usually advanced for this observation is that the speed of the impact test effectively raises the temperature of this secondary transition by this amount (time/temperature equivalence); therefore, secondary transitions occurring near the test temperature at low frequency are effectively shifted into the glassy region at testing frequency. Hartmann and Lee, in a detailed study of the impact resistance of polycarbonate as a function of temperature [17], showed that the maximum in the impact resistance of polycarbonate correlated with the position of the maximum in the secondary transition, which occurs at $-105°C$ at low frequency. This transition is shifted to $-80°C$ at the estimated frequency of the test (roughly 1 kHz), which is very close to the observed maximum in impact strength.

It has been observed further by Heijboer [15] that such secondary transitions must be associated with motion of the polymer backbone, not pendant sidechain groups, for them to be effective at improving the impact resistance. This observation was studied in more detail by Schaefer et al., who characterized these relaxation modes by solid-state nuclear magnetic resonance (NMR) techniques [18,19]. A correlation was found between the ^{13}C relaxation times of the polymer backbone and the impact strength of the unmodified polymers, such as polycarbonate, polyphenylene oxide, and PVC. A more detailed correlation between the impact strength and a parameter representing the spectral density of the relaxation motion was proposed by Schaefer et al. [18]. These results parallel the earlier observations, obtained by dynamic–mechanical measurements, that the allowed motions of the polymer main-chain atoms must be essentially in resonance with the frequency of the impact test.

Because our definition of impact strength is the ability of the material to undergo massive yielding at impact speeds, it is easy to see that main-chain motions that can be activated at impact speed would be expected to correlate with impact strength. Thus, a secondary transition at low temperature is an indication that chain segments possess some degree of mobility at impact speed; it is hoped that this mobility can be translated into large-scale deformation of the polymer chains. In fact, the correlations are surprisingly good, considering that the transitions are measured at very low strains, and large strains are encountered in an impact test.

B. Interaction Between the Additive Impact Modifier and the Matrix Polymer

There are relatively few polymers, such as polycarbonate, that manifest secondary transitions significant enough to yield high impact strength while maintaining a sufficiently high primary T_g for acceptable engineering properties. Even polymers that do manifest a secondary transition at low temperature, such as PVC and polyphenylene oxide, require impact modifiers to yield high impact. In the case of many polymers, such as polystyrene, no secondary transition of any significance is present. Consequently, this concept of providing a secondary transition in the material for impact strength is achieved in practice by addition of rubber impact modifiers having a low T_g.

Because such systems are now, by definition, phase separated to some degree, with an interface between the glassy matrix and the rubber that may be difficult to describe, elucidation of the principles behind effective impact modification by rubber toughening is necessarily more complicated than the study of inherently tough homopolymers. The most well-characterized systems are those that have been in existence the longest, the impact-modified styrenics such as HIPS and acrylonitrile–butadiene–styrene (ABS). The factors that appear to govern the toughening mechanisms in these polymers can provide a good basis for a qualitative understanding of the role of impact modifiers in toughening the more recent examples of impact-modified engineering resins.

The commercial examples of toughened thermoplastics appear to fall into two broad classes of matrix failure mechanism, namely crazing and shear banding, sometimes referred to as type I and type II behavior, respectively, after Bucknall [4] and Wu [20]. Crazes are formed by microfibrillation of the matrix polymer into tendrils that span the growing microscopic voids in the matrix. The voided regions may be a few micrometers across; the fibrils spanning the void are 10–100 times thinner. Deformation by crazing is by definition, accompanied by an increase in sample volume resulting from this void formation. The energy is absorbed by the drawing of the matrix polymer into microfibrils and by the need to create a fresh surface area within the voids. The fibrils bridging the voids appear to be load bearing, insofar as crazed specimens display a mechanical integrity that is absent from materials having macroscopic voids and cracks. Polystyrene is the best example of essentially pure type I behavior; the hallmark volumetric increase under strain that accompanies craze formation has been used by Bucknall to quantify the contribution from a type I mechanism [21]. Crazes are usually found to propagate across the sample at right angles to the applied stress.

The type II or shear band mechanism is better identified as a matrix yielding process, in which the specimen deforms with relatively little volumetric change. The ''shear bands'' are regions of high shear strain that tend to form along the planes of highest resolved shear stress, which is typically at a 45° angle to the applied stress. Although many polymers can be thought of as displaying substantially type II behavior, such as polycarbonate, thermoplastic polyesters, and ABS, most of the polymers other than PS appear to deform by a combination of shear banding and crazing. For example, impact modified Nylon 6/6 has been characterized by Wu as having contributions from 75% shear banding and 25% crazing during energy absorption in a notched Izod impact test [20]. Blends of PS and polyxylenyl ethers have also been shown by Maxwell and Yee [22] to fail by a mixture of shear banding and crazing; the onset of the two mechanisms can be influenced by the applied strain rate, as well as by the PS content in the blend.

In general, rubbery inclusions are thought to function as impact modifiers due to the shear stresses that arise at the rubber–matrix boundaries. Bucknall [4] presents graphically the stress fields operating at the periphery of the rubbery inclusions via the application of Goodier's equations. These locally enhanced stress fields are believed to induce the local matrix deformation, by either the type I or type II process, necessary to keep the effective stress below the critical value at which a catastrophic crack will propagate through the sample. When large numbers of microscopic rubbery inclusions are distributed throughout the matrix, this matrix deformation is delocalized throughout the sample and large amounts of energy can be absorbed.

The significance of knowing the preferred failure mechanism of the polymer that is to be modified lies in the apparent difference between the requirements for the additive modifier for types I and II matrix behavior. In the case of a predominantly crazing mechanism, such as in polystyrene, the modifier is thought to perform two functions: to initiate massive amounts of crazes and to effectively terminate these crazes so that large flaws do not grow in the sample. These apparently conflicting requirements are further complicated by the particle size dependence of the two effects. It is generally held that the modifier particles must be roughly the same size as the crazes, or several micrometers in diameter, to function effectively as craze terminators; see Bucknall's review of this mechanism, as it is thought to operate in HIPS [4]. From this consideration, large modifier particles should be favored. There is still some disagreement on the matter, as summarized by Bucknall et al. [23], as there does not appear to be as strong an influence of particle size on the ability of modifier particles to *initiate* crazes. Because craze delocalization is favored by a large *number* of particles, a small modifier particle size would be favored. As a result, commercial HIPS represents a compromise between these two considerations, with a modifier particle size of 2–5 μm being common. It is also believed that to ensure the efficient initiation and termination of crazes, the inclusion must adhere strongly to the matrix [24]. As a corollary observation, the modulus of the rubber must be sufficiently lower than that of the matrix to set up the dilational shear field that is an apparent prerequisite for craze initiation [25].

In the case of deformation via shear band formation, the primary requirement would appear to be the efficient delocalization of matrix shear yielding. Because this would be favored by a large number of particles, an impact modifier of small particle size should yield the greatest number of particles for a given loading and so should be most efficient. However, because most polymers deform via contributions from both a crazing and a shear banding mechanism, there appears to be an optimum particle size required for efficient interaction. In the case of Nylon 6/6, Wu [26] argues that there is a maximum interparticle distance for efficient initiation of matrix deformation at impact speeds. This interparticle distance will be strongly affected by the modifier particle size at a constant volume fraction of modifier, with smaller particle size leading to increased interaction (smaller interparticle distance). Due to the contribution of crazing to nylon matrix deformation, we would expect there to be a definite minimum to the effective particle size.

Probably the most sophisticated approach to choosing an effective particle size for the impact modifier is to use a mixture of two or more particle size modes. In a polymer such as ABS, where both failure mechanisms are thought to contribute, advantage has been claimed by Monsanto [27] for impact modifier technology that combines a relatively large particle size modifier component with one substantially smaller. In this way, large modifier particles can be present to control crazing while efficient shear band delocalization is also obtained. The detrimental effect of the large particle size modifier on surface gloss is also minimized. Mixed particle size modifier systems are an elegant solution to the compromise in performance due to use of unimodal modifier systems in matrices that show both type I and II behaviors. The proper tailoring of such systems obviously demands a thorough knowledge of the failure mechanism of the matrix in question, as well as a flexible and well-controlled process for the synthesis of the impact-modifier particles.

The question of the need to obtain the appropriate degree of adhesion, or "compatibilization," between the modifier and the matrix is a difficult one to answer, even to this day. The systems that have been the best characterized are the vinyl polymers such as HIPS and ABS, wherein the modifier is grafted with a polymer of generally the same composition as the matrix polymer. This grafting is known to be necessary to obtain good impact strength because simple mechanical blends of polybutadiene rubber and styrenic plastic are not very effective, as reviewed by Bucknall [4]. As mentioned earlier, it is also thought that good adhesion is vital to the operation of the crazing mechanism.

However, what of the case of the condensation polymers, such as polycarbonate, thermoplastic polyesters, and nylon? These polymers are not readily impact modified by the same grafting technology as the vinyl polymers, owing to the different polymerization chemistry. Also, they tend to favor a type II mechanism, which may not be as sensitive to adhesion effects. Recent work by Wu [26] challenges the need to specifically promote adhesion when modifying these polymers; the investigator claims that van der Waals attraction between the modifier and the matrix polymer may be sufficient in these cases. The surface energy between the modifier and matrix may exert a profound effect on the dispersion of the modifier in the matrix and attainment of the proper particle size, but this is really a secondary effect of interfacial adhesion on the properties of the system. Work referenced below, which claims the effective use of polyolefins such as polyethylene (PE) as impact modifiers for engineering resins, tends to minimize concern for the need to maximize adhesion between the modifier and matrix for these systems.

In summary, considerations of the choice of the appropriate particle size for the rubbery inclusions, and methods for retaining their integrity during melt processing via techniques such as cross-linking, appear to take precedence over concerns of interfacial adhesion. As a general rule, a unimodal particle size that lies toward the minimum effective particle size in the matrix will likely result in the most effective system because this size will maximize the number of modifier particles at a given volume fraction of modifier. Multiple particle size systems may be resorted to in attempts to obtain the maximum impact efficiency in systems that show a combination of type I and II behaviors. Tailoring of the modifier–matrix interface should be undertaken only in conjunction with an understanding of the general particle size and overall volume fraction of modifier required by the matrix under consideration.

IV. COMMERCIAL APPLICATIONS OF ADDITIVE IMPACT-MODIFIER TECHNOLOGY

The best way to illustrate the practical aspects of impact modification via the use of additive modifiers is to review the technologies that either are currently in commercial usage or have been disclosed in the patent literature as promising avenues to the production of toughened polymers. The presentation below is by no means intended to be exhaustive, but merely to illustrate the growing variety and sophistication of approaches.

A. Examples of Available Additive Impact-Modifier Polymers

A variety of rubbery polymers are available and in use as additive impact modifiers for thermoplastics. As an overview, representatives of each of the major classes of polymeric additives and their applications are presented in Table 2. The polymers listed span the gamut from simple elastomeric, melt-processible homopolymers such as low density polyethylene to structured, crosslinked multistage copolymers.

Many of the elastomers are block copolymers wherein a minimum of two dissimilar polymer chains, typically a "hard" and a "soft" segment, are covalently joined together during their synthesis. Figure 7 represents this morphology schematically, and for a recent review, consult the work of Meier et al. [28]. The self-association of the "hard" segments, which are below their T_g at room temperature, effectively acts as a physical crosslink, and so these materials have properties similar to thermoset crosslinked rubbers at room temperature. However, at elevated processing temperatures (sufficient to melt the "hard" polymer domains serving as the physical crosslinks), the chains are able to disentangle and flow, although some self-association of the two dissimilar domains remains even in the melt. The melt processibility, morphology, and ultimate mechanical properties of the block thermoplastic elastomers (TPEs) have been shown by Agarwal [29] to be very sensitive to the molecular weights of the various segments, as well as to the shear stress and residence time in the processing equipment. When such materials are used as additive modifiers in a glassy matrix, the domain size of the elastomeric phase derived from the TPE is typically broad and ill-defined. Because the morphology of these blends is dependent on the shear stresses in the melt and because there is typically a distribution of shear stresses in the processing equipment, the broadness of the particle size distribution

Table 2 Commercially Available Additive Impact Modifiers and Representative Applications

Generic designation	Product®/ supplier	Composition/ morphology	Typical applications
Polyethylene	Numerous	Semicrystalline polyolefin	PC, PP
EPDM	TPE (Uniroyal)	Non-crosslinked as supplied, but contains reactive olefin sites	PP
Acid-modified polyethylene	Surlyn, Vamac (DuPont)	Semicrystalline backbone, ionomeric "crosslinks"	Polyamides, PC
Thermoplastic polyurethane	Texin (Mobay), Pellethane (Dow)	Segmented block copolymer, physical "crosslinks"	POM thermoplastic polyesters
Thermoplastic polyester elastomer	Hytrel (DuPont)	Segmented block copolymer, soft polyether/hard polyester, physical "crosslinks"	Thermoplastic polyesters
Thermoplastic styrenic elastomer	Kraton (Shell), Solprene (Phillips), Stereon (Firestone)	Segmented block copolymer, soft P (Bd) or (ethylene/ butylene)/hard PS; physical "crosslinks"	Crystal and impact PS, PP, PPO
Core/shell, acrylic rubber	Paraloid KM 300 series, EXL 3300 series (Rohm and Haas)	Crosslinked P (acrylic) "core," P(methacrylic) "shell"	PVC, PC, PBT, PET, blends, polyesters, crystallized PET
"Modified"	Durastrength (Elf Atochem), Kane Ace FM series (Kaneka Texas Corp)	Additional monomer combinations	Same as above
Core/shell MBS	Paraloid BTA, EXL 3600 series (Rohm and Haas), Metablen series (Elf Atochem), Kane Ace Series B and M (Kaneka Texas Corp.)	Crosslinked P(Bd or Bd/Sty) "shell," P(methacrylic/ styrene) "shell"	PVC, PC, PET, PBT, blends
ABS	Blendex series (GE Specialty Chems), Lustran (Monsanto), Magnum (Dow), Kane Ace j series (Kaneka Texas Corp)	Acrylonitrile/butadiene/styrene block terpolymers	PVC, PC, PET, PU
EVA	Elvaloy (DuPont), Baymod (Mobay)	Ethylene/vinyl acetate/carbon monoxide copolymers	PVC
SBR	K-Resin series (Phillips)	Styrene/butadiene rubbers	Styrenic polymers and copolymers
Nitrile rubber	Nipol series (Zeon Chemicals)		PVC
SAN	Blendex HPP (GE Specialties)	Styrene/acrylonitrile block copolymer	PC
CPE	Tyrin series (Dupont–Dow)	Chlorinated polyethylene	PVC
Polybutylene	Polybutylene L, H series, Indopol (Amoco)		ABS, PP, EVA, EPDM/PP blends, styrenic polymers
E–O copols	Engage (Du Pont–Dow)	Ethylene/octene copolymers	PP, PE, TPO

Self-Association of Hard and Elastomeric Domains

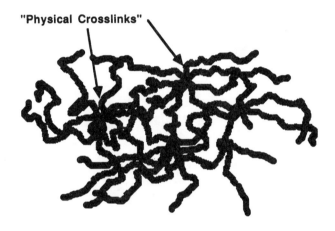

Figure 7 Schematic of block copolymer morphology.

of the TPE blends is a reflection of the uniformity of the shear stresses in the particular piece of processing equipment used to prepare the blends.

The class of "core/shell" polymers is well known in the art of PVC impact modification and can be divided into two subsets: those based on acrylic elastomers, typically polybutyl acrylate, and those based on polybutadiene elastomers, often referred to as "methacrylate–butadiene–styrene" or MBS polymers. These structures are most conveniently synthesized via emulsion polymerization techniques; by use of "seeded" polymerization processes, it is possible to obtain good control over the particle size of the resulting particles, as reviewed by Morgan [30].

The core/shell polymers are typically covalently crosslinked to retain their core/shell integrity even after extensive melt processing. This relieves the blend processor of the necessity to control closely the processing conditions for the sake of morphological control, as is so often required when using unstructured elastomers or even the block copolymers. This attribute of the core/shell polymers

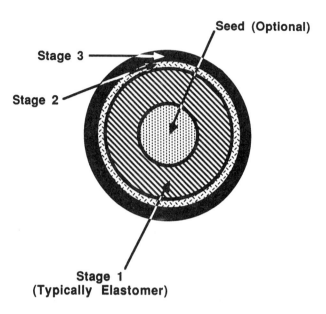

Figure 8 Schematic of core/shell impact modifier morphology.

is particularly valuable in the processing of impact-modified polymer blends, as will be described in Section IV.B. Also, multiple-layered structures can be constructed by following a sequential polymerization scheme. As shown in Fig. 8, the elastomeric "core" of the polymer can be encapsulated by an outer stage of another polymer, which can be chosen to provide good adhesion to the target matrix polymer. It is also possible to have at least one interlayer of polymer, to serve as a transition layer between the inner core and the outer shell. Core/shell polymers can thus be finely tailored to suit the particular matrix to be modified. However, the emulsion polymerization chemistry obviously must be well controlled and well understood, so that such optimization can be carried out in a predictable and repeatable fashion on a commercial scale.

B. Applications of Additive Impact Modifiers in Toughened Plastics

We now review the state of the art regarding the use of additive impact modifiers in preparing toughened thermoplastic blends. We begin with the toughening of the major thermoplastic homopolymers and close with the toughening of engineering resin blends, which is the field generating the most excitement and activity today in the realm of toughened plastics.

1. Polypropylene

The case of polypropylene (PP) well illustrates the choice open to those seeking to produce toughened polymers. One approach, best suited to the manufacturers of PP, is to chemically modify the PP backbone to confer the appropriate amount of toughening. Thus, manufacturers such as Himont incorporate olefinic comonomers, such as ethylene, to produce the toughened copolymer grades of PP. The ethylene is apparently incorporated in a manner that leads to some aspect of block PE character, allowing the formation of microdomains of rubbery inclusion required for toughening.

The alternative approach is to incorporate a polyolefin impact modifier via melt extrusion. To be competitive in physical properties, PP must be stiffened by the addition of substantial quantities of fillers such as calcium carbonate or talc. The accompanying reduction in impact strength can be effectively overcome by the addition of a polyolefin impact modifier. Dao and Hatem [31] have demonstrated that with the appropriate choice of filler and modifier loading, blends that offer an attractive balance of impact strength, stiffness, and price relative to copolymer systems can be obtained.

Ethylene–propylene–diene terpolymer rubber (EPDM) is the polyolefin most often used in toughening PP; a most convenient form is that offered by Uniroyal's TPE, which is a blend of about 75% EPDM in PP. This product is available in a pellet form, as opposed to balestock EDPM, and so is much more convenient for use in a continuous extrusion operation. However, the use of a non-cross-linked elastomer requires careful control of the compounding conditions to achieve the optimum particle size and dispersion of the elastomer in PP. Dao [32] has demonstrated that the use of intermeshing, corotating extruders such as the Werner Pfliederer compounding machines yields the most consistent properties.

Ramsteiner et al. [33] have shown that PE can also be blended with PP to yield toughened blends comparable to the PP/PE copolymers. Again, the formation of phase-separated domains of low-T_g elastomer is required for toughening to be obtained.

It is also possible to effectively impact modify PP by the addition of styrene–butadiene–styrene (SBS) triblock copolymers. Bull and Holden [34] claim that the notched Izod impact strength of PP can be improved significantly at SBS loadings of 20% or more, despite the apparent incompatibility of PP and the PS end blocks of the TPE. However, the SBS triblocks may be at a cost disadvantage compared to the simple polyolefins, which can also be used to toughen PP.

2. Styrenics: HIPS, ABS, and PS/PPO Blends

The use of additive impact modifiers in the commodity styrenic resins is rather limited relative to the *in situ* modifier route wherein the modifier is generated during the mass polymerization of the matrix polymer; this has been well reviewed by Bucknall [4]. This is no doubt due to the long and successful development of the *in situ* polymerization technologies for production of both HIPS and ABS. The added cost of the compounding step required to incorporate a postpolymerization impact modifier is prohibitive in most of the cost-competitive commodity applications of styrenics. However, in the higher-value-added products such as ignition-resistant grades of HIPS, where a compounding step is required for the flame-retarding additives anyway, additive modifiers such as the Kraton SBS triblock polymers are employed to offset the reduction in impact strength caused by the addition of the flame-retardant package.

The Noryl family of impact-modified engineering resin blends produced by General Electric are based on blends of conventional mass-polymerized HIPS and poly(2,6-dimethyl-*p*-phenylene ether) (PPO) and, as such, can be considered an extension of the conventional technology. Shultz and Beach [35] claim that the butadiene–styrene block coploymers are effective at impact modification of these styrenic blends in a manner similar to their use in PS alone. Again, the use of additive modifier here is thought to be small in comparison to that from the one-step HIPS modifier process, primarily due to the higher cost of the block copolymer systems.

3. Poly(Vinyl Chloride)

Rigid PVC is the classic example of the use of an additive modifier to provide impact resistance. The subject has been extensively reviewed, and the interested reader is referred to several leading references on the subject by Ryan and Jalbert [36], Petrich [37], and Aarts et al. [38]. Blends to be used in outdoor applications where good weathering performance is essential, such as extruded siding and window profiles, are typically modified with acrylic core/shell rubbers, although there is also usage of chlorinated polyethylene and ethylene/vinyl acetate copolymers. Blends in which good clarity and optics are required, such as in calendered packaging film and blow-molded bottles, are typically modified with the MBS type of polymer to obtain a refractive index match with the PVC compound.

An interesting trend in weatherable PVC compounds, which is being observed particularly in Europe, amounts to a reversal of the trend seen in other polymers in evolving from *in situ* modifier generation to the use of additive impact modifiers. PVC blends, containing acrylic rubber onto which vinyl chloride has reputedly been grafted during the mass polymerization of the PVC, have been introduced by Chemische Werke Huls and Wacker Chemie. As seen earlier in the case of PP/PE blends, as well as in polystyrene, incorporation of the modifier during the polymerization of the matrix is primarily advantageous to the producers of the base resin, such as Huls and Wacker, in that it will increase the volumetric throughput in their PVC reactors.

The process can apparently be carried out in several ways. One potential route is to carry out a graft polymerization of a rubbery monomer, such as acrylic esters or butadiene, onto a substrate of PVC latex polymer, as claimed by Buna [39,40]. Alternatively, the vinyl chloride monomer can be grafted onto a rubbery spine polymer, such as pololefins, as claimed by Dynamit Nobel [41] and Buna [42], or acrylic esters, as claimed by Goodrich [43]. However, this approach denies the extruder of PVC profiles the ability to retain the complete flexibility offered by the use of additive impact modifiers in custom tailoring the price/performance balance of their formulations. In addition, there does not appear to be a significant advantage in impact efficiency to be offered by the vinyl graft approach in PVC versus the additive modifier approach; that is, the adhesion of high-rubber-content acrylic impact modifiers, such as Paraloid KM-334, is sufficiently high that direct chemical grafting of vinyl chloride onto the rubber does not pose a significant performance advantage.

4. Nylons

The family of crystalline 6 and 6/6 nylons constitutes the first examples of what are now commonly referred to as "engineering resins." It has been suggested by Fahnler and Merten [44] that the development of the engineering polyamide resins has occurred in essentially three stages. First, the nucleation technology developed in the 1950s allowed production of fully crystalline parts via the injection-molding process, opening up the market for melt-processible crystalline engineering resins. Many of today's engineering resins are based on semicrystalline polymer matrices, as this type of morphology yields a better balance of stiffness, strength, and dimensional stability than analogous amorphous thermoplastics. In the 1960s, the concept of glass–fiber reinforcement of polyamides was developed; the combination of high-modulus, high-aspect-ratio fillers and a semicrystalline nylon matrix resulted in stiffness-to-weight ratios that are very favorable compared to those of metals, and allowed penetration of these nylon blends into load-bearing applications formerly dominated by the use of metals. The industry is now divided into two broad segments: the use of Nylon 6 predominates in Western Europe, whereas the 6/6 polymer is dominant in North America.

However, it was not until the mid-1970s that the final stage of nylon blend development was realized. That step was the development of methods for the effective rubber modification of Nylon 6 and Nylon 6/6 to yield ductile impact behavior in notched impact tests while preserving the mechanical properties of the base resin. In general, the modified Nylon 6/6 blends display a better overall physical property balance relative to those derived from Nylon 6, owing to the

Table 3 Properties of Impact Modified Zytel Nylon 6/6

Property	Test temperature (°C)	Samples[a] Zytel 101 (unmodified)	Zytel 408	Zytel ST-801
Notched Izod	−40	0.6	1.3	3.0
impact strength (ft-lb/in.)	23	1.0	4.3	17
Tensile strength (psi)	23	12,000	8,800	7,500
Flexural modulus (psi)	23	410,000	285,000	245,000

[a] Sample conditioning: dry as molded.
Source: Ref. 46.

different crystallization behavior and higher melting temperature of the 6/6 structure. However, this also results in the general trend that the 6/6 nylons are more difficult to impact modify than 6 nylons, particularly at the low molecular weights necessary to produce high-flow, injection-moldable resins. Thus, it was a significant breakthrough when du Pont announced the development of the ''supertough'' Zytel ST-801 technology around 1975. As described by Flexman [45], the use of the appropriate rubber toughening technology results in blends that provide ductile failure even in the presence of razor-sharp notches. The properties of these ''supertough'' 6/6 nylons are compared to those of the corresponding base resins in Table 3. Toughened Nylon 6 blends having similar impact resistances have been developed by manufacturers such as Allied-Signal.

As can be seen in Table 3, a price must be paid in terms of the physical properties for the reduced notch sensitivity of these modified blends. Nylons derive their useful engineering properties in large part from their semicrystalline-phase structure. When high toughness is obtained by the addition of an amorphous rubbery phase, on the order of 25–30% by volume, the stiffness and strength of the material are substantially degraded as a result of the reduction in the crystalline content of the blend. Thus, in the case of nylons, as is with most impact-modified engineering resins, special emphasis is placed on developing efficient impact-modifier systems so that the volume fraction of modifier, hence degradation of physical properties, can be minimized. The processibility of the resin may also be adversely affected by the presence of the dispersed elastomeric phase. Highly modified resins may not fill thin-walled parts as readily during injection molding—again, a good reason to minimize the volume fraction of the additive impact modifier.

Because the matrix of nylon 6 and 6/6 is both polar and semicrystalline, it is difficult to disperse simple hydrocarbon elastomers in such a matrix and also obtain adhesion sufficient to promote effective impact modification. Consequently, the general approach to toughening of nylons is to employ elastomers that have been modified with the addition of small amounts of carboxylic acid functional groups. As mentioned earlier, there are some acid-modified PE polymers available on the market. Blends of this class of polymers with nylons have been studied, and the acid modification has been shown by Chang and Han to be effective in improving the dispersion of elastomer in the matrix and improving adhesion [47]. However, the preferred method for obtaining the maximum impact efficiency appears to be the use of polyolefins that have been modified by the addition of a few percent of a dicarboxylic acid or anhydride, preferable maleic anhydride, as claimed by Epstein [48].

Because the impact modification of semicrystalline polymers such as Nylon 6 and Nylon 6/6 via acid-modified polyolefin represents quite a departure in chemistry from the classical grafted vinyl polymers such as HIPS and ABS, the impact reinforcement mechanism of toughened nylon has received a good deal of attention. The results of extensive studies, most notably by Wu of du Pont, have shown that the behavior of toughened nylons is quite similar to that of the rubber-toughened vinyls, although the chemical routes to providing the modified systems are certainly different.

The predominant mode of matrix deformation appears to be matrix shear yielding, with about 25% of the total being ascribed to a crazing or voiding mechanism; see the work of Wu [20], Ramsteiner and Heckmann [49], and Hobbs et al. [50]. This is very reminiscent of the failure mode of ABS, although the higher energy required for shear yielding of a semicrystalline matrix results in notched Izod impact energies of 20–25 ft-lb/in. for the toughened polyamides versus only 8–10 ft-lb/in. for ABS. There is a very strong dependence of impact efficiency on the particle size and rubber volume fraction: The best balance of properties appears to occur at a rubber volume loading of 25%, with the particles being dispersed as domains of a submicrometer particle size. This dependence has been interpreted by Hobbs as arising from the need to achieve a critical stress–field overlap between the rubber inclusions [50], which then induces the massive matrix yielding observed for the cases in which high-impact strength is achieved. Because the number of modifier particles varies geometrically with the particle diameter at a constant total volume fraction of the modifier, the most effective modifier systems are those in which the modifier particle size is at the minimum required for effective interaction with the matrix, as this provides the maximum *number* of particles per unit volume. In the case of Nylon 6/6, it appears that high-impact efficiency can be achieved when size is reduced to about 0.3 μm. However, it may also be feasible to use a modifier system of mixed particle sizes to maximize the synergism between the elastomer and the polyamide. Electron photomicrographs obtained by Fahnler and Merten [44] of a commercial Nylon 6 system, Durethan BC402, suggest that modifiers of two distinct particle sizes, approximately 0.5 and 0.1 μm, are being used. It is not apparent from the literature whether the critical particle size constraints for Nylon 6 are the same as those for the more heavily studied Nylon 6/6 systems.

Much attention has been focused on the need to employ reactive acid functionality to generate effective impact-modifier systems for polyamides. In the case of the maleate functionalized polyolefin systems, the formation of true nylon/olefin graft polymers via reaction of nylon amino end groups with the maleic anhydride to form a covalent imide bond has been postulated to occur by Chang and Han [47b] and by Hobbs et al. [50] In the application of core/shell impact-modifier technology to toughening of polyamides, the use of carboxylic acid functionality has similarly been claimed to be effective at promoting high-impact efficiency by Baer [51].

However, recent work by Wu [26] calls into question the degree of adhesion required for the purposes of impact modification of polyamides. When comparing melt blends of Nylon 6/6 with polyolefins, with or without acid modification, Wu finds that the acid modification may be more important to stabilization of the small particle size required for toughening of nylon. He argues that because nylon is an example of a type II polymer that fails by matrix yielding, the adhesion between modifier and matrix is less critical than achieving the appropriate delocalization of elastomer particles in the matrix, to promote this yielding. This is in agreement with the work of Hobbs in Nylon 6/6, in which he showed [50] that there was a definite (low) minimum to the amount of adhesion required for impact modification of nylon and that the stress–field overlap effect was dominant. Although a certain amount of adhesion is certainly required to prevent matrix debonding during impact, Wu suggests that this surface energy lies closer to that of simple van der Waals attractions at the modifier–matrix interface, not as high as the covalent type of attachment previously postulated.

If the primary requirement for toughening in type II polymers (PVC, nylons, PC) is the dispersion of small particle size rubbery inclusions, this suggests that the emulsion-prepared core/shell polymers should be particularly effective at impact modification of this class of polymers. Because the appropriate morphology and particle size can be set by cross-linking during the synthesis of the modifier, it should be possible to achieve high impact simply by ensuring proper extrusion compounding of the core/shell modifier to achieve dispersion back to the primary particles, and good "wet-out" of the particles by the matrix polymer to achieve full van der Waals contact. Although there has apparently been comparatively little activity with this class of modifier in the nylons, they have been shown to be very effective in the other type II polymers, as will be described in Section IV.B.5, below.

5. Thermoplastic Polyesters

Although thermoplastic polyesters such as polyethylene terephthalate (PET) have been known for some time from the fibers trade, it was not until Celanese introduced the Celanex polybutylene terephthalate (PBT) resin family that injection-moldable polyesters became a factor in the engineering resin marketplace. Although higher in cost, PBT resins display a much more rapid crystallization chemistry relative to that of PET, which, in the absence of additives, is too slow to crystallize to allow practical molding cycles. As in the case of the nylons, the crystallizable thermoplastic polyesters were initially available only in glass and mineral-reinforced grades for engineering applications; now, rubber-toughened blends with greatly reduced notch sensitivity are available.

Paraloid core/shell impact modifiers can be effectively used for impact modification of thermoplastic polyesters, as described by Farnham and Goldman [52]. The impact efficiency of the modifiers is strongly influenced by the molecular weight of the polyester (Fig. 9), which can be estimated from the intrinsic viscosity (IV). This is the same effect as was seen for PVC and PC, and it further illustrates the general nature of the effect of matrix molecular weight on impact modification.

In addition to modification of PBT, the core/shell modifiers are also effective at impact modification of PET and copolyesters derived from it (Fig. 10).

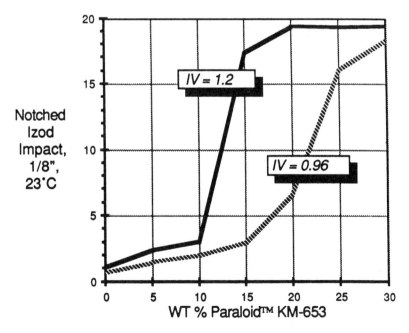

Figure 9   Impact modification of thermoplastic polybutylene terephthalate with core/shell polymers.

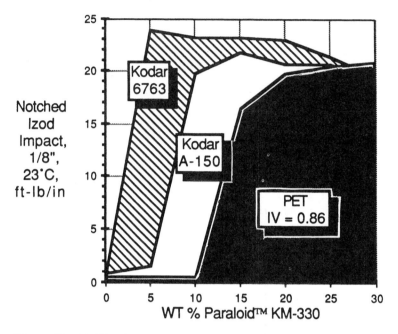

Figure 10   Modification of PET and related copolyesters, with core/shell polymers.

As noted earlier, the molecular weight of the polyesters, most conveniently approximated by IV, is crucial to determining the loading of impact modifier required for ductile notched impact behavior. In addition, substitution of comonomers such as isophthalate for terephthalate in the PET and alternative glycols for ethylene glycol results in a more easily modified matrix polymer, perhaps owing to the enhanced amorphous character of injection-molded test specimens prepared from the copolyesters.

There are currently several families of impact-modified polyester blends on the market. The Valox resins manufactured by General Electric are based primarily on PBT; recently, Celanese announced the introduction of their Duraloy 2000 impact-modified PBT blends. In addition, Bayer is now importing through Mobay their Pocan impact-modified PBT blends: Although these materials span a variety of blend compositions, some of them have been described by Binsack et al. [53] as being based on an ABS-modified PBT.

Advances in the nucleation chemistry of PET have allowed du Pont to commercialize blends based on PET that can be injection molded in cycles competitive with those of PBT. This Rynite family of resins had previously been available only in the traditional glass- and mineral-reinforced grades; however, the Rynite SST grade was recently announced, which heralds the commercial introduction of rubber-toughening technology to this family of filled polyester resins.

6. Acetals

The prototypical acetal polymer is polyoxymethylene (POM). Two general classes of acetal polymers are in commercial production: the Delrin resins of du Pont, which are essentially straight POM polymers, and the Celcon resins manufactured by Celanese, which are copolymers of POM and ethylene oxide; similar copolymers are manufactured by Hoechst. The copolymers offer greater thermal stability, but at the price of reduced physical properties owing to reduced crystallinity. The valued chemical resistance of acetal, as well as its stiffness with glass or mineral reinforcement, are due to the very high degree of crystallinity of this polymer relative to others such as the nylons or polyesters. Acetal enjoys considerable use as a metals substitute in automotive applications, particularly those requiring excellent gasoline resistance.

Unfortunately, this high degree of crystallinity and chemical inertness make acetal a difficult polymer to effectively impact modify. The current approaches favored by acetal manufactures revolve around the use of thermoplastic polyurethanes (TPUs) as the impact modifier. The use of TPUs as impact modifiers has recently been reviewed by Bonk et al. [54]. Within the past year, two "supertough" acetal grades have been introduced. First was the Delrin 100ST, which claims an notched impact strength of essentially "no break" at $\frac{1}{8}$ in. thickness at room temperature. This system is reputedly modified with Texin TPU from Mobay. Recently, Celanese announced the Duraloy 1000 family of supertough acetals, which claim to have "no-break" notched impact resistance at room temperature, as described by Dolce [55]. Because this system is based on a melt-extruded mixture of acetal and (apparently) non-crosslinked TPU, the domain structure and mechanical properties of the blend are very dependent on the processing conditions. The domain size of the "modifier" in POM/polyurethane systems can be manipulated by controlling the melt viscosity of the POM continuous phase, as described by Kloos and Wolters [56]. The best impact appears to be obtained when the modifier domains are small and finely dispersed, which is consistent with the findings of Kloos [57] that shear banding plays a dominant role in toughening of POM. The Delrin 100ST material is reputed to suffer from a high degree of sensitivity to processing conditions and is based on a TPU modifier system similar to that of Duraloy.

The levels of elastomer required to achieve this toughness cause the modulus and the tensile strength of the modified blends to fall drastically below the levels of the base polymers: These "supertough" acetals are more accurately described as chemically resistant flexible thermoplastics rather than rigid engi-

Table 4   Properties of Duraloy 1000 Series of Impact-Modified Acetal

Property	Test temperature (°F)	Celon M25-04 (unmodified)	Duraloy 1000	Duraloy 1200	Duraloy 1100
Notched Izod	−20	1.0	2–4	1.0	No break
impact strength	73	1.5	No break	No break	No break
(ft-lb/in.)					
Dynat up total	−20	—	11	—	33
energy (ft-lb)	73	4	33	—	27
Tensile					
strength (psi)	73	8,800	5,400	6,300	5,200
Tensile					
modulus (psi)	73	410,000	170,000	90,000	95,000

Source: Ref. 55.

neering plastics (Table 4). Indeed, the most likely immediate application for the Duraloy 1000 system is in extruded, chemically resistant flexible tubing.

7. Polycarbonate

Although PC is considered a tough polymer when compared to other unmodified systems such as nylon or thermoplastic polyesters, it is still somewhat notch sensitive, particularly in thicker sections and at low temperatures. Consequently, low levels of additive modifiers are added to provide an extra measure of toughness for demanding applications. As was shown in Fig. 3, the Paraloid core/ shell modifiers are effective in reducing the sensitivity of PC impact strength to specimen geometry during notched impact testing.

In fact, a wide variety of elastomeric additives has been claimed to be effective as modifiers for PC, including the core/shell polymers (both MBS and acrylic), polyolefins of all varieties, olefin/acrylic rubbers, nitrile and ABS rubbers, and styrenic block copolymers, as claimed by Liu et al. [58]. The apparent effectiveness of such a wide variety of elastomers in modification of PC lends credence to the hypothesis of Wu described earlier stating that specific polymer–modifier interactions are less important to type II polymers such as PC. The choice of the appropriate modifier then revolves around more practical considerations:

How easily can the modifier be dispersed in the matrix?
How low is the T_g of the elastomer with regard to potential low-temperature impact applications?

What is the thermal and oxidative stability of the modifier to the conditions
of processing and the end-use application?
Last but not least, what is the cost of an effective loading of modifier?

It would appear that the manufacturers of impact-modified PC blends have
a wide latitude of formulation options within which to optimize the blend properties for specific applications.

8. Impact-Modified Engineering Resin Blends

The area of toughened plastics receiving the most attention at present is impact
modification of polymer blends; the resulting materials are sometimes referred
to as "alloys." With the wide variety of matrix polymers already on the market,
it is becoming advantageous to generate new molding compounds by melt blending two or more polymers together, rather than seeking a chemical solution to
the problem such as copolymerization or the formation of block and/or graft
copolymers. The use of polymer blends allows the rapid targeting of specific
applications with relative ease, provided the initial outlay in terms of compounding equipment has been made and a thorough understanding of the melt-blending
behavior of the polymers in question has been developed.

The subject of formation of polymer blends or alloys has been comprehensively reviewed in the literature; a thorough discussion of the fundamentals of
blend behavior lies outside the scope of this work. An excellent, relatively short
review on the subject has been put forth by Paul and Barlow [59]. More detailed
treatments of the subject are available as well by Paul and Newman [12], Manson
and Sperling [13], and Klempner and Frisch [60]. Although the concept of creating
new commercial polymer blend products by simply combining currently available
homopolymers is appealingly simple on the surface, new problems and complexities tend to offset the benefits of polymer alloying. Control of the various phase
morphologies during their initial melt blending and subsequent melt processing
and fabrication can present a formidable task; the weld line strength of polymer
blends can be seriously diminished where flow fronts meet in the injection-molding tool. It is true, though, that progress is being made in understanding how
polymers adhere to one another, so that problems of interfacial delamination can
perhaps be minimized. The subject of characterizing and modifying the polymer–polymer interface has recently been comprehensively reviewed by Wu [61].

Even after prospective manufacturers of polymer blends have successfully
dealt with the foregoing considerations, they are confronted by the same problem
that has occupied us in the preceding work: the need to provide for sufficient
impact strength. The development of a thorough understanding of an impact-
reinforcement mechanism in alloys is made a good deal more complicated by
the need to extend our theories to systems in which the additive polymer is but

one of a number of dispersed polymer phases. The fracture mechanics of rubber-toughened polymer alloys is a subject still in its infancy because toughened polymer alloys themselves are only now becoming generally available on a commercial scale. The next few years should provide some exciting new insights into the mechanisms of rubber toughening as our theories are tested by extrapolation to these new systems.

To review the state of the art with regard to toughened alloys, we will use as a working definition of "toughened alloy," a system in which there are at least two phase-separated polymeric domains *excluding* the rubbery modifier. There is a tendency for some to label as "alloys" such resin–elastomer blends as the acetal/TPU or EPDM/PP systems, as described in a review by Jalbert [62]. We believe those systems are more accurately described as simple impact modifier/resin blends. Most of the Noryl family of resins can thus be excluded from our definition, as the PS/PPO blends are completely miscible, and hence present only one phase for impact modification. However, the recently introduced alloy of PPO and nylon, Noryl GTX, appears to be the first member of the Noryl family that can truly be considered to be an impact-modified multiphase polymer blend.

The presence of more than one phase presents an interesting morphological problem to those seeking to modify such a blend with an additive modifier. In the previous discussions, we have been mainly concerned with dispersing a rubbery modifier polymer uniformly and to a discrete particle size within a single, homogeneous matrix. Let us assume that this same modifier is now to be melt blended into a matrix that itself is already made up of two phase-separated polymers. The possible morphologies that might result from this operation are shown schematically in Fig. 11.

In configuration A of Fig. 11, the additive modifier tends to associate, for thermodynamic or kinetic reasons, with the polymer that comprises the continuous matrix phase. Alternatively, the polymer compatibilities may be such that the additive polymer has a strong association with the polymer forming the discontinuous phase and is essentially encapsulated within these domains, as depicted in configuration B. Finally, the polymer interactions may be such that the additive polymer will tend to bridge the boundary of the two phases, as in configuration C. The mode of interaction between the additive modifier and the blend matrix in terms of impact reinforcement is likely to be very different in cases A and B, and it may be difficult to predict in case C.

Consequently, interpretation of the fracture mechanics of these ternary polymer blends on molecular terms cannot be undertaken without knowledge of the distribution of the impact modifier between the matrix phases. Because these phase distributions are often strongly affected by the melt processing conditions used to prepare the samples, the blend morphology seldom can be described by

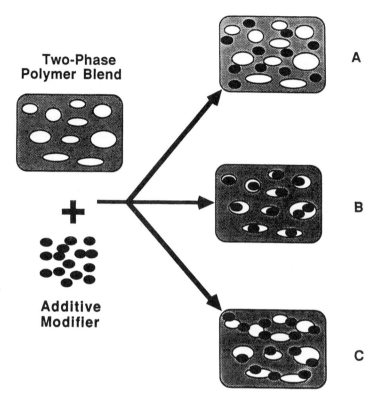

Figure 11   Possible morphologies for impact modification of multiphase polymer blends or alloys.

a singular configuration. The investigator is instead faced with a *range* of morphologies, which must be characterized and may often differ between different sections of the same test specimen because of orientation imposed during the fabrication step.

By and large, the commercially introduced impact-modified alloys to date are dominated by toughened alloys of PC. Not surprisingly, the first examples of this class of modified blends, ABS/PC, were based on an alloying component in which the rubbery modifier was chemically grafted to one of the phases, namely the styrenic phase, thus simplifying the maintenance of the correct blend morphology. These materials were introduced by Borg–Warner as the Cycoloy family of resins in 1964 [63]; they are now competing with similar blends introduced by Mobay (Bayblend). The polar acrylonitrile comonomer is thought to confer some compatibility between the styrenic phase and the PC when present in appropriate amounts, as shown by Paul et al. [64].

Even so, these blends have not enjoyed the amount of growth predicted for them at their introduction because of problems with controlling the blend morphology, hence the properties, during processing. Even with the use of a rubbery modifier grafted to one of the alloy matrix components, the processing of a multicomponent allow is significantly more complicated than the processing of a typical rubber-modified plastic. As a result, manufacturers of ABS/PC blends such as Mobay are still pursuing optimization of their compositions with regard to processing stability and consistency [65]. The shear sensitivity and poor color stability of the ABS/PC blends have served to largely offset the modest price advantage for these blends over PC alone. The advent of the impact-modified PC has also served to address some of the drawbacks of unmodified PC, such as thickness sensitivity, on which the ABS/PC blends had hoped to capitalize.

Despite this early experience with the added complexities of producing impact-modified polymer blends, economic considerations have encouraged the major resin manufacturers to press forward in developing even more complicated impact-modified alloys. The development costs associated with bringing to commercial production a polymeric system based on new monomer(s) are very high. If the technology of blending can be sufficiently developed and understood, the ability to tailor resin properties exactly to individual specifications via blending allows a rapid response to customers' changing needs. In addition, a resin supplier can provide a stable of blended products, covering a wide range of property profiles, by the clever manipulation of a relatively small number of polymer blend components; the commercial attractiveness of such a blending strategy has been reviewed [66].

The most versatile method of preparing impact-modified polymer blends or alloys is one in which the additive impact modifier can be added as a phase separate from those comprising the matrix polymers. Thus, instead of an ABS/PC impact-modified blend, one could prepare a blend of SAN, PC, and a diene rubber. As we shall see below, the impact-modified alloy field appears to be evolving toward this sort of arrangement.

One particular family of impact-modified engineering resin blends is generating a good deal of interest, from both commercial and technological points of view. This is the Xenoy family of engineering resin blends being marketed by General Electric. The basic technology is the blending of (typically) crystallizable thermoplastic polyesters, such as PET or PBT, with amorphous PC. These blends well illustrate the synergism that is sought when preparing polymer alloys. The crystalline polyester phase provides the resistance to chemical attack and creep, as well as the stiffness required for engineering applications, whereas the PC phase acts to help toughen the matrix polymer. However, the final ingredient is the use of a suitable elastomeric impact modifier to provide toughness at low temperatures and at high-impact speeds; in fact, sufficient toughness has been obtained that Xenoy is finding acceptance as a replacement for metals in automotive bumper applications and other exterior body parts as described by Bertolucci and DeLaney [67]. The Ford Motor Company has been most active in introducing this technology on a commercial scale—first in Europe on the Sierra, and now in the United States on Ford's (recently introduced) Taurus/Sable line of family cars. The use of tough thermoplastics as a bumper material gives the consumer the benefit of bumpers able to withstand 5-mph collisions without visible damage. The ability to fashion complex curvilinear surfaces from an injection-moldable thermoplastic allows Ford to pursue the aerodynamic styling that has come to typify their entire product line.

As might be expected, the Xenoy blends present a complicated morphology in terms of understanding the structure–function relationships of impact resistance. Because the PC and PBT are not cross-linked, their phase structures depend heavily on their respective melt viscosities and their sensitivity to shear, as well as the relative volume fractions of the two phases. The proper control of the PC/PBT blend morphology has been shown by Bartosiewicz and Kelly [68] and by Birley and Chen [69] to be critical to obtaining the desired toughness and chemical resistance from the blend.

The need to properly disperse an impact modifier adds another dimension to the problem of morphology control in modified alloys. Suppose that an non-crosslinked elastomer is to be used, as in the case of the toughened acetal polymers described earlier, to modify a polymer blend instead. How are we going to be able to adjust the shear rate and residence time in the processing equipment so that the proper modifier morphology *and* the proper matrix morphology can be simultaneously obtained? The simplified morphological scheme of Fig. 11 now becomes the superposition of a *range* of impact-modifier particle sizes and distribution on top of the distribution of the matrix polymer phases.

For these reasons, the use of crosslinked, core/shell elastomers should offer a distinct advantage over the non-crosslinked elastomers and block copolymers when it comes to preparation of impact-modified, multiphase polymer alloys. The ability to fix the particle size and chemical compatibility of the core/shell elastomer independent of the matrix polymer allows the resin compounder to concentrate on generating the proper phase distribution of the non-crosslinked matrix polymers. Thus, although there are some claims made by Dieck [70,71] for the effectiveness of thermoplastic elastomers, such as the SBS, as modifiers for PC/polyester blends, the use of crosslinked elastomers such as the core/shell variety predominates in the patent literature for impact-modified PC/polyester blends. Modifiers based on acrylic rubber have been claimed to be effective as

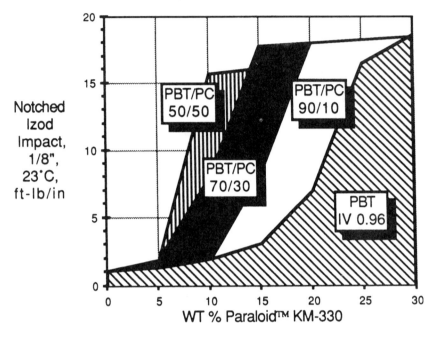

Figure 12 Impact modification of PC/PBT blends with core/shell polymers.

modifiers for PC/polyester blends by Cohen and Dieck [72], Fromuth and Shell [73], and Stix et al. [74], the utility of diene rubbers in this application have been claimed as well by Dieck [75] Agarwal [76], Fromuth and Shell [77], and Nakamura et al. [78]. An alternative and somewhat related technology is the use of ABS-type terpolymers as the impact modifiers for these polyester blends as claimed by Binsack et al. [79].

The effective and synergistic combination of PC/polyester blends and core/ shell impact-modifier technology is illustrated in Fig. 12 for the system PC/PBT/ Paraloid KM-330. The simple addition of PC to the PBT, even at loadings of 50%, does not appreciably improve the notched Izod impact strength of the blend, which remains at 1–2 ft-lb/in. (although the toughness to a GVHIT-type impact is improved). However, in the presence of the elastomeric core/shell impact modi-fier, the PC is seen to act effectively as a secondary ''impact modifier'' for the blend. As increasing amounts of PC are added to the blend, the loading of core/ shell impact modifier required to achieve ductile Izod impact behavior can be dramatically reduced. This allows a wide range of blended products to be pro-duced, all with excellent toughness, but with differing cost/mechanical property balances depending on the relative content of PC and impact modifier. The substitution of an MBS type of impact modifier for the all-acrylic resin would likely allow enhanced low-temperature impact performance to be realized.

We might expect that the use of core/shell impact modification in PC blends will increase as the demand grows for blends having highly tailored balances among impact, mechanical properties, and cost. For example, a related technology practiced by Arco Polymers is the blending of their Dylark impact-modified styrene/maleic anhydride resins with PC to produce the Arloy family of resins, a lower-cost alternative to ABS/PC blends. A recent patent to Bourland [80] claims that the usage of MBS-type core/shell impact modifiers in Dylark/PC blends offers advantage over reliance on the bulk polymerized diene rubber of the Dylark phase alone.

We also would expect that manufacturers will find ways to employ polymers other than PC with which to fashion impact-modified, high-performance polymer alloys. The combination of an amorphous and a crystalline resin as the two primary matrix components, as in the case of Xenoy, is especially attractive in that the good chemical resistance and stiffness of the crystalline component can be balanced against the reduced shrinkage and warpage of the amorphous phase. A recently introduced continuation of this trend is the Noryl GTX resin blends of General Electric described by Chambers [81]. In this family of resins, crystalline Nylon 6/6 forms the continuous phase, whereas amorphous PPO is ''compatibilized'' (by a proprietary process) and dispersed as the occluded phase. The high heat resistance of PPO and the chemical resistance of nylon are thus claimed to be advantageously combined. It is not clear at this time whether any impact modifier is employed to toughen these blends. However, given the characteristics of the two matrix materials, it would be surprising if some sort of rubber toughening to reduce the notch sensitivity of the blend were not found to be necessary.

V. EFFECT OF ADDITIVE IMPACT MODIFIERS ON BLEND PROPERTIES OTHER THAN TOUGHNESS

The preceding discussion has centered primarily on the attributes of additive impact modifiers that exert their primary effect on the thermoplastic matrix polymer, namely impact resistance. However, numerous other factors must be considered when setting out to design an impact-modified blend. There can be a tendency to focus primarily on the type and level of impact modifier required to achieve maximum toughness while disregarding the effect of such an additive modifier on the other properties of the matrix polymer. In concluding, we present some generalizations on the effect of impact modifiers on base resin properties other

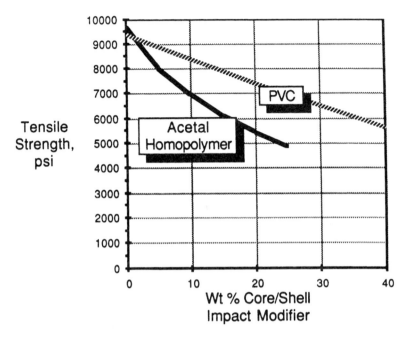

Figure 13   Effect of additive impact modifier on tensile strength.

than toughness. It should be stressed that most of these considerations apply to the use of rubber-toughening agents *regardless* of their method of incorporation, whether by postpolymerization extrusion blending or by *in situ* graft polymerization.

Because rubbery tougheners by definition must be above their glass transition temperature if they are to be effective in service, they typically have mechanical properties much inferior to those of the matrix polymer. Consequently, the most noticeable effect of an impact modifier is on the mechanical properties of the polymer being toughened. The reduction in key physical properties such as tensile strength tends to follow the volume fraction of dispersed rubbery phase in the matrix, which is conveniently approximated by the weight fraction (Fig. 13). Here, because the two polymers we are comparing, PVC and acetal homopolymer, have similar (high) densities, data calculated according to weight can be compared. If we wished to compare these plots to data for blends prepared from lower density resins such as polystyrene, correction of the data to the volume fraction would be necessary.

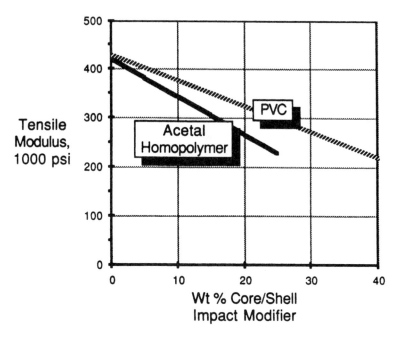

Figure 14 Effect of additive impact modifier on tensile modulus.

In the case of amorphous matrix polymers such as PVC, a simple rule-of-mixtures plot typically results, wherein the tensile strength of the blend lies along a line drawn between the tensile strength of the matrix polymer and that of the impact modifier. In the case of matrix polymers that rely on crystalline domains for their mechanical strength, such as the highly crystalline acetal homopolymer, we find that the tensile strength appears to fall off more rapidly. This is probably due to the reduction in overall crystalline content of the blends resulting from the addition of the amorphous modifier phase. The data for toughened nylon in Table 3 shows that a substantial loss of tensile strength accompanies the addition of the impact modifier used to produce the ''supertough'' Zytel ST-801. A similar effect is obtained when the tensile modulus is measured (Fig. 14). Here, the discrepancy between the data for PVC and POM is not as large, although the modulus of the POM blends still appears to fall off more rapidly as the modifier content is increased. Again, these general trends are seen no matter how the impact modifier phase is incorporated.

The gain in toughness obtained by the use of impact modification must therefore offset the reduction in mechanical properties. The need for the designer to compensate for the reduction in physical properties by increasing slightly the cross section of parts prepared from toughened blends is usually far outweighed by the safety of knowing that the part will be able to withstand loadings at impact speed. Improvements in the ductility of the material also usually result in savings due to less breakage during fabrication.

The addition of an impact modifier does not always result in a detrimental effect on physical properties. In addition to improving the toughness of the material in the traditional sense, additive modifiers can improve the "chemical toughness" of plastics as well. Some polymers are quite susceptible to stress cracking when a load is applied in the presence of hydrocarbon solvents, this behavior limits their application in environments in which lubricating greases or food oils are encountered. For example, the addition of the appropriate impact modifier can greatly improve the chemical stress cracking resistance of plastics, as Russell [82] has shown in the case of polystyrene and as demonstrated by Liu [83] in the case of PC.

Additive impact modifiers, particularly those that are covalently cross-linked, generally tend to increase the melt viscosity of the blend over that of the base polymer. This effect must be considered if the intended application requires fabrication by injection molding. The incorporation of lubricants or other processing aids, as described elsewhere in this volume (see Chapter 9), is often desirable if high loadings of impact modifier are required. On the other hand, the added melt strength afforded by the additive impact modifier may be beneficial in blow molding or thermoforming operations.

The particle size and structure of the impact modifier may affect the surface appearance of the blend. In the case of HIPS, the conventional bulk polymerization process yields an impact-modifier particle size of 2–5 μm, which is large enough to cause a flatting effect on the surface of the fabricated part. Relaxation of rubber particles at the surface causes the gloss of the material to be reduced. This gloss is variable because it is affected by the molding pressure and mold temperature used during fabrication. If a smaller modifier particle size is employed (≤0.5 μm), the rubber particles on the surface of the molded part are too small to effectively interact with light and better gloss is obtained.

Some considerations apply in particular to the use of the additive-impact-modifier approach. Because the general method for postpolymerization incorporation of additives is melt blending, provision must be made for the need to achieve complete dispersion of the phase-separated impact modifier domains in the matrix polymer. This may entail the use of sophisticated extrusion compounding equipment. The ease of dispersing the additive modifier in the matrix of choice can be influenced to some extent by the structure of the modifier. The modifier must also be designed so that it can withstand the degree of shearing required to obtain adequate dispersion in the matrix polymer.

Because a melt processing step is required for the incorporation of an additive impact modifier, care must be taken to assure that the additive polymer is stable enough to withstand the processing conditions. It is often necessary to incorporate stabilizers against thermal or free-radical decomposition pathways, or against oxidative decomposition, as in the case of modifiers based on diene rubbers. These stabilizers may be incorporated into the modifier itself before blending, or they may be added during the melt-blending step. If outdoor weathering resistance is required, it may be necessary to add UV stabilizers. Again, it should be noted that many of these concerns also apply to modified blends that arise from *in situ* grafting reactions. If unexpectedly poor results are obtained when screening an additive impact modifier in matrices such as high-temperature engineering resins, care should be taken to demonstrate that the additive polymer has not suffered decomposition during processing before concluding that the fundamental modifier chemistry is not suitable for effective impact modification. Substantial progress is being made in terms of stabilizing elastomers for use as impact modifiers in high-temperature engineering resins and in aggressive environments. The fundamentals of additive stabilizers for thermoplastics is discussed elsewhere in this book (see Chapter 7).

For these reasons, the designers of impact-modified blends must always seek to develop systems that use the additive modifier most efficiently. By using the minimum level of additive modifier sufficient to provide controlled failure over the range of impact conditions likely to be encountered, the best balance of impact strength and mechanical properties is likely to be attained. Hopefully, the principles and examples reviewed in this work will serve as a useful starting point for the successful design of toughened polymers via the use of impact-modifying additive polymers.

VI. APPENDIX

Table A1   Impact-Modifier Product Designations/Recommended Applications

Type	Supplier	Trade names/grades
ABS	General Electric	Blendex 101, 131, 201, 310, 336, 337, 338, 340, 405, 424, 467 (1,2)[a]
Acrylics, mod-acrylics	Elf Atochem	Durastrength 200, D638 (3,2); 300 (2); mod-acr.
	General Electric	Blendex 975 AS polymer (3)
	Kaneka Texas	Kane Ace FM 10, 20, 25 (3,2); FT 80 (2) mod-acr.
	Rohm and Haas	Paraloid KM 334, 346 (3,2); KM 388 (3,2,4); KM 350 (3); KM 346 (3,2); KM 390, 940, (3,4); KM 318F (6); EXL 3330,377, 3361 (5).
CPE	DuPont–Dow	Tyrin 3615, 2614A (3,2).
EVA	DuPont	Elvaloy 838G,K-EP 4924 (3,2)
	Mobay	Baymod 450 (2)
MBS	Elf Atochem	Metablen C 201, 202, 301, 301P, 320 (1,2); 223 (2)
	Kaneka Texas	Kane Ace B 31, 51, 52, 56, 58A, 382, 513, 522 (1); M511,521 (2)
	Rohm and Haas	Paraloid BTA 702, 715, 730 (1); 733 (1,2); 751 (2); 753 (2); EXL 3691 (5)
Nitrile	Zeon Chemicals	Nipol 1422; DP 5123P, 5125P, 5128P. (2) (5).
Inorganic Filler-Impact Modifiers		
Calcium carbonate	Specialty Minerals	Ultrapflex 200 (2,3)
(fine particle size, coated)	OMYA	Omyacarb UFT (2,3)
Alumina trihydrate	Solem Industries	Micral 800, 900 series (2,3)
(fine p.s., coated)	ALCOA	Lubral, Hyral series (2,3).

[a] Recommended primary use: (1) clear PVC; (2) opaque, general-purpose PVC; (3) weatherable, opaque PVC; (4) dual purpose (e.g. impact modifier plus processing aid); (5) use in engineering polymers and blends; (6) opaque, cellular PVC.

Table A2 Impact-Modifier Suppliers Reference

Company/address	Telephone
Aluminum Company of America 151 ALCOA Bldg. Pittsburgh, PA 15219	412-553-2354
Amoco Chemical Co. 200 E. Randolph Drive Chicago, IL 60601-7125	1-800-621-4567
Dow Chemical Co. 2020 Dow Center Midland, MI 48674	517-636-5411
DuPont–Dow Co. 1007 Market St. Wilmington, DE 19898	302-774-1000
Elf Atochem 3 Parkway Philadelphia, PA 19102	215-587-7118
Firestone Synthetic Rubber and Latex Co. P.O. Box 2611 Akron, OH 44319-0006	1-800-282-0222
Kaneka Texas Corp 17 S. Briar Hollow Lane Suite 307 Houston TX	713-840-1751
Mobay Corp. Mobay Rd. Pittsburgh, PA 15205-9741	412-777-2000
Omya, Inc. 61 Main St. Proctor, VT 05765	1-800-451-4468
Phillips Chemical Co. 101 ARB Plastics Technical Center Bartlesville, OK 74004	1-800-537-3746
Rohm and Haas Company Independence Mall West Philadelphia, PA 19105	215-592-3000
Salem Industries 5824-D Peachtree Corners E. Norcross, GA 30092	404-441-1301

Table A2 Continued

Company/address	Telephone
Shell Chemical Company 910 Louisiana Houston, TX 77002	800-874-3532
Speciality Minerals Industries, Inc. 640 N. 13th Street Easton, PA 18042	215-250-3307
Uniroyal TPE Benson Road Middlebury, CT 06749	1-800-322-3243
Zeon Chemicals, Inc. 4111 Bells Lane Louisville, KY 40211	1-800-735-3388

ACKNOWLEDGMENTS

Data on the performance of impact modifiers in PVC were provided by Dr. Richard Clikeman of the Rohm and Haas Company. Test speciments were prepared by milling powder blends, then compression molding the fluxed material into $\frac{1}{8}$-in. sheets.

Data on the performance of impact modifiers in polyesters, polycarbonate, and polyacetal were provided by Dr. Evan Crook of Rohm and Haas. The blends were melt compounded via single-screw extrusion, and ASTM test specimens were prepared via injection molding.

Data on the performance of impact modifiers in HIPS were generated by the author. Blends were melt compounded via twin-screw extrusion, and ASTM test specimens were prepared via injection molding. The instrumented impact tests were conducted in conjunction with Dr. Steven Havriliak of Rohm and Haas, using a Dynatup instrumented drop tower manufactured by Effects Technology, Inc.

The author acknowledges numerous other co-workers at Rohm and Haas for their helpful comments and insight on the subject of impact modification, particularly Richard Weese. The author also thanks Rohm and Haas for permission to publish this work.

REFERENCES

1. JL Amos, OR McIntire, JL McCurdy. U.S. Patent 2,694,692 (1948) (to Dow).
2. S. Havriliak. Rohm and Haas Co., personal communication.

3. JD Ferry. Viscolelastic Properties of Polymers, 2nd ed., New York: Wiley, 1970.
4. CB Bucknall. Toughened Plastics. London: Applied Sciences Publishers, 1977.
5. PL Fernando, JG Williams. Polym Eng Sci 20:215–220, 1983.
6. WJ Moritz. Mod Plast 52:60, 1975.
7. AJ Winuk, TC Ward, JE McGrath. Polym Eng Sci. 21:313, 1981.
8. KC Dao. J Elastomers Plast 15:227, 1983.
9. S Turner. Pure Appl Chem. 52:2739, 1980.
10. CB Bucknall. J Elastomers Plast 14:204, 1982.
11. H Kekkula, ed. Toughened Plastics, Vol. 2. Nashua, NH: MMI Press, 1982.
12. DR Paul, S Newman, eds. Polymer Blends, Vols. 1 and 2. New York: Academic Press, 1978.
13. JA Manson, LH Sperling, eds. Polymer Blends and Composites. New York: Plenum Press, 1976.
14. RD Deanin, AM Crugnola, eds. Toughness and Brittleness of Plastics, ACS Advances in Chemistry Series No. 154. Washington, DC: American Chemical Society, 1976.
15. J Heijboer. J Sci. 16:3755, 1968.
16. RF Boyer. Polymer 17:996, 1976.
17. B Hartmann, GF Lee. J Appl Polym Sci. 23:3639, 1979.
18. J Schaefer, EO Stejskal, R Buchdahl. Macromolecules 10:384, 1977.
19. TR Steger, J Schaefer, EO Stejskal, RA McKay. Macromolecules 13:1127, 1980.
20. S Wu. J Polym Sci 21:699, 1983.
21. CB Bucknall, WW Stevens. J Mater Sci 15:2950, 1980.
22. MA Maxwell, AF Yee. Polym Eng Sci 21:205, 1981.
23. CB Bucknall, D Clayton, WE Keast. J Mater Sci 7:1443, 1972.
24. S Newman. Polym Plast Technol Eng 2:67, 1974.
25. ER Wagner, LM Robeson. Rubber Chem Technol 43:1129, 1970.
26. S Wu. Polymer 26:1855, 1985.
27. Monsanto. Belgian Patent 886,223 (1979).
28. DJ Meier, ed., Block and Graft Copolymers. Nashua, NH: MMI Press, 1983.
29. SL Agarwal. Polym Eng Sci 17:497, 1977.
30. LW Morgan. J Appl Polym Sci 27:2033, 1982.
31. KC Dao, RA Hatem. Properties of blends of rubber/talc/polypropylene. SPE ANTEC, 1984.
32. KC Dao. J Appl Polym Sci 27:4799, 1982.
33. F Ramsteiner, G Kanig, W Heckmann, W Gruber. Polymer 24:365, 1983.
34. AL Bull, G Holden. Rubber Chem Technol 49:1351, 1976.
35. AR Schultz, BM Beach. J Appl Polym Sci 21:2305, 1977.
36. CF Ryan, RL Jalbert. In: LI Nass, ed. Encyclopedia of PVC, Vol. 2, New York: Marcel Dekker, 1977.
37. RP Petrich. Polym Eng Sci 13:248, 1973.
38. M Aarts, PD Hall, AJ Meier. Kunstsoffe 75:224, 1985.
39. Chem Werk Buna VEB. East German Patent 137, 309 (1970).
40. Chem Werk Buna VEB. East German Patent 132,072 (1978).
41. Dynamit Nobel AG. Belgian Patent 855,015 (1976).
42. Chem Werk Buna VEB. East German Patent 144,924 (1979).

43. BF Goodrich. Belgian Patent 783,191 (1971).
44. F Fahnler, J Merten. Kunststoffe 75:157, 1985.
45. EA Flexman. Polym Eng Sci 19:564, 1979.
46. Du Pont Engineering Plastics Sales Guide E-32986, 1984.
47. (a) HK Chang, CD Han. J Appl Polym Sci 30:165, 1985; (b) HK Chang, CD Han, J Appl Polym Sci 30:2431, 1985; (c) HK Chang, CD Hans. J Appl Polym Sci 30: 2457, 1985.
48. BN Epstein, U.S. Patent 4,174,358 (1979) (to du Pont).
49. F Ramsteiner, W Heckmann. Polym Commun 26:199, 1985.
50. SY Hobbs, RC Bopp, VH Watkins. Polym Eng Sci 23:380, 1983.
51. M Baer. U.S. Patent 4,306,040 (1981) (to Monsanto).
52. SB Farnham, TD Goldman. U.S. Patent 4,096,202 (1978) (to Rohm and Haas).
53. R Binsack, D Rempel, C Lindner, L Morbitzer. European Patent Application EP 131,202 (to Bayer).
54. HW Bonk, R Drzal, C Georgacopoulos, TM Shah. SPE ANTEC, Washington, DC, 1985.
55. T Dolce. High performance toughened polyacetal copolymer alloys, SPE RETEC, Athens, OH, 1985.
56. F Kloos, E Wolters. Kunststoffe 75:735, 1985.
57. F Kloos. Angew Makromol Chem. 133:1, 1985.
58. (a) PY Liu, NR Rosenquist. U.S. Patent 4,456,725 (1984) (to General Electric); (b) M Witman. U.S. Patent 4,378,449 (1983) (to Mobay); (c) FF Holub, PS Wilson. U.S. Patent 4,226,950 (1980) (to General Electric); (d) AJ Yu, J Silberberg. U.S. Patent 4,148,842 (1979) (to Stauffer); (e) WJJ O'Connell. World Patent 8,000,154 (1980) (to General Electric); (f) PY Liu, EJ Goedde. German Patent DE 3,031,539 (1981) (to General Electric); (g) K Kishida, A Hasegawa. German Patent DE 2,728,618 (1977) (to Mitsubishi Rayon); (h) HL Heiss. German Patent DE 1,930,262 (1969) (to Mobay).
59. DR Paul, JW Barlow. J Macromol Sci—Rev Macromol Chem 18:109, 1980.
60. D Klempner, KC Frisch, eds. Polymer Alloys. New York: Plenum Press, 1980.
61. S Wu. Polymer Interface and Adhesion. New York: Marcel Dekker, 1982.
62. RL Jalbert. In: Modern Plastics Encyclopedia. New York: McGraw-Hill, 1983–84, p 98.
63. TS Grabowski. U.S. Patent 3,864,428 (1963) (to Borg–Warner).
64. (a) JD Keitz, JW Barlow, DR Paul. J Appl Polym Sci 29:3131, 1984; (b) H Suarez, JW Barlow, DR Paul. J Appl Polym Sci 29:3253, 1984.
65. DA Folajtar. Recent improvements in the performance of Polycarbonate/ABS Blends, SPE RETEC, Athens, OH, 1985.
66. G Forser. Plast World 41:45, 1983.
67. MD Bertolucci, DE DeLaney. Xenoy 1000 series—Polymer blends for exterior automotive applications, SPEVANTEC, 1983.
68. L Bartosiewicz, CJ Kelly. Res Dev 80, August 1985.
69. AW Birley, XY Chen. Br Polym J 17:297, 1985.
70. RL Dieck. U.S. Patent 4,271,064 (1981) (to General Electric).
71. RL Dieck, AL Wambaugh. U.S. Patent 4,220,735 (1980) (to General Electric).

72. SC Cohen, RL Dieck. U.S. Patent 4,257,937 (1981) (to General Electric).

73. HC Fromuth, KM Shell. U.S. Patent 4,264,487 (1981) (to Rohm and Haas).

74. W Stix, W Novertne, H Buding. German Patent DE 3,302,124 (to Bayer).

75. RL Dieck. U.S. Patent 4,280,949 (1981) (to General Electric).

76. SH Agarwal. European Patent Application 79,477 (1983) (to General Electric).

77. HC Fromuth, KM Shell. U.S. Patent 4,180,494 (1979) (to Rohm and Haas).

78. Y Nakamura, R Hasegawa, H Kubota, U.S. Patent 3,864,428 (1975) (to Teijin Limited).

79. R Binsack, D Rempel, C Lindner, L Morbitzer. European Patent Application 131,202 (to Bayer).

80. LG Bourland. U.S. Patent 4,530,965 (1985) (to Atlantic Richfield).

81. GR Chambers. Noryl GTX: A new modified PPO blend, SPE RETEC, Athens, OH, 1985.

82. RJ Russell. U.S. Patent 4,371,663 (to Dow).

83. PY Liu. U.S. Patent 4,481,331 (to General Electric).

9
Lubricants

Richard F. Grossman
Halstab Division, The Hammond Group, Hammond, Indiana

I. BACKGROUND AND MECHANISM OF LUBRICATION

A. Lubricants in Compounding

Polymer formulation with additives appears to have begun 4000 years ago with use of metal carboxylates, in the form of wood ash, as lubricants for naturally occurring or derived lipids used to coat chariot axles [1, p. 2]. A dramatization of this event has recently been reconstructed [2]. Despite this long history of empirical lubricant usage, the absence of an underlying rationale should not be assumed. Quite the contrary, it now seems that considerations of polarity and mobility can be used to predict lubricant behavior in a variety of polymeric environments. One possible approach will be presented in Section III.

B. What Is a Lubricant?

All proposed definitions have been stated in terms of function, stressing comprehensiveness at the expense of brevity. For the purposes of the following discussion, a lubricant is a substance that, when added in small quantities, provides a considerable decrease in resistance to the movement of chains or segments of a polymer of at least partly amorphous structure, without disproportionate change in observable properties. The definition has meaning when it is considered how a polymer molecule moves; in particular, how it unwinds from a tangle with its neighbors so as to pass through a runner, gate, or roll nip. Two factors contribute to the resistance to flow (i.e., to process viscosity): interactions with nearby polymer chains and other ingredients, and the bulk of the molecule itself (i.e., its molecular weight). Disassociation from its neighbors, other ingredients, or parts of itself requires, in addition to movement by translation, rotation around (typically) carbon–carbon bonds in the polymer chain. Ease of internal rotation is the key to polymer flow. Polymers whose structure permits facile internal rotation, such as poly(dimethyl) siloxanes, are known for low viscosity even at high molecular weight. A lubricant is an additive whose characteristics permit it to assist the internal rotation of a polymer chain under strain. In other terms, it adds to the free volume associated with that segment of the polymer.

C. A Lubricant Is Not a Plasticizer

Is this more than a semantic distinction? A plasticizer is also an additive that promotes movement of chains and segments of an at least a partly amorphous polymer. Normally, however, plasticizers are used in more than trace amounts and are expected to provide significant changes in observable properties (e.g., increased flexibility and extensibility). In addition, the addition of plasticizer provides improved processability and may be selected partly on that basis. Further, the same species may be used as a plasticizer in one system [dioctyl phthalate (DOP) in poly(vinyl chloride) (PVC)] or in a trace quantity as a lubricant in another [DOP in low-density polyethylene (LDPE) or ethylene vinyl acetate (EVA)].

Mechanistic differences between the action of plasticizers and lubricants are suggested by the difference in the levels at which they are used. Recent views of the mechanism of a plasticization center on the equilibrium replacement of dipolar, dispersion, and hydrogen-bond interactions between polymer chains (the forces inhibiting internal rotation) by similar interactions between the polymer and functional groups on the plasticizer molecule [3]. The remainder of the plasticizer molecule supplies substantial free molar volume (i.e., has a weak interaction with its neighbors) [4]. A plasticizer is often considered a nonvolatile solvent [5]. In this context, a lubricant becomes an analog of a surfactant. The distinction is that a lubricant has a balance of polarity that makes it very mobile in the polymer phase. Its function is to repeat its facilitation of internal rotation at different sites on the polymer. This is not to say that a plasticizer is completely bound to one segment or domain, merely that its balance of polarity is consistent with lower mobility in the polymer phase than that of a lubricant. In summary, a lubricant provides a plasticizer function at a series of particular sites on the polymer, sites that are under strain from processing, and, therefore, function at a low additive level. Why (it should be asked) does the lubricant show up at the right place at the right time? The driving force is applied strain. The effect of deforming strain is to increase polarization; polar groups on the polymer become more polar. The lubricant must be selected to prefer the polarity of the strained section over those at rest.

As with many surfactants, part of the lubricant molecule is often a saturated hydrocarbon chain. This will be referred to as the nonpolar or noninteractive section of the molecule. The other end, the interactive section of the lubricant, may be varied among different functional groups to achieve a range of polarity, enabling interaction with various polymers and surfaces. Both sections may be characterized (approximately) by estimates of their solubility parameters. That of the noninteractive section should be thermodynamically incompatible with, and less polar than, the polymer [i.e., have a solubility parameter two to three SI units (megapascal, MPa) lower]. For a hydrocarbon such as polyethylene, with a solubility parameter around 16, this would ideally consist of a fluorocarbon or silicone side chain, with a solubility parameter of 12–14. More polar polymers, polystyrene, acrylonitrile–butadiene–styrene (ABS), and PVC, are accommodated by saturated aliphatic hydrocarbon side chains [6].

D. What Is Being Lubricated?

Historically, a division has been made between lubricants that are presumed to facilitate movement of one polymer chain with regard to another (internal lubricants) and those intended to increase flow in the vicinity of objects such as the surfaces of process equipment (external lubricants). It has been pointed out that this division is often not clear [7]. Further, what might be classed as an external lubricant in one system (paraffin in PVC) might well be an internal lubricant in another (paraffin in PE) and vice versa (calcium stearate in PVC versus in PE). What is meant, therefore, is that in a specific thermoplastic composition, the data at hand suggest that a given additive functions mainly, but probably not exclusively, as an internal or external lubricant.

The above model is particularly opaque when concerned with polar polymers such as PVC, where flow before and during fusion of powdered compounds involves movement of one domain relative to another. In this case, the most helpful model is one in which the domains are visualized as being coated with lubricant (most often a metal stearate) [8]. Some technologists consider such action to be "external" [9]; traditionalists maintain that it is "internal" [10]. As long as the model is kept in mind, the semantics are not important. In general, if a lubricant appears to be effective in improving flow but does not have much effect on surface tension, it is considered internal. If it is found on the surface or on adjacent surfaces, or if it modifies observables associated with the surface, it is considered external. If it behaves in one way under one set of conditions and in another way under a different set of circumstances, it is considered both an internal and external lubricant. These are referred to as "balanced," "combination," or "multifunctional" lubicants in the literature. This, again, has significance only in terms of a specific composition in a specific environment.

E. Measurements of Lubricity

The distinguishing characteristic of internal lubrication is that bulk viscosity is reduced. Unequivocal determination requires apparatus for measurement of bulk viscosity under conditions where perturbation of the composition by the apparatus is controlled and reproducible. This may be done in the laboratory with a variety of instruments: cone plate and parallel plate viscometers, capillary rheometers, and a number of devices measuring a reflection of bulk viscosity (torque, power draw) in a miniaturization of an actual process such as extrusion or mixing. What is of particular interest is whether viscosity reduction is found under conditions approximating a particular process. Therefore, the apparatus selected should provide a shear rate of the same order of magnitude as the process at similar temperatures. The ability to vary temperature and shear rate is found in several commercially available instruments [11].

Ideally, the composition exhibits a bulk viscosity stable with time at a particular temperature and shear rate. As the molecular weight of the polymer is increased to the point where viscoelastic behavior is found, such stability is almost never the case at shear rates high enough to be of practical interest. Nonetheless, the composition should contain typical levels of antioxidants or stabilizers so as to minimize the drift of viscosity with time. This is particularly true with polymers subject to rapid chain scission with shear [polymethyl methacrylate (PMMA), polyvinyl dichloride (PVDC)] and those whose decomposition temperature when unstabilized is in the same range as the melting point (PVC). The viscosity of a typical polypropylene resin versus time is shown in Fig. 1, where the Brabender rheometer torque is plotted at two temperatures.

The experiment of interest usually is the relation of bulk viscosity of a specific composition to the concentration of lubricant. This is combined with background knowledge of what level of viscosity at a given shear rate is required for successful operation of a particular process. The conclusion is the level of internal lubricant to employ or the suitability of a particular species for such use. Translation into the pilot plant or factory then provides information as to whether the lubricant has affected behavior associated with surfaces (i.e., has also functioned as an external lubricant) or bulk properties of the composition in more than a trivial way (i.e., has functioned as more than a lubricant).

Ideally, the addition of a lubricant will have essentially no effect on the glass temperature (T_g) of the base polymer; that is it will result in no lowering of maximum service temperature. This is often the case with modest levels of lubricants that act primarily externally; less often so with internal lubricants.

Although experiments regarding behavior as an internal lubricant are, from a practical viewpoint, referenced to the conditions of a specific mixing or fabrication process, they may be considered independently as simply measurements of the properties of a composition. This may be meaningful to a supplier considering a range of new materials with potential as lubricants. Behavior as an external lubricant must, on the other hand, always refer to a particular process.

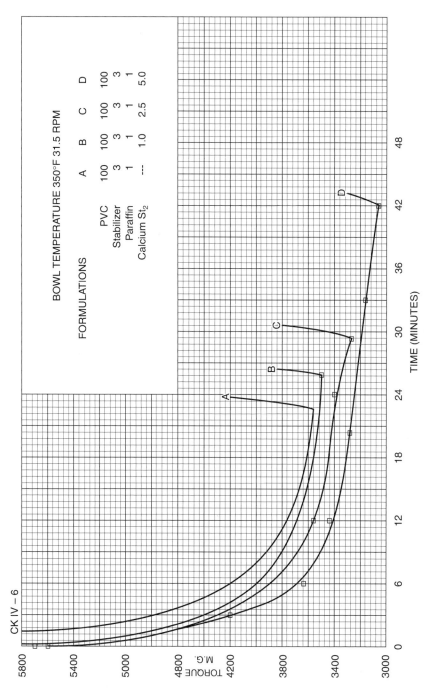

Figure 1 Rheometer viscosity of polypropylene versus time. (Data supplied by C.W. Brabender Instruments, Inc.)

A common situation for external lubrication relates to the interface between process equipment and the polymeric composition. The metal surface of the equipment bears a polar oxidized layer and usually a pattern of microcracks and other discontinuities. Polar ends of lubricants are adsorbed at such sites. Movement of polymer molecules in the vicinity of the surface then involves interaction between nonpolar ends of lubricant molecules associated with the polymer and those adsorbed on the metal surface. The effects of such lubrication (increased ease of internal rotation in the vicinity of the metal surface) include faster flow through restrictions used to shape the product (higher output, improved fill), lower heat buildup in processes with high shear rates, and improved surface appearance from easier release and a more streamlined flow. Effectiveness of external lubricants should be determined using laboratory or pilot equipment analogous to the actual process (e.g., a small extruder or calender, or a spiral-flow mold in the case of injection molding). Then, lubricant concentration can be related to extruder or calender output, to appearance and equipment power draw over a range of process temperature and machine speed, or to mold fill and part appearance versus injection pressure, time, and temperature. As has been pointed out, these procedures actually evaluate a combination of internal and external lubrication [12]. In practice, it is often convenient to proceed with an assessment of both functions at once (will this additive work?) when faced with the need for practical problem solving. Most studies in the literature are of this type; viscosity of compositions of interest is followed by means of equipment sensitive to both internal and external lubrication, commonly a Brabender torque rheometer, and the data used to infer the mechanism of lubrication [13, p. 838].

The use of laboratory models of production equipment also facilitates the evaluation of undesired side effects which may accompany external lubrication. One such is as a participant in the deposition of materials on metal surfaces, variously known as plateout and mold fouling (*inter alia,* all perjorative.) In an ideal situation, a monolayer of lubricant is adsorbed on the metal surface and each polymer chain has the exact number of lubricant groups needed to help it move about in the region proximate to the lubricated surface. In reality, those molecules adsorbed on metal surfaces are constantly swept away and replaced. The factor that permits efficient external lubrication, decreasing solubility of partially compatible additives in polymers with increasing physical strain, is usually also a factor in deposition. In the ideal situation, shear strain has decreased the compatibility of the lubricant to the point where the surface of the compound and matching metal surface are suitably coated. A point of flow restriction may then further reduce compatibility, causing the deposition of more than a monolayer. This may or may not be followed by a zone in which the excess is resolvated. If a multiple layer is deposited, the likelihood of exposed polar groups greatly increases (from the second layer tending to be deposited in the opposite polar orientation from the first). These are, of course, the groups specifically introduced to interact with polar species passing by. In passing the deposit, shear strain on the composition will now tend to expel whatever polar inclusions are present: fillers, pigments, stabilizers, and so forth. Thus, aggregation is promoted and chances for degradation and resinification are maximized [14].

To minimize such side effects, overlubrication must be avoided, emphasizing the importance of compound development using test equipment comparable to plant analogs. In addition, the incompatible part of the lubricant should have low surface energy (i.e., be incompatible because its solubility parameter is significantly lower, not higher than that of the polymer). Dipolar materials in which the polymer-incompatible section is more polar than the polymer *can* function to improve processability in thermoplastics (e.g., polyterpene resins in LDPE or EVA) but are also apt to be tackifiers. Low surface energy will usually be reflected by the pure lubricant having a relatively low viscosity at compound processing temperatures. Simple laboratory screening tests for the tendency to accumulate deposits have been proposed [15].

Often, exactly the opposite situation is found. The introduction of new or rebuilt machinery having newly finished surfaces is commonly accompanied by a temporary loss in output because of the absence of sites for external lubricants. A similar phenomenon is the temporary drop in output associated with a major formulation change (as one lubricant attempts to displace another, particularly in mature equipment having a high microsurface area). This often mitigates against the adoption of a new recipe which has shown sterling performance in the laboratory (where the small size of the equipment minimizes the effect).

Experienced formulators are well aware that additives which ameliorate problems bring hazards of their own. The reduction in bulk viscosity resulting from internal lubrication invariably extracts a price. The glass transition temperature will be lowered to some extent; properties dependent on a thermoplastic having a high glass temperature (heat distortion, impact strength) are likely to be affected. Adsorption on polymer chains may desorb other additives, lessening their effect and increasing chances of their exudation. The hazards of external lubrication are obvious: the quantity needed for best processing at one stage may be an obnoxious presence at some other point; lower surface energy may interfere with subsequent processing (painting, coating, adhesion) [16].

There are, of course, occasions where permanent modification of surface energy is desired. As defined earlier, additives in this category are not properly process lubricants, but slip or antiblocking agents. These may also serve as external lubricants during processing and vice versa. In cases where the surface of the thermoplastic composition is used to store or transmit information (tapes, disks, phonograph records), such agents are carefully chosen and used at very low levels.

F. What Is an Incompatible Additive?

Experimental observations that an additive exudes or deposits do not imply that the equilibrium concentration in the polymer phase is zero. No additive may be considered totally incompatible. This simplest of observations underlies the recent use of materials normally classed as totally incompatible species in the role of external lubricants. To date, this approach appears to be limited to use of fluorocarbon and silicone additives in polyolefins [high density polyethylene (HDPE), linear low density polyethylene (LLDPE), and polypropylene (PP)] as extrusion aids [17,18]. Additive levels range from parts per million to parts per thousand, usually added as dilute concentrates. It is intended that, under shear strain, these additives will deposit and that their low surface energy will permit the displacement of previous deposits and repel the deposition of other materials. The same additives are, of course, internal or combination lubricants when used in silicone or fluoropolymers.

G. What Surfaces Are External?

Polymer flow in regions adjacent to solid insoluble inclusions presents special problems. It is normally not desired that polymer molecules transport themselves through nips and dies or down runners, leaving particles of filler or pigment behind. (This is often the case if turbulent flow becomes excessive.) Occasionally, in service it is useful if polymeric compositions chalk a layer of solid particles to the surface, but, normally, it is expected that dispersed solids will follow the polymer in processing. In fact, a considerable effort is made to choose fillers and pigments and to disperse them such that extreme magnification is needed to disclose the existence of multiple phases. That fractionation of polymer–filler mixtures can nonetheless occur during processing, often at points of sudden pressure change, is well known to the experienced technologist.

At rest, the solid particle functions as a physical crosslink, similar to a point of dipolar or dispersion force interaction between polymer chains. For the composition to flow, polymer–filler bonds must continually break and re-form [19]. This is particularly true of particles with high aspect ratios; the consequence of strong interaction (high reinforcement value) is high viscosity [20]. Fillers vary greatly in their ability to adsorb dipolar mobile species, but, regardless of their surface area or structure, they are almost always more polar than their polymeric binder. Therefore, as the interactive section of the lubricant is made more polar, lubricant–filler interaction becomes more prevalent. This tends to lower process viscosity of a filled composition but may increase overall lubricant requirements. If the lubricant has other functions (e.g., is also a stabilizer), filler competition with the polymer may be detrimental unless taken into account.

The lubricants effective at the polymer–filler interface are those which are strongly dipolar. They may therefore be classed as internal if the polymer is polar (with regard to their polymer–polymer interactive behavior) or external if used in nonpolar thermoplastics. This may be altered considerably by the use of fillers or pigments that have been previously coated or which may be treated during mixing. Such coatings include reagents that bind to the filler surface by ionic, dipolar, or hydrogen-bond forces and those which react with surface groups to form covalent links. The net effect is to generate a particle of lower surface energy than existed previously. The ability of lubricant-bearing polymer chains to flow in the vicinity of the particle is thereby enhanced.

A number of factors, other than effectiveness in lowering viscosity, influence the choice of filler surface lubricant (e.g., temperature stability, resistance to various media the composition may encounter in service, potential toxicity, etc.). If free to choose, the first principle of compounding should be followed (put things that like each other together.) Thus, if the filler at hand contains Si—O linkages (silicas, silicates, clay), experiments should include reactive silane treatments [21]. In the case of Ti—O linkages (titanium dioxide, metal titanates), certainly reactive titanates should be investigated [22]. Where the filler is a metal salt (calcium or magnesium carbonate or silicate), often metal carboxylates are highly effective.

In this category, there is some evidence that choosing the same cation (e.g., using calcium stearate to coat calcium carbonate) or a cation whose chemistry is very similar provides the greatest utility in processing [23]. It is often preferable to treat the filler with a lubricant having a high specific attraction for that particular surface (i.e., one that will only provide filler–polymer lubrication) and use other materials for internal and external lubrication. This is tedious in terms of expanding raw material inventory, but convenient in practice. In many actual processes, thermodynamic equilibrium is never reached; these are path dependent (observable properties of the composition vary with process conditions). The use of one additive to achieve a succession of different effects will generally increase path dependency (the change in properties associated with process variation will increase).

II. SPECIFIC LUBRICANTS

A. Hydrocarbons, the Simplest Lubricants

Mixtures of saturated aliphatic hydrocarbons with molecular weights ranging from several hundred to several thousand are obtained from refined petroleum as paraffin wax or, where economics dictate, via Fischer–Tropsch synthesis. They are used extensively because of low cost, good color, relative nontoxicity, and reasonable stability to heat, light, and oxidation. They are nonpolar but not nonpolarizable. Unbranched segments of greater than 12 methylene groups form crystallites; branch points lie outside in amorphous domains. The latter also contain traces of unsaturation and oxidized groups (to a lesser extent in synthetic paraffin). Amorphous domains in paraffin are sufficiently polarizable to coordinate with polar groups of polymers. The crystalline sections are incompatible with almost all thermoplastics except for low molecular weight (MW) polymers and copolymers of ethylene. In such cases, paraffin becomes a plasticizer (e.g., in hot-melt coating compositions).

Low levels of paraffin are often used as internal lubricants in polyolefins. There is little point in considering them for external lubricants in such surroundings, as their solubility parameter is close to that of the hydrocarbon polymer [6]. In high-value-added products (e.g., insulated wire or automotive components), processability is normally optimized by selection of the most suitable MW distribution (often broad for extrusion and calendering applications; narrow for injection molding).

This may be done by resin selection or blending or the use of copolymers, or elastomer-modified grades. A considerable volume of polyolefins is, however, directed to lower-end applications where economics forces the use of grades that can be obtained at lowest cost. In these cases, the addition of paraffin may serve as the most suitable avenue to improved processing.

The scenario is similar for styrenic polymers [PS, ABS, styrene–acrylonitrile (SAN), etc.]. Although a great range in processability can be obtained by grade selection, there is some use of paraffin at very low levels, often in conjunction with other lubricants. As these polymers are not totally nonpolar, even at low levels the action of paraffin is likely to be both internal and external.

The greatest volume of paraffin, as a lubricant, is in unplasticized PVC compositions, where its function is largely external. At levels usually below 1 part per 100 of polymer, it is an effective external lubricant in extrusion, calendering, and molding formulations. Paraffin is of little value in modifying filler surfaces or providing internal lubrication in compositions this polar. Suspension polymerized PVC yields the complex granules typical of this method [8,24] and is almost unique in that often the initial stages of compounding (incorporation of stabilizers) are carried out below the melting point (at which temperature the unstabilized polymer rapidly decomposes) by absorption and adsorption processes. Many fabrication routes employ blends of PVC plus compounding ingredients; intimate mixing, fusion, and shaping of the product occur sequentially in line. As with all continuous compounding–fabricating operations, it is essential that each stage occur at the proper moment. For example, it is desired that the stabilizer and internal lubricant permeate and that filler particles adhere to the resin granule prior to fusion. It may or may not be desired (depending on the process) that external lubricants delay some or all of these by coating the polymer. Therefore, investigations of the levels of external lubricant in general and paraffin in particular must consider effects on fusion as well as on subsequent processing. The level of paraffin chosen for external lubrication must be appropriate to the time for fusion needed in a given operation. The need for compromise frequently dictates blending with a secondary lubricant having some external capability.

Some PVC copolymers and blends of PVC with other polymers [acrylics and chlorinated polyethylene (CPE)] are sufficiently less polar that paraffin can function both externally and internally [10]. As a result, somewhat higher levels are used, often one part or more in PVC/CPE blends. Similarly, one to two parts may be used as a combination lubricant in thermoplastic CPE formulations, such as for pond liners, although blends of lubricants are more common [25].

B. What Is a Wax? What Is a Polymer?

The mixtures of saturated hydrocarbons refined from petroleum are available in grades with melting (or softening) points from 55°C to 75°C. Similar species made by low-MW polymerization of ethylene or propylene range from softening points of 85–160°C, indicating greater regularity and higher crystallinity as well as higher average MW. Compared to paraffin wax, these additives also exhibit good color and stability; they are not, however, as economical. Their usage is usually more specific: PE wax is more compatible with standard PE resins than is paraffin, and less compatible with polar polymers. Similarly, atactic PP wax has improved compatibility with PP resins.

Where these materials function as internal lubricants (i.e., in polyolefins), increased MW provides improved heat stability. Commonly, polyethylene waxes with softening points in the range of 90–110°C are used in LDPE, EVA, and low-MW HDPE waxes and with higher softening points in HDPE. In polypropylene, HDPE waxes are only partly compatible (ethylene and propylene crystallites preferring separate domains) and are used as external or combination lubricants, the internal lubrication function more appropriately assigned to low-MW polypropylene waxes. Amorphous PP is, conversely, only partly compatible with PE and may improve processing, but with a tackifying action. Blends of low-MW PE and amorphous PP waxes are highly compatible in themselves and serve as bases for hot-melt compositions. This is a further reminder that all comments on compatibility must refer to a specific range of MW.

Low-MW PE and PP waxes are common color concentrate binders for the respective resins and therefore occur more often than obvious at first glance. When added as lubricants, levels as high as one to two parts are routine, and are at times higher when regrind, particularly of a miscellaneous nature, is employed. In these cases, the function may go beyond process lubrication; improvements in environmental stress crack resistance are also found.

Low-MW PE waxes are also used at low levels (0.1–0.5 part) in styrenics, particularly in ABS, as combination lubricants. This is also common in thermoplastic CPE and CSM (chlorosulfonated PE), where they may function internally at one stage and then as mill or calender release agents at another. A consequence of a high MW (as compared to paraffin) is increased incompatibility at elevated temperatures. With paraffin, a poor solvent for other than hydrocarbon polymers, decreasing compatibility competes with temperature-enhanced mobility. With PE wax, mobility is lower, and decreased compatibility with increasing temperature prevails. As a result, such waxes can provide release from mill and calender rolls well above their softening temperatures.

These properties underlie the use of PE waxes as external lubricants in PVC. Low levels, 0.1–0.2 part, are effective in unplasticized formulations; generally, high-softening-point grades are used to minimize the effects on physical properties. These levels provide good surfaces and increased output in extrusion and calendering without impairing further surface finishing (e.g., printing) as much as comparable levels of paraffin. Generally, they are used in combination with other lubricants that are dual function.

In plasticized PVC formulations, levels of 0.2–0.6 part are used, principally in calendering and extrusion. Low-MW PE is particularly useful in wire insulation because of its excellent electrical properties and inertness to water. At correct levels, the resultant exterior surface can be easily printed, yet the interior surface does not develop undue adhesion to the conductor.

Low-MW PE waxes are readily oxidized under controlled conditions at chain ends (usually points of unsaturation) to yield carboxylic acid groups without MW reduction. This can be carried out without notable degradation or color formation (i.e., without substantial oxidative scission or cross-linking at points within the chain). This is not true of low-MW PP, which should be protected with an antioxidant.

Low-MW PE waxes have solubility parameters in the same range as standard PE resins. Oxidized low-MW PE, depending on the extent of oxidation, may have a solubility parameter nearly identical to PVC [6]. Therefore, a particular grade may be used in PVC as an internal or combination lubricant. Oxidized grades also function as internal lubricants in thermoplastic CPE and CSM formulations. Good synergism is often found in these cases with combinations of oxidized PE wax as an internal lubricant interacting with low-MW polyethylene glycol wax as a filler coating.

Similar products also result from end-group oxidation of paraffinic waxes made by the Fischer–Tropsch process. These have softening points in the range of 90–110°C and should be useful as alternates to lower-MW grades of oxidized PE wax. Both oxidized PE and paraffin waxes are also available as partially neutralized for use as additives in acrylic and vinyl/acrylic coatings.

Increased polarity may be added to the PE wax by copolymerization as well. Copolymers of ethylene with unsaturated acids can have acid numbers as high as 120. Such additives have sufficient polarity to function as lubricants for cellulosic polymers.

At more modest acid numbers (15–50), use of these additives in PVC parallels that of oxidized homopolymers. Other useful copolymers are low-MW EVA waxes. These have very broad solubility parameters (wide compatibility) and are excellent internal lubricants in polyvinylchloride–acetate (PVC–Ac) copolymers, EVA/CPE and EVA/PVC–Ac blends. In compounding with low-MW polymers (paraffin, PE wax) as lubricating additives, attention must be paid to several structural factors: mass (MW), the ratio of polar to nonpolar composition, and the solubility parameters of each. The effect of high mass is generally to keep the additive in place. For example, in a polar polymer such as PVC, where hydrocarbon lubricants function mostly externally, increasing the MW, as in going from paraffin to PE wax, will keep a low-energy hydrocarbon layer on the surface throughout a broader range of process conditions [24]; or, in a nonpolar polymer such as PE, where the same lubricants are mainly internal, increasing the MW tends to maximize the internal rather than external function. As the MW of the additive is decreased to a few hundred, the mass effects disappear. Therefore, with monomeric lubricants, there is increased reliance on the balance of polar and nonpolar functionality and the increased variation in the types of polar groups in use.

C. Small Molecules as Lubricants

Many interesting applications involve saturated aliphatic acids. Two major types predominate: the ubiquitous "stearic acid" (a blend of C14–C18 even-numbered straight-chain aliphatic acids from animal fat or vegetable oil) and "montanic acid" (a blend of C26–C32 analogs obtained from the fossil fuel lignite). The carboxylic acid group (see Fig. 2) never suffers from lack of coordination—if not with other groups, then as a dimer. Although even-numbered saturated straight-chain acids crystallize with melting points (MPs) above room temperature, at C14 and above, the carboxylic acid groups lie outside the crystalline domains. An equilibrium with monomer exists, but is normally in the direction of the dimer, with the acid hydrogen of each carboxyl group hydrogen bonded to the carbonyl oxygen of the other. This interaction is clear in the O—H and C=O regions of the infrared spectra of such acids [1, pp. 48–55]. Thus, even the C14 member of this series, myristic acid, in its dimeric form is almost as large as a low-MP paraffin wax. The acid may remain a dimer during processing

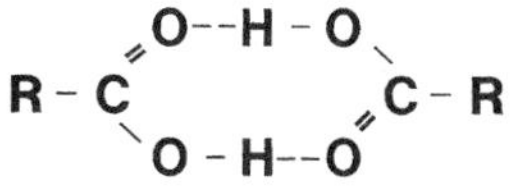

Figure 2 Structure of aliphatic carboxylic acids.

or may dissociate through coordination of the carboxylic acid group to some other site, depending on circumstances. For example, in a polyolefin, a stearic acid dimer acts as an internal lubricant, lowering viscosity comparably to paraffin. The acid group is attracted to oxide-bearing metal surfaces, and the dimer dissociates to coat such process surfaces. It is similarly attracted to the surface of polar fillers, particularly if surface salt formation is possible. After processing, the same additive may be of value as a release agent. In a polyolefin compound, stearic acid, therefore, may serve as an internal and external lubricant, filler coating, and release agent simultaneously, or, if used judiciously, sequentially instead. What could be more useful? (Many alternatives, it turns out.)

The ease with which such acids equilibrate between self-coordinated dimer and other-coordinated multipurpose additive overlies a remarkable migratory ability. In service, they are likely to be found in whatever adjoins the polymer into which they were introduced. They show a preference for interfaces. In applications involving adhesion, straight-chain organic acids must be used with the same caution reserved for silicone fluids. Such acids are often synergists for the migration of other ingredients with which they may coordinate. Even when not found in deposits (e.g., mold fouling), a reduction in level will frequently lower the deposition tendency of other species.

In polar polymers, stearic acid may also function in a complex manner. In unplasticized PVC compounds, it is principally an internal lubricant. This may include the acid's participation in coating fillers; it may also improve the mobility of metal-containing stabilizers to which it can coordinate. Similarly, it may improve pigment dispersion. The reduction in viscosity through internal lubrication probably involves acid dimers interacting with the polymer via dipolar attraction; flow in the vicinity of solid particles no doubt reflects the surface adsorption of monomeric acid molecules interacting with lubricated polymer. Common usage is of the order of 0.1–0.3 part, assuming that another lubricant (e.g., paraffin) is present at an equivalent or higher level to provide external lubrication. In cases where another lubricant capable of both internal and external lubrication is present, stearic acid may assist both functions.

As the polarity of the polymer is decreased, stearic acid becomes an all-purpose lubricant and is used at higher levels. Levels of 0.5–1.5 parts are common in filled CPE and in acrylonitrile-butadiene rubber (NBR) elastomer-modified PVC compositions. Some carbon-black-filled CPE/PVC blends contain as much as 5–7 parts stearic acid, particularly in conjunction with lead stabilizers, on whose surface conversion to lead stearate may occur [25]. It appears to be useful in improving blend compatibility, or in promoting the compatibilizing effect of other additives. This may be considered a special case of internal lubrication—the enhancement of movement of one polymer in the vicinity of another. In the case of thermoplastic elastomers or thermoplastics toughened with elastomers, where phase-separation persists during processing and where in the final state, one phase may act as a reinforcement, the distinction between internal lubrication and surface treatment blurs.

In plasticized PVC formulations, where much of the mobility needed for flow has been furnished by the plasticizer, stearic and montanic acids act as external lubricants, the longer-chain montanic acid being more efficient (at correspondingly increased cost). Here, they may be used at levels of 0.2–0.6 part with some caution in applications where migration is intolerable. In applications where critical color (or the lack of color) must be maintained, grades of stearic acid of low unsaturation (from traces of unsaturated acid components) must be used. Montanic acid, derived from an oxidative treatment of its naturally occurring precursor, has very little residual unsaturation.

In almost all of the above applications, a blend of stearic acid and oxidized PE wax could be used. This is advantageous in cases where stearic acid is intended as a multipurpose lubricant and in situations where need for decreased migration is indicated.

D. Truly Monomeric Lubricants

Lower aliphatic alcohols show evidence of association through hydrogen-bonding in the liquid state (in the lowest member of the series, H—OH, to an extent great enough to be considered a structural feature). In the higher members of the series, hydrogen-bonded dimers predominate [26]. The effect is not as significant with straight-chain alcohols as with the corresponding acids, particularly as the chain length increases. Thus, caprylic acid (C8) has about the same melting point as n-hexadecane (C16); n-octanol is intermediate between a C8 alkane and C8 acid [27]. Use as lubricants in thermoplastics is mainly limited to C16–C18 fractions, although effects of the use of a C22 fraction have been reported [16]. If C16 predominates, the additive is referred to as cetyl alcohol; if C18, it is referred to as stearyl alcohol.

The balance of polarity is such that long-chain alcohols could be used as internal lubricants in polar polymers, external lubricants in nonpolar polymers, and combination lubricants in semipolar polymers. In addition, they are relatively economical species of good color and thermal stability, the terminal primary alcohol functionality being comparatively inert in service (presuming absence of oxidizing acids). Yet, in comparison with other lubricants, they are currently limited to special situations. Perhaps the most common use is in conjunction with lead stabilizers in PVC. Stearyl or cetyl alcohol will coordinate via the hydroxyl group with stabilizers such as tribasic lead sulfate or dibasic lead phosphite. This is another case of the promotion of an internal lubrication in the region of a solid particle by coating the latter. In this case, the lead compound's function as a stabilizer is also improved. Levels of 10–20% of the stabilizer content are used. Higher levels may be used in unplasticized PVC, where decreased fusion times are desired. In this context, such use tends to amplify the effect of and decrease the need for other lubricants, both external and internal.

By analogy with the above use, long-chain alcohols should be investigated in cases where the composition contains items likely to react with —OH groups. For example, silica fillers are commonly treated with low-MW polyethylene glycol to lower compound viscosity. This surface treatment could be combined with stearyl alcohol to increase the contribution to internal lubrication.

Another special-purpose lubricant, whose use illustrates the same compounding approach, is stearone (distearyl ketone). This is commonly used as a lubricant in EVA and EVA/PVC–vinyl acetate (VA) blends containing tackifying resins. Here, the central carbonyl group of stearone can coordinate with carbonyl groups of the resin as well as with the ester groups in these semipolar polymers and thereby act as a combination lubricant. The release agent function is also provided at relatively low levels (one part or less) without plasticization. Further investigation in other compositions containing ingredients with carbonyl groups is worth considering.

Long-chain acid amides consitute a diverse class of specialty lubricants. The simplest members, normal primary amides (e.g., stearamide), are more polar than the corresponding acids, with higher melting points, reflecting association but not simple dimerization. A major use is as an external lubricant in polyolefins, where slip, antiblock, and mold-release functions are also provided. Freedom from isomers, leading to a sharp melting point, has made erucamide the most popular straight-chain amide used in polyolefins. Migration to adjacent layers is reduced in comparison with stearic acid, permitting use as an adhesion modifier in hot-melt applications.

In polar polymers such as PVC, primary amides are mainly internal lubricants but here appear to offer little advantage over stearic acid, except as slip agents. Oleamide, stearamide, and mixtures are used. This also is the case with regard to function as a combination lubricant in semipolar thermoplastics. The situation changes drastically when such amides are ethoxylated and converted to alkanolamides. Although a reduction in the polarity of the end group, this modification results in an increase in specific attraction for water. As a result, such species find use in polar and semipolar polymers as external lubricants that are also antistatic agents and synergists for other antistatic agents (long-chain amine oxides and quaternary salts). Caution must be exercized when using amide lubricants or, for that matter, any nitrogen compound with potential nucleophilic tendency in halogen-containing polymers because of the possibility of promoting degradation at elevated temperatures [28]. This does not mean that there are not cases where amides can be used successfully in PVC or CPE, nor that there are not instances where certain nitrogen-containing compounds may actually be heat stabilizers. It is further worth noting that in at least one case of PVC impact-modified with acrylics, the usual situation was reversed; the generally innocuous

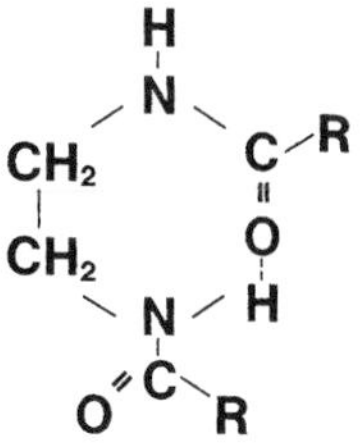

Figure 3 Possible internal coordination in EBS.

stearic and montanic acids led to a reduction in heat stability, whereas an amide lubricant proved harmless [29].

A more important category is that of diamides, principally ethylene bis-stearamide (EBS); see Fig. 3. The latter melts at 140°C, compared to 90–109°C for stearamide (MP dependent on the level of pure C18 amide present), and has increased thermal stability. This enables its use as a combination (although primarily external) lubricant in polyamides, polyesters, ionomers, and other thermoplastics with high process temperatures. Similar considerations prompt its use in higher-melting polyolefins (HDPE, PP), where its function is mainly external. Again, at low levels (0.1–0.5 part), EBS is often used as an external lubricant in styrenic polymers. With polyolefins and other thermoplastics sold in pellet form, EBS can be added by simply tumbling or blending with the resin pellets prior to fabrication without danger of lubricant agglomeration. This is no great advantage with polymers sold not as pellets but in powder form (PVC, CPE, etc.), which are easily blended with additives having a variety of physical forms.

Unlike stearamide and other primary amides, EBS provides external as well as internal lubrication in polymers as polar as PVC. In fact, it is of value as an external lubricant throughout a broader range of polymers than any other additive yet reported; in all cases, there seems to be an internal contribution as well. This suggests an equilibrium between configurations analogous to stearic acid.

Another area for consideration of amide lubricants is that of blends of polar polymers (polyamides, polyesters) with less polar polymeric modifiers (e.g., various elastomers added for purposes of impact toughening), particularly in situations involving high process or service temperatures. In effect, such lubricants are monomeric analogs of block-copolymer compatibilizing agents and, in some applications, may be used in polymer blends in place of the latter, with desirable economy.

E. Different Approaches: Inappropriate Plasticizers

Almost all the above-cited lubricants could be used in some circumstances as active plasticizers (e.g., paraffin in EVA coating compositions). With ingenuity, materials normally classed as plasticizers may be used at lower levels as lubricants in environments where the balance of polarity is such that they would be poor (insufficiently compatible) plasticizers. It was discovered serendipitously that levels of 0.1–0.2 part DOP (leaking into a Banbury® mixer from lubricated dust seals) improved processability in filled LDPE compositions [P. Smith, personal communication]. Whether the use of similar levels of dioctyl sebacate (DOS) in polyamides to improve monofilament output resulted by chance or design is unknown. These agents have two characteristics in common: They are known as excellent plasticizers in other systems, but are only slightly compatible with the above polymers. This suggests that plasticizers should be investigated as lubricants in polymers that they do not plasticize. In fact, many already have.

Stearate and related esters are excellent plasticizers in certain systems (e.g., glycerol monostearate in cellulosics and butyl or iso-octyl stearate in chlorosulfonated polyethylene (CSM) and semipolar elastomers). The former is not surprising given the affinity of cellulose derivatives for hydroxyl-bearing additives of all types; nor is the latter, considering the broadness of plasticizer compatibility shown by many elastomers. The relevant unrelated systems in which such esters should be (and, of course, are) useful lubricants are various thermoplastics.

Highly nonvolatile esters such as stearyl stearate and pentaerythritol tetrastearate are common lubricants in polycarbonate, possibly an extension of an older use in acrylics. In these polymers, compatibility is sufficient so as not to affect clarity, yet combination external–internal effects are gained. Effects on impact strength and stress cracking are minor at the low levels (below 0.5 part) used. Lower congeners such as butyl and iso-octyl stearates are common in styrenics where their action is primarily internal. In such systems they are usually used in conjunction with other lubricants. Combination effects are observed in thermoplastic urethanes and in a variety of thermoset resins. Glycerol monostearate is an external lubricant for polyolefins and finds use in blown film production (where it is also an antistat).

In PVC, the situation with ester lubricants is characteristically complex. In unplasticized compounds, relatively nonpolar esters such as stearyl stearate are primarily external in function and retard fusion time. As the parent alcohol is made smaller and the polar ester group thus more available, internal lubrication takes over. This is the case with butyl stearate (which tends to decrease fusion time). In between cases (e.g., iso-octyl stearate) appear to be combination effect lubricants [30]. Glycerol monostearate will function as a combination lubricant, mainly internal, when used in conjunction with paraffin, PE wax, or EBS. When used with other clearly internal lubricants, it will contribute significantly to external lubrication and be relatively neutral with regard to the effect on fusion time. The hydroxyl functionality is of additional value when used with lead stabilizers (which coordinate with —OH groups).

Combination lubricants also result from esterification of montanic acid with long-chain alcohols. Generally known as "montan ester wax," these have much broader use than, for example, stearyl stearate. In unplasticized PVC, a combination of external and internal lubrication invariably results. They are particularly useful when combined with a clearly internal lubricant (e.g., stearic acid) and an external lubricant (e.g., paraffin), appear to enhance the action of both, and permit use of low levels of each, consistent with resistance to migration and deposition. In plasticized formulations, low levels of esters provide external lubrication and are synergistic with other external lubricants.

Comparable behavior is found with proprietary blends of glycerol mono-, di-, and tri-esters. In unplasticized compounds, with and without acrylic modifiers, combination (or balanced) lubrication is obtained without sacrifice in heat stability [28]. In plasticized compounds, somewhat higher levels provide useful external lubrication, again with good stability [31].

In semipolar polymers (thermoplastic CPE, EVA, PVC–VA) most esters are sufficiently compatible to function efficiently as plasticizers. This is also true with most elastomers; therefore, typical lubricant effects should not be expected of esters used in rubber-modified thermoplastics (e.g., NBR/PVC blends).

Many lubricants not primarily intended as lubricants are also partially compatible esters. These include the widely used antioxidant, stearyl di-*t*-butyl-hydroxyphenyl propionate, and the synergists dilauryl and, especially, distearyl dithiopropionate. Popular in polyolefins, styrenics, and thermoplastic elastomers, combinations of these ingredients contribute to internal lubrication. It is commonly found that these additives contribute to processibility not only by increasing process safety but also by lowering viscosity, often synergistically, with more typical internal lubricants. The domain of an antioxidant should be regarded as simply one more location where some degree of polymer motion is desirable, analogous to the useful coordination of stabilizers with lubricants.

F. Metallic Stearates, Versatile Surfactants

Considering the range of additives which function as lubricants, no group has a broader range than that of the metallic salts of organic acids. These are principally metal stearates, although there has been some use of analogous montanates, particularly calcium. In addition, alkali metal and ammonium salts of oxidized waxes are frequently used in emulsion coatings of thermoplastics, and metal salts of the lower acids (octoates, laurates, etc.) find widespread use as additives in many polymeric systems. No doubt, some of the various aspects of lubrication are involved in these applications, at least secondarily. Nonetheless, systematic investigation as lubricants has centered on the use of metallic stearates.

In polyolefins, it is expected that stearate salts would function externally; clearly, the use of calcium stearate in HDPE and PP provides external lubrication in extrusion and injection molding. At times, minor amounts of sodium stearate are included to boost the external function [32]. Surprisingly, calcium stearate also helps internal lubrication in polyolefins. Two factors are involved: First, the synergism between stearates and other lubricants. These may be added specifically for internal lubrication (e.g., waxes) or may be present for other functions such as the lubricating antioxidants discussed earlier. The second is the ease with which stearates associate with a variety of pigments and fillers, providing internal lubrication by coating the surface of such particles, because of the equilibrium between tautomeric forms shown in Fig. 4.

With styrenics, a similar situation obtains. In PS, zinc stearate is commonly used as an external lubricant, again, at times, bolstered by sodium stearate. Enhancement of internal lubrication of additives such as stearate esters is typical of both zinc and barium stearates in PS. In ABS, calcium and magnesium stearates serve as combination lubricants, lubricating internally through filler surface treatment. It would be anticipated that calcium stearate would be preferable in this regard with calcium carbonate filler, magnesium stearate with fillers which are magnesium salts (talc) [23]. Reversing this (i.e., decreasing the filler–lubricant association, or pretreating the filler) would maximize external function.

In the above systems, metal stearates may also function as secondary stabilizers. This may be true as well in polycarbonates, where calcium stearate is used as a combination lubricant, at times in concert with stearate esters. Zinc stearate provides balanced lubrication in cellulosics with similar combinations. In polyamide formulations, aluminum stearate is often used with silicate fillers (for which it has strong affinity), and zinc or cadmium stearate is used for balanced lubrication.

Where external function is important, this may be promoted by the addition of sodium or lithium stearate [33]. Thermoplastic polyesters also derive balanced lubrication from zinc stearate, as do thermoplastic elastomers.

The use of stearates in PVC is, as usual, complex. In plasticized wire insulation compounds, where lead stabilizers are used to minimize moisture absorption, lead stearate is a common secondary stabilizer, also providing internal lubrication through coordination with the lead stabilizer. The external lubrication provided

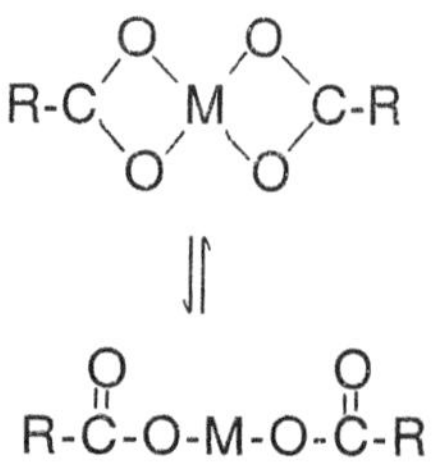

Figure 4   Divalent metallic stearates.

by ester or amide additives may, in addition, be enhanced. In some insulation compounds, the only lubricant may be the stabilizer; most often a barium stearate–lead stearate blend; less often a barium stearate–cadmium stearate blend. In plasticized, lead-stabilized formulations other than in wire and cable, blends of lead stearate and calcium stearate are used as lubricating costabilizers. This is also common in unplasticized PVC formulations which use lead stabilizers, as in pipe compounds in some parts of the world. In these cases, the blend of lead and calcium stearates functions internally, and paraffin or other wax is used for external lubrication. Unless added as a lubricant–stabilizer or a lubricant used with conventional lead stabilizers, there is little point in using lead stearate. Some use has also been found in thermoplastic CPE and CSM compounds, also in conjunction with lead stabilizers and in situations where low moisture absorption is desired. In plasticized compounds not containing lead, blends of barium and cadmium stearates, used as the primary stabilization system in conjunction with organic phosphites or epoxides, also contribute to lubrication.

Calcium stearate probably appears more often than any other ingredient (except PVC) in unplasticized PVC recipes. It functions as a secondary stabilizer (e.g., in tin-stabilized compounds), in some cases as the primary stabilizer, but it is always also a lubricant. If it is the principal lubricant (other internal lubricants not present) and used, for example, with EBS as external lubricant and if added early (to promote its rapid adsorption), its function is classically internal. Time to fusion is decreased and consequent process viscosity high [28]. If other internal or balanced lubricants are present, particularly if they are added early in the blending cycle, and calcium stearate added later with external lubricants, the reverse behavior is found. In such compounds, it has been shown that many stearates may show behavior usually attributed either to external or internal lubrication by varying the other lubricants in the system and the relative proportions [34]. Several discussions of the role of calcium stearate in rigid PVC explore this behavior in detail [35,36]. Despite this potential for complexity, it has been shown that in simple systems using paraffin as the external lubricant and no other internal lubricant, calcium stearate exhibits a completely internal function and can compensate for external overlubrication [13, p. 839]. Contradictory reports are found only when stearates are used in concert with multifunctional lubricants. [See Section I.D. regarding whether calcium stearate should be considered internal in PVC (the traditional view) or as an external lubricant acting internally.]

Lubricants may become multipurpose through the combination of ingredients or by the design of a unique additive. An interesting example is the combination of ester and metal salt functionality obtained by partial saponification of a montanic acid ester. Such products, offered commercially by Hoechst, provide a range of polarity in the functional group. In PVC, styrenics, and thermoplastic urethanes, a combination of internal and external lubrication invariably results. This may be shifted in one direction or the other by further addition of a specific internal or external lubricant, because the range of solubility parameter provided by the ester/salt offers compatibility with a variety of other lubricants. This useful approach could be extended, for example, to ester–amide combinations, with potentially wide applications.

III. ANALYSIS AND SELECTION OF LUBRICANTS

A. Classifying Lubricants Semiquantitatively

The factors in selecting a lubricant are as follows:

1. The polarity of different sections of a polymer and the volume fractions of these sections
2. The same considerations with regard to the lubricant molecule.
3. Other species to be encountered: the surface of fillers, pigments and stabilizers; equipment surfaces
4. Compound behavior without the additive (i.e., what modification is desired)

The first two factors suggest a donor–acceptor scale for correlating lubricants with polymers. The feasibility of such scales has been discussed in detail by Jensen [37]. In order to be generally useful to the applied plastics technologist, a scale should be based on readily available parameters. At this time, the most reasonable choice is the solubility parameter. What is needed is not only the solubility parameter of the polymer and lubricant overall but also estimates of the contributions of the interactive and noninteractive parts of each. These may be calculated by several methods (e.g., from group contributions to the heat of vaporization and to the molar volume) [38]. Compilation of values from a number of methods is given in the monograph by Barton.[39], and some of interest regarding lubricants are given in Tables 1–4. Systeme International (SI) units

Table 1 Solubility Parameters (MPa^{-1}) of Functional Groups

—CH—CH	14.3
—CH—	16.8
—CH—CH(CH—)CH— (branch point)	18.6
—CH—COOH	26.5
—CH—COOH dimerized	21.8
—CH—COCH—	19.7
—CH—CONH	34.2
—CH—CONH— hydrogen-bonded	22.5
—CHCl—	19.9
—CH—COOCH—	19.3

Source: Calculated from data from Ref. 38.

Table 2 Solubility Parameters (MPa^{-1}) of
Thermoplastics

Polyolefins	16–17.5
Styrenics	18–18.8
PVC homopolymers	19–19.5
PVC copolymers	18.5–19
Cellulosics	20.5–23

Source: Data from Ref. 40.

Table 3 Solubility Parameters (MPa^{-1}) of
Lubricants

Stearic acid monomer	19.2
Stearic acid dimer	17.8
Glycerol monostearate	17.5
Stearamide	19.8
EBS (hydrogen-bonded)	17.5
Divalent stearates	18
Alkali metal stearates	19
PE wax, paraffin	17
Oxidized waxes	18–19
Long-chain esters	17.5–18.5

Source: Data compiled from Ref. 39.

Table 4 Interaction Parameters (MPa^{-1}) of
Surfaces

Titanium dioxide	34
Carbon black	25–35
Carbonate fillers	26–28
Sulfate fillers	26–28
Silicate fillers	30–40
Iron oxide	28

Source: Data from Ref. 41.

are used. Many older tables (e.g., Ref. 6) use units previously in vogue (based on calories instead of joules). These solubility parameters may be converted to SI units by multiplying by 2.05 (the square root of the conversion factor from calories to joules). This can be a source of confusion at first glance because 15–16 Hildebrands (based on calories) would denote a very polar material, but 15–16 MPa$^{-1/2}$ (SI units), a nonpolar one.

B. Principles of Lubricant Selection

Because function as a lubricant depends on partial compatibility, the first principle (perhaps a Zeroth Law) is that the solubility parameter of the lubricant must not be the same as that of the polymer (it is then a plasticizer.) At the other extreme, partial compatibility implies that these values be no more than 3–4 MPa$^{-1/2}$ different. The top of this range, if the additive has the lower value, is appropriate to the case of an external lubricant whose exudation is desired (see incompatible lubricants Sec. II.F). If the additive has a solubility parameter 3–4 MPa$^{-1/2}$ higher than that of the polymer, it may be a filler treatment, a tackifier, a nucleating agent, or a partly soluble inclusion of other (or perhaps no) utility.

For internal lubrication, the following conditions must be met. The overall solubility parameter of the lubricant must be within 3 MPa$^{-1/2}$ of the polymer. The solubility parameter of the interactive functional group should be a good match for the polar part of the polymer (e.g., ester, ketone, or carboxylate groups for PVC; hydrocarbon branch points for polyolefins). The solubility parameter of the noninteractive tail of the lubricant should be a poor match for the polymer (≥ 3 MPa$^{-1/2}$ lower); for semipolar polymers, a 12-carbon-atom chain normally suffices.

The combination of the above requirements places limits on the fractions of polar and nonpolar components. For instance, propionic acid contains both a terminal methyl group and reactive acid dimer. The molar volumes involved, however, generate an overall solubility parameter (21.5–22) which would be useful only in extremely polar polymers. Internal lubrication in common thermoplastics with additives having typical functional groups constrains hydrocarbon side chains to the C14–C22 range. If the lubricant has another function (e.g., a lubricating stabilizer or antioxidant), the success of this function is usually dependent on the criteria for internal lubrication being met.

For effective filler surface treatment, the functional group of the lubricant should have a higher solubility parameter than that of the polar part of the polymer. This permits the highly polar filler surface to compete with polymer for lubricant. If a filler is chosen that is less polar than the polymer, the above does not apply. A normally internal lubricant may be of some value if the polymer-noninteractive section is a good match for the nonpolar filler. Filler pretreatment with additives usually does not lower surface polarity drastically enough to put the resultant product in the category of nonpolar fillers. (This is intentional; interfering with the ability of the polymer to wet polar fillers usually is not a goal.)

C. Multifunctional Lubricants

In most processes, a balance of lubricating functions is sought. In high-shear mixing, either with an internal mixer or twin-screw extruder, internal lubrication promotes flow of the polymer around other additives and rapid recombination (i.e., enhances distribution). Filler treatment with lubricant (within limits, see Section III.B) increases the rate of wetting by the polymer. The combination of these effects should not, however, lower the shear stress to the point where dispersion (the reduction of particle agglomerates) is hindered. In such cases, it is preferable to pursue the needed level of dispersion before lubricant addition. Sufficient external lubrication is required for the compound to remain more cohesive than adhesive and exit cleanly. With high-shear mixing, it is therefore desirable that internal lubrication and filler surface treatment occur either very early in the cycle or immediately after distribution of a filler which requires strong dispersive action, and that external lubrication take place toward the end of the cycle. This calls for a lubricant or blend having a polar group of relatively broad solubility parameter and overall size large enough so that total compatibility decreases strongly with increasing temperature. If, for example, low-MW PE wax is used as the internal lubricant for an unfilled polyolefin, at the dump temperature of an internal mixer, say 150–165°C, it will also function as an external lubricant [despite being of little use externally at low temperatures, (e.g., as a mold-release agent)]. Using this approach, it is feasible to include process release agents in highly adhesive compositions and deliver them more or less cleanly from mixing equipment without compromising their later function. With high-shear mixing, the choice lies between a lubricant (or blend) providing sequential function and the addition of an internal lubricant early in the cycle and an external lubricant near the end.

With low-shear mixing of powder blends, there are two distinct cases. Powdered polyolefins, CPE, and some specialty thermoplastics are normally fluxed completely before shaping the product and are usually not heated during blending. In this case, it is desired to adhere any lubricant which must later function internally to the polymer surface and any surface treatment to the filler particles. This may be accomplished simply by the order of addition. This is the prevalent practice with polyolefins; therefore, combinations of lubricants rather than multifunctional lubricants predominate.

Other powdered thermoplastics (PVC, chlorinated PVC, some fluoropolymers) are fluxed or fused at temperatures below the melt point during shaping. Here, the blends are commonly heated during mixing. The choice of lubricants in these cases is governed by the nature of the subsequent fabrication process.

The goals of early permeation of internal lubricant into the polymer and adsorption of surface treatment on the filler remain, but the choice and point of addition of external lubricant is dictated by the fusion time found to be optimum. Short fusion times are consistent with the early addition of combination lubricants of primarily internal function and with the late addition of low levels of specific external agents. Long fusion times are reached by the early addition of lubricants having significant external function. This is often possible through use of a high-MW lubricant (e.g., oxidized PE wax in PVC). Although the overall solubility parameter match is very good, suggesting normal function as an internal lubricant, its high MW results in a low rate of permeation during blending. The high MW also reduces high-temperature compatibility. Therefore, the additive functions first externally to delay fusion, then internally during fabrication, and later externally at die, nip, and runner surfaces. This type of behavior has prompted considerable contradiction in the literature with regard to the classification of lubricants [7].

Sequential function has an important presumption: sufficient time to complete one process before starting the next. Processes demanding a short fusion time are generally output driven; time is short. As a result, such processes are usually served by the multiple addition of specific function lubricants. Time being short refers to the volume of compound processed per time, not the line speed. Extrusion of PVC pipe is a process in which time is short; the separate early addition of calcium stearate and late addition of paraffin during blending is common. Extrusion of small-gauge plasticized PVC wire insulation from a dry blend may involve very fast line speeds, but extruder dwell is not short (in comparison). Here, multifunctional lubricants are widely used.

The balance between internal and external lubrication is unique in the case of molding. In almost all other processes, external lubrication is intended to facilitate flow in the region of a hot process surface. In extrusion, for example, streamlined die flow is desired, but undue loss of friction at the cooler screw surface is not. In injection molding, the need for this balance of properties during the extrusion phase is followed rapidly by the urgent requirement that the compound release from a cold metal surface. Lubrication invariably plays a part in this. There are two distinct cases. As the shaped compound cools and shrinks, it prefers thermodynamically to dewet the mold surface or it does not (i.e., is an adhesive for that surface). If the latter is the case, improved mold release follows with external lubrication. Decreasing polymer compatibility with decreasing temperature (the most common situation) calls for a low-MW additive. These are naturally prone to deposit and should be chosen so as to facilitate cleaning. The operation of this type of release may be checked by analysis of deposits and of the final concentration of lubricant at the surface versus the interior of the part.

The molded compound, on the other hand, may not be an adhesive but not dewet anyway, simply because the rate is too slow to be useful; that is, the compound viscosity at demolding temperatures may be so high that the rate of movement of chain segments from the surface vanishes. In this case, internal lubricants promote mold release by increasing the segment mobility (internal rotation). If this mechanism operates, usually deposits will not be found and analysis will fail to indicate a difference in lubricant level at various points in the part cross section. In extremely viscous compounds (e.g., thermosets), the release properties of internal lubricants have been shown to be very significant [42].

Balanced lubricants are available which are blends having a single functional group, but different sizes of the noninteractive section (e.g., proprietary ester blends) and also the reverse [different functional groups, single-sized noninteractive tail (e.g., partly saponified montan esters)]. The pattern of commercial use suggests that both approaches are effective. Then, there are those that are serendipitously multifunctional: stearic acid with, in effect, two functional groups, monomeric and dimerized —COOH; and stearates, with polarity of the functional group influenced by that of the environment (one versus both oxygens coordinated to the cation) [1, pp. 48–55].

Whether one uses a commercial blend or prepares one's own is a matter of convenience and economics. In doing the latter, it must be appreciated that not all blends will be multifunctional (paraffin wax and silicone fluid being, in themselves, complex blends). A blend to provide balanced lubrication in a given polymer should contain components with the solubility parameter characteristics discussed earlier for the individual functions. If simultaneous function is desired, the components should have closely similar overall solubility parameters. If sequential function is needed, the components should meet the above criteria for specific behavior but be less compatible among themselves.

It can be seen, it is hoped, from the above analysis that the use and selection of lubricants differs little from other areas of formulation, that what is required is principally an accurate statement of the requirements of process and service conditions, that level and choice of lubricant cannot be separated from selection of other ingredients, and that a logical rationale for experimentation, if not perfected, is at least developed sufficiently to help in most cases.

D. Guide to Lubricant Types

Table 5 is a summary of current reported usage of lubricants in thermoplastic compounds. It is intended to supplement rather than replace consideration of the properties which must be achieved, the processing which must be accommodated, and the cost framework which must be met by a given set of ingredients. The compounder is encouraged to use the principles of partial compatibility to extend or to deviate from current practice as his or her analysis may indicate.

Table 5 Guide to Lubricant Types

Lubricant	Olefins	Styrenics	PVC	Others
Paraffin	Int.	Comb.	Ext.	Comb. CPE, blends
PE wax	Int.PE Ext.PP	Comb.	Ext.	Comb. CPE, CSM
PP wax	Int.PP			Tackifier PE
Oxid. PE wax			Comb.	Int. CPE, CSM
Fatty acids	Comb.	Comb.	Int.	Comb. plasticized PVC
Long-chain alcohols		Int.		
Stearone	Comb. (EVA)			Comb. (PVC–VA)
Stearamide	Ext.	Comb.	Int.	
EBS	Comb.	Ext.	Ext.	Comb. polyamide
Stearate esters	Ext.	Int.	Int./Comb.	Comb. PC
Complex esters	Ext.	Comb.	Comb.	Ext. plast. PVC
Stearate salts	Ext.	Ext./Comb.	Int./Comb.	Comb. TPE

Note: Ext., external; Int., internal; Comb., combination.

REFERENCES

1. RC Mehrotra and J Nehva. Metal Carboxylates. New York: Academic Press, 1983.
2. RF Grossman. The origins of compounding. SPE Palisades Section, 1994.
3. AK Doolittle. J Polym Sci 2:121, 1947.
4. JD Ferry. Viscoelastic Properties of Polymers. New York: New York, 1961, p 365.
5. AK Doolittle. The Technology of Solvents and Plasticizers. New York: Wiley, 1954.
6. Allied Chemical, "Solubility Parameters." Specialty Chemicals Div. Technical Data Bulletin G-7, 1978, p 4.
7. MC McMurrer. Plast Compound 74, July/Aug. Vol. 5, 1982.
8. EB Rabinovitch, E Lacatos, JW Summers. J Vinyl Tech. 6:98, 1984.
9. RA Lindner. Calcium stearate, internal or external lubricant? SPE Palisades Section, 1995.
10. EL White. In: LI Nass, ed. Encyclopedia of PVC, Vol. 2. New York: Marcel Dekker, 1977.
11. T Hawkins, J Vinyl Tech 4(3):110, 1982.
12. G Ullman. SPE J 23(6):71, 1967.
13. AJ Yu, P Boulier and A Sandhu. 1984 SPE ANTEC Proceedings, 1984, p 838.
14. RF Lippoldt. Plastics Engineering, p 37 Sept. 1978.
15. T Jennings. J Vinyl Tech 2(1):67, 1980.
16. A Bohaczuk. 1985 SPE Vinyl Div. RETEC PVC Primer, 1985, p 216.
17. GR Chapman and D Priester. Polyolefins V. SPE, Brookfield, CT, 1987, p 271.
18. DF Klein. Polyolefins V. SPE, 1987, p 285.

19. H Green. Industrial Rheology and Rheological Structures. New York: Wiley, 1949.
20. HS Katz, and JV Milewski. Handbook of Fillers and Reinforcements for Plastics. New York: Van Nostrand Reinhold, 1978, p 55.
21. B Arkles. Silane Coupling Agents. Bristol, PA: Petrarch Systems, 1984.
22. SJ Monte and G. Sugarman. J Elastomers Plast 8(1):30, 1986.
23. RF Grossman, FW McKane. Rubber Plast News, 23 March 1987, p 12.
24. JW Summers. 1986 SPE Vinyl Div. RETEC Proceedings, 1986, p 229.
25. Dow Chemical Co. Technical Bulletin Nos. 305-895-281 and 305-896-281, 1981.
26. LF Fieser, M Fieser. Organic Chemistry. New York: Reinhold, 1950, p 113.
27. RL Shriner and N Fuson. The Systematic Identification of Organic Compounds. New York: Wiley, 1965, pp 312–325.
28. RA Lindner. Plast Compound July/Aug. Vol. 6, 1983, p 134.
29. DM Detweiler, and MT Purvis. Soc Plast Eng Tech Pap 19:647, 1973.
30. PJ Davis, SJ Fraser. Soc Plast Eng Tech Pap 19:477, 1973.
31. RA Lindner, R.A. Bohaczuk. Plast Compound Jan./Feb. Vol. 6, 1983, p 90.
32. DJ Eckert, S Meinstein. Plast Compound July/Aug. Vol. 5, 1982, p 65.
33. TE Breuer. Modern Plastics Encycopedia. New York: McGraw-Hill, 1985–1986, p 157.
34. JE Hartitz. Polym Eng Sci 14:392, 1974.
35. RA Lindner. Plast Compound Sept./Oct. Vol. 4, 1981, p 35.
36. JD Bower. Plastics Compounding 2:64, 1979.
37. WB Jensen. Educational Symposium No. 6, Am. Chem. Soc. Rubber Div., Akron, OH, 1981.
38. RF Fedors. Polym Eng Sci 14(2):147, 1974.
39. AFM Barton. Handbook of Solubility Parameters. Boca Raton, FL: CRC Press, 1983.
40. RF Blanks, JM Prausnitz. Ind Eng Chem 3:1, 1983.
41. CM Hansen. J Paint Technol 39:505, 1967.
42. TC Jennings and CW Fletcher. 1972 SPI Proceedings, 27th Technical Conference, 1972.

10
Optical Brighteners

Thomas Martini
Hoechst AG, Frankfurt, Germany

I. INTRODUCTION

Even in ancient times, attempts seem to have been made to improve the yellowish appearance of ''white'' products, principally textiles. The first bleaching process put into practice may well have been bleaching in the sun (bleaching on grass). In the literature, it is reported that toward the end of the eighteenth century, extracts made from chestnuts were used to strengthen the action of bleach liquors. In 1919, Krais isolated esculin (Structure I) as the fluorescent substance from horse-chestnut extract and demonstrated that it made possible a marked improvement in the whiteness of bleached linen yarn and viscose artificial silk, though, unfortunately, the fastness properties to washing and light were unacceptable. Confirmation was thus afforded of the principal enunciated in 1921 by von Lagorio [1] that fluorescent colorations.reflect *more* visible light than they receive as radiation. Krais' article, entitled ''Uber ein neues Schwarz und ein neues Weiss'' [2] contains the historic proposition that even the whitest white can be made whiter.

I

The development of optical brighteners, which began in the 1940s, attempts to provide the wide spectrum of products required by industry. These products

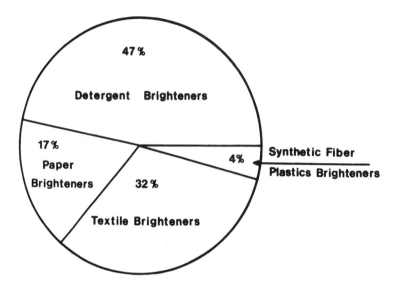

Figure 1   Breakdown of optical brighteners by percentages.

are used in detergents paper, and textiles and also in the manufacture and process-ing of fibers and plastics [3].

The total market is estimated at about $450 million on a worldwide basis, the principal products being brighteners for detergents and paper. It can be seen from Fig. 1 that the proportion of brighteners for plastics and synthetic fibers is relatively low, at 4%. However, increasing demand for high-quality products enables optimistic market forecasts to be made.

II.   MECHANISM OF OPTICAL BRIGHTENING

A.   How Optical Brighteners Act

The presence of ultraviolet (UV) radiation in light is essential for the action of optical brighteners.

Visible light denotes electromagnetic ₒradiation in the wavelength range between 400 and 700 nm (1 nm = 10 AA = 10^{-9} m). The portion of the spectrum between 400 and 430 nm appears violet, from 430 to 485 nm, it appears blue, from 485 to 570 nm, green, from 570 to 585 nm, yellow, from 585 to 610 nm, orange, and above 610 nm, red. When the light rays of daylight fall on an object, they can pass through it, be absorbed, or reflected. If all the light radiation is absorbed, the object is black, whereas total reflection makes the object appear

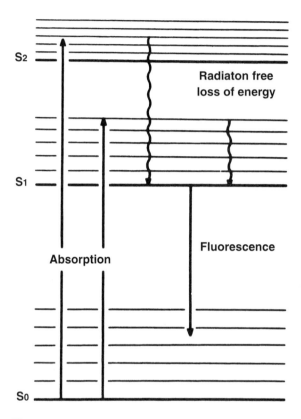

Figure 2 Emission of UV radiation as visible light.

white. As a result of their chemical structure, optical brighteners absorb in the ultraviolet region of the light spectrum and reflect in the visible region at about 440 nm. Accordingly, they transform invisible UV light into a bluish-violet fluorescent light. The phenomenon of the transformation of UV light can be understood more easily by means of a simplified diagram (Fig. 2). From the ground state, the electron passes, as the result of absorbing light, into an excited state (S_1 or S_2 in Fig. 2). From this excited state, the electron then decays into the lowest vibrational level of the first excited state S_1. This transition takes place without radiation and is brought about by loss of energy as heat within the excited molecule. The transition is relatively short-lived and has a life of 10^{-12}–10^{-15} s. On the other hand, the life of the lowest excited state S_1 is longer by approximately 10^{-8} s. The electron then drops back into the ground state. The radiation emitted is recorded as fluorescent radiation. The fluorescence is displaced toward longer wavelengths because of the loss of energy while in the excited state.

For an optical brightener, this means absorption in the near-UV range (350–380 nm) and emission (fluorescence) within the visible range (430–450 nm). In the ideal case, an optical brightener should emit one quantum of light for every quantum of light that is absorbed; the quantum yield should be 1. However, energy loss in the excited state results, in some cases, in quenching of fluorescence (caused, e.g., by small amounts of extraneous substances), so that the quantum yield can be less than 1.

A high degree of purity and a solution or excellent dispersion in the substrate to be brightened are fundamental requirements for the effectiveness of a brightener.

The ratio between the light reflected by an object at a specific wavelength and the component of light reflected by an ideal white surface is known as *diffuse reflectance*. If the various values of diffuse reflectance at different wavelengths within the region of the visible spectrum from 400 to 700 nm are plotted for a specific body color, a curve of diffuse reflectance is obtained which is characteristic of that body color. Finally, if the reflection is different in different regions of the visible spectrum, an impression of color will be obtained. The curves of diffuse reflectance will also vary correspondingly.

Diffuse reflectance is quoted as a percentage, a diffuse reflectance of 100% indicating the reflection of all the radiated light. The standard of measurement used for this purpose is magnesium oxide, which reflects light radiation almost completely.

If the curve of diffuse reflectance of a plastic [e.g., polystyrene (containing 0.5% of anatase titanium dioxide)] is recorded against magnesium oxide, the following differences will be observed (Fig. 3, curves A and B).

1. The diffuse reflectance curve of the plastic is below that of magnesium oxide (i.e., polystyrene is less white than magnesium oxide).
2. Curve B shows a decrease in passing from longer to shorter wavelengths, the greatest decrease of diffuse reflectance being found in the blue region. This blue defect is the cause of the yellow color.
3. The curve can be extended and the maximum in the yellow–red region can thereby be weakened by adding a blue or blue–violet dye, which does not reflect in the region of longer wavelength, but rather absorbs.

Although this reduces the yellow shade of the material, at the same time the total diffuse reflectance is decreased (curve C in Fig. 3). Because of the deficiency in yellow, the article appears whiter to the observer; this psychological effect plays an important part. The tinted plastic is, however, less bright, because its emission of light is reduced.

The action of optical brighteners is completely different. As mentioned previously, these substances absorb in the ultraviolet portion of the spectrum and reflect at about 440 nm in the visible region. If the diffuse reflectance curve of

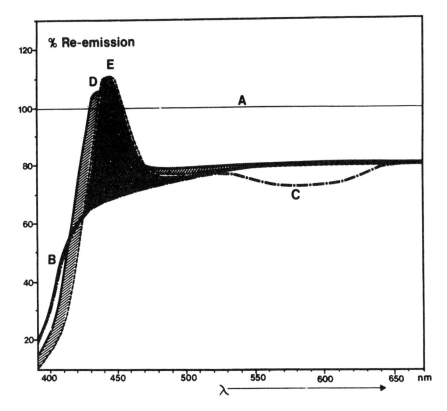

Figure 3 Diffuse reflectance curves: A = magnesium oxide, B = polystyrene, C = polystyrene with a blue dye, D = polystyrene brightened with a reddish-tinged brightener, and E = polystyrene brightened with a greenish-tinged brightener.

a material treated with an optical brightener is recorded, a curve shaped like D or E in Fig. 3 is obtained. It will be seen that the curves have risen to the extent indicated by the hatched section. As a result, the diffuse reflectance in the blue–violet region of shorter wavelength greatly exceeds the diffuse reflectance of magnesium oxide, hence the 100% limit. The diffuse reflectance is thus no longer predominantly in the yellow to red region of longer wavelength.

Optical brighteners of different chemical structures differ in their diffuse reflectance value, hence in effectiveness, and in the position of the diffuse reflectance maximum. The latter property makes it possible to identify a first class of optical brighteners: compounds whose maximum is in the decidedly shortwave region from about 430 to 440 nm. Products of this type are also often described as red–violet or reddish-tinged brighteners (Fig. 3, curve D).

In contrast, bluish tinted products have a fluorescence maximum at 440–445 nm, whereas optical brighteners whose diffuse reflectance maximum extends to about 450 nm appear as decidedly greenish (Fig. 3, curve E).

B. Degree of Whiteness

Although white is not formally a color, such as blue, green, or red, it can be determined by colorimetric methods in terms of defined standards of whiteness. By calculating the standard tristimulus values X, Y, and Z by the spectral method or by direct measurement using the tristimulus method [4], numerical values can be obtained from whiteness formulas; these values are comparable. A whiteness formula proposed by Ganz [4] has the general form

$$W_{\text{whiteness}} = (DY) + (Px) + (Qy) + C$$

in which Y, x, and y are the appropriate numerical color values for brightness and trichromatic coefficients, and D, P, Q, and C are parameters for the magnitude of the color shade preference. A further formula enables the shade of a brightening to be characterized colorimetrically [5]. In spite of the problems in assessing white visually (physiologically induced individual preference for a specific shade), in practice, visual color matching of a white article is still of prime importance because none of the known whiteness formulas can satisfy all requirements.

C. Chemistry of Brighteners

A monomolecular distribution in the substrate (solution or excellent dispersion) is a fundamental requirement for an optical brightener to be effective in a plastic. The variation in solubility in different plastics is one of the causes of whiteness effects being dependent on the substrate. The fundamental chemical structures II, III, and IV are the most important and effective brighteners currently marketed for plastics and synthetic fibers.

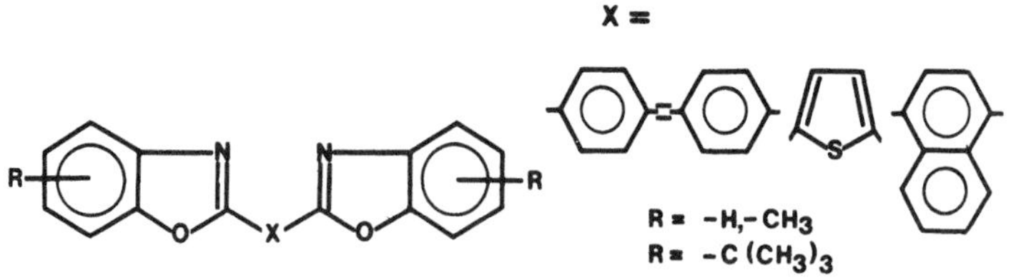

II

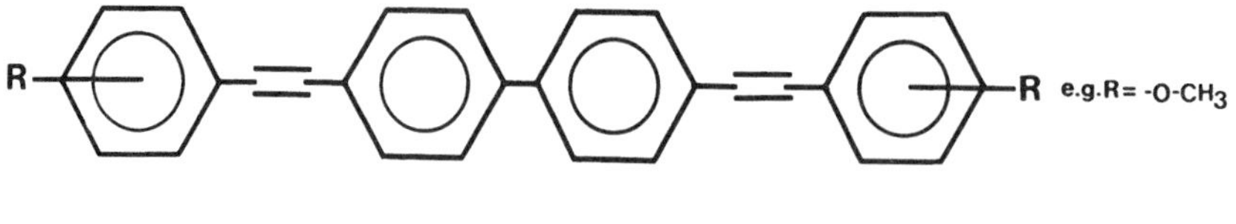

III

IV

III. USE OF OPTICAL BRIGHTENERS IN PLASTICS

A. Areas of Use

As mentioned in Section II.A in comparison with "bluing," it is advantageous to employ brighteners for plastics to compensate for the yellow tinge often noticed in these materials. Very low amounts (1–10 ppm) of an effective brightener are sufficient to produce a distinct increase in the whiteness of a pigment-free plastic. Particularly good brightening effects can be achieved using a combination of brightener and dyes. Brighteners for plastics are therefore also sold in the form of mixtures with dyes or colored pigments. Adequate concentrations of dye or colored pigment are as low as 0.1–0.5% of the optical brightener.

In virtually all fields in which plastics are used, there is a demand for very white finished articles of everyday use: handbags, raincoats, refrigerators, cables, and so on. These requirements are often not met using white pigments such as titanium dioxide (rutile or anatase), chalk, barium sulfate, or zinc sulfide alone. The use of 50–500 ppm of optical brighteners results in improved effects that are technically and commercially satisfactory.

Optical brighteners coloristic effects are not limited to white or colorless plastics. Interesting effects can also be achieved by adding them to pastel colorations. The resulting color shades are clearer and more brilliant and are shifted toward blue.

B. Heat Resistance

Plastics are commonly processed at sufficiently high temperatures that the resistance to heat of a brightener in contact with the plastic melt becomes of particular importance. Resistance to sublimation under these extreme conditions must also be ensured, as nonreproducible whitening effects can result. Experience shows that the customary brighteners for polyester spinning compositions satisfy the requirements of the plastics manufacturing and processing industries in this respect.

C. Lightfastness

High requirements are, of course, needed for optical brighteners in respect of fastness to light. In general, it can be said that modern brighteners for plastics have the same high level of fastness as high-grade organic pigments.

If the data are examined, it will be seen that the fastness values exhibit a wide range of scatter over various types of plastics. It is known not only that the fastness values can differ according to the origin of the plastics concerned but also that additives can have a decisive effect on the stability values. As in the case of dyes or pigments, it is possible to determine fastness values only on test specimens taken from actual practice.

Three methods of testing will be outlined briefly, representing approaches used in the United States and in Europe.

1. AATCC Test Method 16C–1977

In the practical method of the American Association of Textile and Color Chemists (AATCC) [6], the brightened plastic is illuminated behind glass and with air circulation, the test cabinet facing south in the northern hemisphere and north in the southern hemisphere. The window glass used has an optimum transmission of 90% at 370–330 nm and is opaque to light of a wavelength below 310 nm. The changes in the plastic under the influence of light are recorded by standard comparisons [blue wool lightfastness standards: 2 (very low)–9 (very high)], relative to a gray scale. In this scale, an increase of one point indicates a doubling of the stability. The relatively long exposure time is a disadvantage of this method. An accelerated natural sunlight test method is given in ASTM D4364, where Fresnel reflecting elements are used to concentrate sunlight.

2. DIN 54,004

In the European standard numbered 54,004 [7], the light source used is a xenon arc lamp with a color temperature of 5500–6500 K. Ultraviolet radiation is reduced by means of special UV-absorbing glass (90% transmission of the 380–750-nm range). Exposure is carried out under precisely defined humidity conditions (climatically controlled conditions). Fastness is measured on a scale of eight blue standard dyeings on wool, graduated in fastness in ratings running from 1 to 8 in ascending order of quality. In the assessment, the alteration in the standards of the sample is compared with the alteration in contrast (yellowing or graying) of the brightened sample and the standard of lightfastness is assigned, which shows a similar difference in contrast with the unexposed part of the dyeing, again relative to a gray scale.

3. Current Methods

Changes in visual appearance in plastic articles are most commonly investigated using artificial sources such as a xenon arc (ASTM D4459) or calibrated fluorescent tubes (ASTM D4329). These have largely replaced carbon arc sources (ASTM D1499). Although visual assessment of changes is still used [8], there is increasing reliance on instrumental methods.

D. Migration and Toxicological Properties

The brightener dissolved in the plastic must neither migrate to the surface nor pass into an article in contact. In the first case, the brightened material acquires an uneven yellow–green appearance and becomes unusable. Long-term-storage tests, including tests at elevated temperatures (e.g., 50°C) and an accurate determination of the saturation concentration of the brightener in the plastic are fundamental requirements in making recommendations concerning applications.

The migration of the brightener into an article or medium in contact is a particularly important factor in the toxicological assessment of plastics that come into contact with food. A determination of the extent to which the brightener migrates into test substances similar to food, such as peanut oil, water, ethanol, and 3% acetic acid, at various temperatures and for various contact times, is essential for official clearance by government agencies concerned with public health. Because of their many possible applications in plastics, sanctioned brighteners for plastics are subjected to very precise toxicological examination. Expensive trials must be carried out covering long-term animal feeding investigations, mutagenicity tests, determination of acute toxicity, toleration by the skin and mucous membranes, biological degradability, and so on.

E. Compatibility with Other Additives

1. White Pigments

White pigments and other additives for plastics can have an adverse effect on the action of optical brighteners. In principle, it is necessary to regard critically any component capable of absorbing UV light. In the case of the pigment titanium dioxide (TiO_2), a distinction is drawn on the basis of crystal structure between the rutile and anatase types. If the absorption properties of the two types of

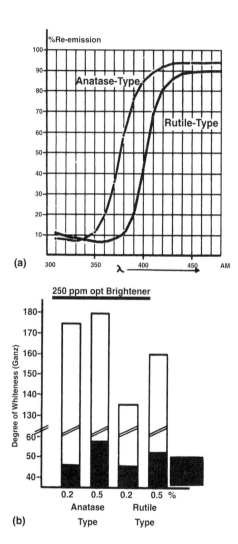

Figure 4   The effect of the pigment TiO$_2$ on the reemission and whiteness of polyethylene terephthalate.

pigment are compared, it can be seen (Fig. 4a) that the rutile pigment has a distinctly stronger UV absorption than the anatase type. As a result, UV light is withheld from the brightener, resulting in lower whitening effects (Fig. 4b).

If the rutile content is high (10%), the brightening action can be reduced to a minimum, so that no essential improvement in whiteness can be observed in comparison with the brightener-free plastic [8]. The best whitening effects are obtained with brighteners that are classifiable chemically as bisbenzoxazolyl derivatives and have an absorption maximum at 370 nm (Structure II).

2. Light Stabilizers and Antioxidants

Like titanium dioxide, the light stabilizers employed in plastics, based on benzophenone, benzotriazole, or oxalic acid diarylamide also have UV light absorption. The overlapping of the regions of absorption of the light stabilizer and the brightener also results in a lack of UV light for the brightener, and this is associated with a reduction in the whiteness of the brightened material. An exception must be made, however, for light stabilizers belonging to the group of sterically hindered amines (HALS). The transparency of these products in the UV range guarantees optimum whiteness in a number of plastics. The whitening effects of the brightener Hostalux KS in polypropylene in the presence of two different light stabilizers are compared in Fig. 5. In comparison with light stabilizer C (a sterically hindered amine), light stabilizer B (a benzophenone derivative) has a very considerable adverse effect on whiteness.

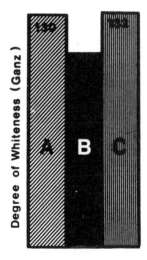

Figure 5 The effect of light stabilizers on the optical brightening of polypropylene (250 ppm Hostalux KS): A = without light stabilizer; B = with 0.5% benzophenone derivative; C = with 0.5% steric-hindered amine.

Light stabilizers are frequently combined with antioxidants. If these additives belong to the groups comprising thiodicarboxylic acid esters, alkyl disul-
fides, or organic phosphites, they lack UV absorption, do not affect the whitening
effect of the brightener, and can be employed without problems.

F. Incorporation into Plastics

In principle, optical brighteners can be incorporated into plastics like other additives by being applied to granules or powders by tumbling or dusting. It is essential
for the effectiveness of the brightener that it be soluble in the plastic (see Section
III.C). The monomolecular distribution (dissolution) in the plastic is affected
by heat treatment when the plastic is injection molded, compression molded,
calendered, extruded, or milled. If the plastic is to be processed from a solvent,
the brightener must be adequately soluble in the latter, which is usually the case
at the low concentrations employed. Depending on the brightener, the whitening
effect desired, and the type of plastic, the amounts added are between 1 and 500
ppm. In most cases, transparent plastics require only 1–10 ppm, whereas up to
500 ppm can be employed in very white grades containing pigment. Higher
concentrations are rarely used.

Because the controlled addition of such small amounts can produce inaccu-
rate results when processing plastics, it is preferable to prepare concentrates,
called masterbatches, containing 50–100 times the amount of optical brightener
in the same polymer. These concentrates are opaque to transparent, yellow–green
to red pellets and are incorporated by being mixed with the material to be brightened, which is then processed further. Formulations including chalk or plasticizers, which generally contain 10% of an active brightener ingredient, afford another
possible means of adding an accurate amount of brightener. Hostalux KS-C (chalk
formulation) and Uvitex OB-P (dicyclohexyl phthalate formulation) are examples
of this.

In the case of polyester, polyamide, and polystyrene plastics, it is also
possible to add the optical brightener during the manufacture of the polymer.
Details are given in connection with the individual plastics.

G. Analysis in Plastics

In many cases, the manufacturers and processors of plastics want to have a quantitative check on the amount of optical brightener employed. If the plastic is adequately soluble, this can be done without excessive analytical effort by extinction
measurements at the absorption maximum of the brightener. The plastic containing the brightener is dissolved in a suitable solvent, any pigment present is filtered
off, and the extinction value is determined at an appropriate dilution (2–10 mg

of brightener per liter). It is advisable to use a brightener-free solution of the plastic of the same concentration as a reference solution.

If the amounts of brightener are accurately measured, they can also be used to construct a calibration curve, hence to establish the content of brightener.

IV. CURRENT USAGE

A. Polyolefins

Among polyolefins, high and low density polyethylene and polypropylene have achieved the greatest commercial importance. Their possible uses are manifold. Domestic articles, packaging material, tool components, pipes, toys, fibers, carpet-backing fabric, and needle felt are a few examples of these. Transparent and white articles are frequently called for, and optical brighteners are employed.

The incorporation of the brighteners causes no difficulties from the point of view of process technology. They can be added and homogeneously distributed together with other additives by dry mixing with granules or powders. Because of the heat stability of the optical brighteners, no problems are encountered in processing by methods such as extrusion, injection molding, or extrusion/blow molding. It is usual to employ master batches. In fiber manufacture, the granules containing brightener are forced from a melting pot into the heated spinning unit.

Two brighteners for plastics marketed by Hoechst AG, Hostalux KS-N and Hostalux KS (stilbene bisbenzoxazolyl derivatives) exert whitening effects on polyethylene and polypropylene that are difficult to achieve with brighteners of different chemical structure. In the case of transparent articles, pronounced improvements in whiteness are achieved (compensation for yellow tinge) using only 1–10 ppm (0.0001–0.001%). If a pigment is present, the concentration of brightener added should be increased to 50–300 ppm. Figure 6 shows the whitening effects of the brighteners Hostalux KS-N, Hostalux KS, and Hostalux KCB expressed as whiteness on the Ganz scale as a function of the concentration employed. The polyethylene and polypropylene grades employed each contain 1.4% titanium dioxide (anatase). It will be seen that using as little as 200 ppm of Hostalux KS-N gives optimum whitening effects that can be achieved only with difficulty if Hostalux KCB is used.

The fact that polyethylene and polypropylene produce different whitening effects with the same quantities of brightener is explained primarily by the difference in basic whiteness between the plastics (Ganz whiteness of PE, 50; of PP, 44).

The lightfastness values shown in Table 1 are taken from the data sheets of the brightener manufacturers and are determined by the blue scale in the Xenotest

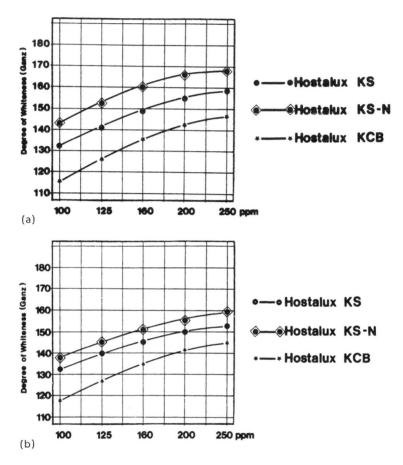

Figure 6 Whitening effects of optical brighteners on polyethylene (a) and polypropylene (b).

Table 1 Lightfastness Values of Various Optical Brighteners for Polyethylene and Polypropylene, Ganz Scale

	Hostalux KS-N	Hostalux KS	Hostalux KCB	Uvitex OB	Leukopur EGM
Polyethylene	4–5	4–5	4–5	2–4	2–3
Polypropylene	4–5	4–5	4–5	1–4	2–3

instrument. Reference should be made to the direct measurements of alteration in whiteness during exposure, discussed in Section III.C.

The fastness to light of brightened polyolefins is considerably improved in combination with light stabilizers, so that the spectrum of use of optically brightened articles is markedly extended thereby. Using 0.5% of Hostavin N20 (HALS), the lightfastness rating of Hostalux KS in polyethylene is increased from 4–5 to 6.

B. Poly(vinyl chloride)

Optically brightened poly(vinyl chloride) (PVC), plasticized or unplasticized, is to be found in many articles of everyday use. Handbags, sponge bags, raincoats, packaging film, household articles, cable sheathing, containers, and bottles are examples.

Together with pigments or other additives in the PVC compound, optical brighteners can be incorporated easily by calendering, extrusion, or injection molding. If the plastic is processed from an organic solvent (e.g., an acetone/carbon disulfide mixture), the brightener must be adequately soluble in this solvent. The concentrations used depend on the grade of PVC and the whitening effect desired. In the case of transparent films, 1–10 ppm of Hostalux KS-N is sufficient to compensate for the yellow tinge. Pigmented material is brightened with not more than 200 ppm of Hostalux KS-N or Hostalux KS. The whitening effects of a PVC compound composed of 92% of PVC powder, 5% of plasticizer, 1.5% of anatase titanium dioxide, and 1.5% of a heat stabilizer as a function of the Hostalux grades employed are shown in Fig. 7.

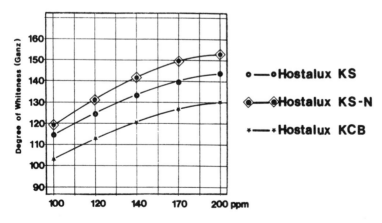

Figure 7 The whitening effects of optical brighteners on rigid PVC.

Table 2 Lightfastness Values of Some Brighteners for PVC, Ganz Scale

	Hostalux KS-N	Hostalux KS	Hostalux KCB	Uvitex OB	Leukopur EGM
Plasticized PVC	5–7	5–6	4–6	4–7	5
Rigid PVC	5–7	5–6	4–6	4–7	6–7

In plasticized PVC containing considerable plasticizer, under certain circumstances, migration to the surface of the plastic can occur, thus transporting the optical brightener (exudation). The shaped article thereby acquires a yellow–green appearance and becomes unusable. It is therefore particularly important to carry out migration tests, including tests at elevated temperatures, and to maintain the recommended concentration of brightener.

The lightfastness values of some important commercial products shown in Table 2 are taken from the data sheets of the manufacturers.

The frequent use of calcium carbonate as a pigment has induced the brightener manufacturers to provide a pigment/brightener blend. For Hostalux KS-C and Uvitex OB-C, this is in each case a chalk blend containing 10% of brightener. As mentioned earlier, this also facilitates controlled addition.

C. Styrenics

Polystyrene and acrylonitrile–butadiene–styrene (ABS) have increased greatly in importance in recent times. Applications for brightened grades are to be found in the packaging industry and for household articles, household machines, electrical appliances, and many others.

As conventional thermoplastics, polystyrene and ABS can be processed by injection molding and extrusion. The optical brightener is added to the granules to be processed on its own or in the form of a master batch, by dry mixing, most suitably together with further additives. If the granules to which powder has been applied are to be transported, it is advisable to use an adhesion promoter (e.g., white oil) to prevent separation. In addition, it is also possible to add the brightener to the monomer or mixture of monomers, which, of course, requires that it be adequately soluble therein. In general, 1–10 ppm is sufficient to compensate for the yellow tinge of a transparent raw material.

In the case of very white grades containing pigment, optimum values are obtained with 200 ppm of Hostalux KS or Hostalux KS-N. The whitening effects achieved using various bisbenzoxazolyl stilbene or naphthalene derivatives are

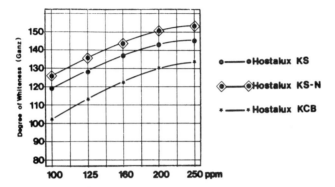

Figure 8 The whitening effects of optical brighteners on polystyrene.

shown in Fig. 8 (impact-resistant polystyrene containing 1.5% of anatase titanium dioxide).

In the case of ABS, it is advisable to increase somewhat the concentration of optical brightener added (e.g., 250 ppm of Hostalux KS or Hostalux KS-N), as this copolymer generally has a lower basic whiteness. The possibility of combining optical brighteners with dyes in pastel transparent products should be borne in mind. Marked improvements in brilliance are achieved, particularly in the case of pastel shades with a blue and red tint.

Table 3 shows the lightfastness ratings published by brightener manufacturers (blue scale, Xenotest).

D. Polyesters

In the United States, optical brighteners are used in about 90% of the white polyester textiles. In other applications, polyester is used for sheeting and as a

Table 3 Lightfastness Values of Optical Brighteners in Polystyrene, Ganz Scale

Brightener	Rating
Hostalux KS-N	6
Hostalux KS	4–5
Hostalux KCB	3–5
Uvitex OB	2–5
Leukopur EGM	6–7

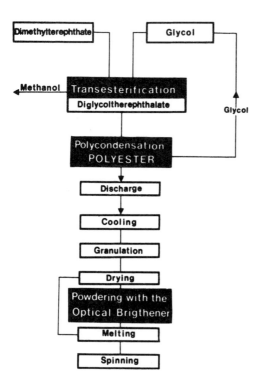

Figure 9   The possible use of brighteners for polyester spinning.

molding material for extrusion and injection molding. Polybutylene terephthalate is mainly employed in the industrial sector (electrical engineering, household appliances, machines, and apparatus engineering).

Using polyethylene terephthalate as an example, Fig. 9 shows the possible uses of optical brighteners in the manufacturing process for polyester (dark areas). The brightener is added during transesterification or polycondensation, or by powdering onto the polyester granules. The latter process is important for the plastics converter, if relatively small amounts of polyester or special grades are to be brightened. In this case, it is, however, advisable to use master batches containing 100–200 times the concentration of brightener, based on an absolute added amount of brightener of 100–300 ppm. Optical brighteners suitable for the manufacture of polymers must be able to survive process temperatures without damage and must have adequate fastness properties for subsequent use.

Properties important for textile materials include fastness to light, fastness to bleaching, and resistance to washing at the boil. Figure 10 shows the whitening

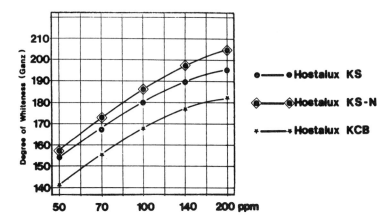

Figure 10 The whitening effects of optical brighteners on polybutylene terephthalate.

effects of Hostalux KS and Hostalux KS-N on polybutylene terephthalate, injection molded at a composition, temperature of 250°C (0.5% of anatase TiO_2). The manufacturers' data on fastness to light and shade are given in Table 4 (blue scale, Xenotest).

E. Polyacrylonitrile

Polyacrylonitrile (PAN) is used virtually only for fiber manufacture, the main use being as textile fibers. Fibers composed of 100% polyacrylonitrile are used for industrial purposes without brightening. Textile fibers are manufactured from PAN that has been chemically modified to achieve better receptivity for dyes, water absorption, and other properties. Styrenesulfonic acid, alkylsulfonates, and vinylsulfonic acid are anionic components that can be copolymerized with acryl-

Table 4 Lightfastness Values and Shades in Polyester, Ganz Scale

Brightener	Rating	Shade
Hostalux KS-N	7	Neutral
Hostalux KS	6–7	Red–neutral
Uvitex OB	7	Neutral–blue
Leukopur EGM	7	Blue–green
Eastman OB-1 (Eastman Kodak)	60–80 h, Fade-O-meter	Red

amide, methacrylamide, vinylidene nitrile, methacrylonitrile, vinyl acetate, and acrylonitrile. The polymer is then converted into fiber by wet or dry spinning, in which optical brighteners are also employed. In dry spinning, the polymer is dissolved, together with the brightener, in a solvent (e.g., dimethylformamide, dimethylacetamide), and the solvent is removed with hot air. Examples of brighteners employed in this process are Hostalux NSM (Hoechst AG), Blankophor MAN (Bayer), and Blankophor MAR (Bayer). Chemically, these products belong to the 1,3-diarylpyrazoline group. They are very readily soluble and can be added in high concentrations (up to 1500 ppm).

F. Polyamides

The textile industry consumes 90% of world production of polyamide; the remainder is used in the plastics industry (sheeting, the electrical industry, and the construction of machines). The following methods of brightening are available to the manufacturer of fibers and plastics:

1. Brightening during polymer manufacture
2. Brightening granules by the exhaustion process
3. Bulk brightening carried out in the melt

In method 1, the brightener is added to the starting materials before polymerization. Brilliant effects are obtained with diaminostilbene acid derivatives at concentrations of 0.1–0.3% [9].

In method 2, granules can be treated with a water-soluble or dispersible polyamide brightener in an autoclave at 120–130°C or at the boil, using the exhaustion process in which the polymer absorbs the additive, in a manner similar to the methods employed for brightening polyamide textiles [10]. Disadvantages experienced are, however, stated to be large quantities of effluent, the need for long-term planning, and increased stock holding [11].

The third process is the one most often used and can also be applied to relatively small batches. As in the case of other thermoplastics, the brightener either is added to the polymer melt (if desired, in the form of a master batch) or is added in the form of powder to the granules in a high-intensity mixer before processing in a thermoplastic manner. Very brilliant brightening is obtained using Hostalux KS or Hostalux KS-N at 1–200 ppm, depending on the type of goods (transparent or pigmented).

Figure 11 shows the whitening effects given by the two brighteners as a function of the concentration in which they are added to polyamide 6. The light-fastness values (blue scale, Xenotest) of some polyamide brighteners are given in Table 5.

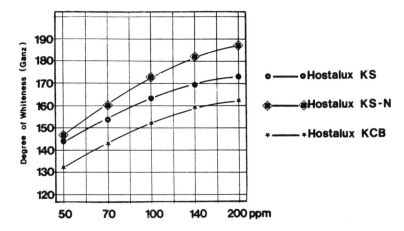

Figure 11 The whitening effects of optical brighteners on polyamide 6.

G. Thermoplastic Polyurethanes

Thermoplastic polyurethane is processed in a manner similar to that used for polyamide. The best procedure is to apply the optical brightener, alone or with other additives, to polyurethane granules by tumbling or powdering and to carry out injection molding or extrusion at the usual processing temperatures of 170–200°C. As in the case of polyamide, brilliant whitening effects are achieved with Hostalux KS-N and Hostalux KS at added concentrations of 1–200 ppm, depending on the type of plastic to be brightened (transparent or pigmented).

An interesting process has been claimed by Bayer for brightening films, sheeting, linings, and coatings [12]. In this process, optical brighteners that are

Table 5 Lightfastness Values of Some Brighteners for Polyamide Compositions, Ganz Scale

Brightener	Rating
Hostalux KS-N	4–6
Hostalux KS	4–5
Uvitex OB	3–5
Uvitex MP	2–3
Leukopur EGM	3
Eastman OB$_1$	No data available

Table 6 Lightfastness Values of Optical
Brighteners in Polyurethane, Ganz Scale

Brightener	Rating
Hostalux KS-N	4–5
Hostalux KS	4–5
Leukopur EGM	2

soluble in organic solvents are added to the polyadduct before the latter is dispersed, and the combination of optical brightener and polyadduct is uniformly dispersed in water. Removal of the solvent leaves stable aqueous dispersions that contain the optical brightener dissolved in the polyurethane and can be employed for the purposes mentioned in Section III.A. Nonionic 1,3-diarylpyrazolines (Hostalux NSM and Blankophor MAN) may be singled out from the large number of brighteners recommended for this purpose.

Some lightfastness values are given in Table 6 (blue scale, Xenotest).

H. Polycarbonates

Because of excellent mechanical and dielectrical properties, polycarbonates are used in the electrical and automotive industries, and for household and sporting articles, safety glazing, office equipment, and other applications.

A large number of optical brighteners are described as suitable for polycarbonate in the patent literature. However, the products Hostalux KS-N, Hostalux KS, Hostalux KCB, Uvitex OB, and Leukopur EGM are of particular interest to the processor.

The materials involved are mainly processed by injection molding, extrusion, and blow molding at 275–350°C, and an additive must therefore meet high requirements in regard to heat stability. Preliminary trials are advisable, particularly for processing temperatures above 300°C. As is usual in the case of thermoplastics, the optical brightener is applied before processing by tumbling or powdering, most suitably in combination with other auxiliaries. The possible use of masterbatches should be borne in mind. Thin films can also be brightened by means of solutions. The concentrations employed vary between 10 and 500 ppm, depending on the type of brightener, the whitening effect desired, and the type of plastic. The whitening effects and the lightfastness rating (blue scale) of Uvitex OB in polycarbonate in the presence of 2% of rutile titanium dioxide are described by Eschle [13].

Table 7 Lightfastness Values of Some Optical
Brighteners in Cellulose Acetate, Ganz Scale

Brightener	Rating
Hostalux KS-N	6
Hostalux KS	6
Uvitex OB	5–7
Leukopur EGM	Not tested

I. Cellulosics

Among the plastics comprising the cellulosics group, such as nitrocellulose, cellulose ethers, and cellulose esters, only the last mentioned are of interest for optical brightening. Cellulose acetate, propionate, and butyrate are used as decorative strips, eyeglass frames, combs, ballpoint pens, and so on, applications in which the aesthetic appearance of the plastic is important. Optical brighteners are incorporated by the dry coloring process, in combination with other additives and in the usual concentrations of 1–500 ppm, before the plastics are processed by injection molding, extrusion, or blow molding. Particularly in the case of cellulose acetate, brightening effects of good fastness to light and not inferior to the fastness values of polyester are achieved. If the cellulose ester is processed by the solvent method, the brighteners employed must have an adequate solubility.

The brightener manufacturers provide the data on fastness to light presented in Table 7 (blue scale, Xenotest). See Ref. 13 for the effectiveness of Uvitex OB as a function of the concentration employed.

J. Polyacetals

The polyacetal plastics belong to the class of condensation thermoplastics, such as polyamide or polycarbonate. Sheeting and thermosetting molding materials composed of polyacetals can be optically brightened without problems. Optical brighteners and other additives are incorporated into the opaque white or natural-colored granules by the dry-coloring process, using slow-running mixers. The final homogeneous distribution of the additives is effected during processing on an extruder or injection molding machine. Controlled addition is facilitated by the use of masterbatches. The processing temperatures are in the range of 180–230°C (composition temperature). Excellent whitening effects are achieved with Hosta-

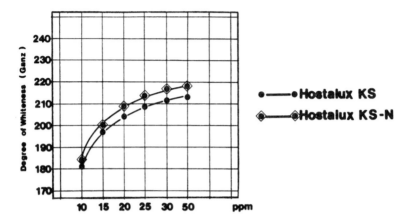

Figure 12   The whitening effects of optical brighteners on polyacetal.

lux KS and Hostalux KS-N at concentrations of 10–25 ppm in an opague starting
material (Fig. 12). The lightfastness values (blue scale, Xenotest) are 3 to 4.

K.   Other Polymers

Polyphenylene oxide (PPO) is important mainly as a mixture with impact-resistant
polystyrene, so that the use of brightened polystyrene also provides a means of
brightening PPO. Possible uses include household equipment and casing components. The applications of polyphenylene sulfides (PPS) are solely industrial and
render optical brightening unnecessary.

More than 90% of epoxide resins are used for the manufacture of lacquers.
Transparent products have a pale yellow shade that makes optical brightening
difficult (low basic whiteness). The lightfastness values and whitening effects
should therefore be checked. For brightening tests, the optical brightener can be
dissolved in a solvent or reactive thinner suitable for the epoxide resin concerned,
and the viscosity can thus be adjusted to the desired figure. It must, of course,
be possible to remove the solvent again. If white fillers, such as powdered porce-
lain, talc, or aerosol, are used, the resin and the fillers are homogenized at an
elevated temperature before the curing agent is added. Here, too, the controlled
addition of brightener is possible.

Because of the processing temperatures, which in some cases are extremely
high (polytetrafluoroethylene), and because the applications of the fluorine containing plastics are entirely industrial, there are virtually no opportunities for the
use of optical brighteners.

Table 8 Physical Properties of Some Important Brighteners for Plastics

| Brand | Fundamental chemical structure | Melting point (°C) | Maxima in DMF[a] (nm) | | | Extinction constant 1% (E_1 cm in DMF) |
			Absorption	Fluorescence	Shade	
Hostalux						
KS	Bisbenzoxazolyl-tilbene derivatives	>300	375	436	Red–eutral	1940
KS-N		>300	375	436	Neutral	1940
KCB	Bisbenzoxazolyl-naphthalene derivative	211	373	438	Neutral–blue	900
Uvitex OB	Bisbenzoxazolyl-thiophene derivative	200–201	375	439	Blue	1080
Leukopur EGM	Coumarin derivative	253–254	375	441	Blue–green	1290
Eastman optical brightener	Bisbenzoxazolyl-stilbene	365–368	375	434	Red	1940

[a] DMF = dimethylformamide.

L. Commercial Products

The physical properties of some important brighteners for plastics are shown in Table 8.

REFERENCES

1. A von Lagorio. Z Angew Chem 34:585, 1921.
2. P Krais. Melliands Textilber 10:468, 1929.
3. HW Zussmann. Modern Plastics Encyclopedia, Vol. 43, 1a, New York: McGraw-Hill, 1966, pp 490–495.
4. R Griesser. Methods of assessing the whiteness of textiles colorimetrically and possible uses of this technique. Ciba-Geigy [White Plains, NY] internal Publication No. 9134, and literature quoted therein.
5. G Rosch. Chemiefasern/Textilindustrie 24(76):373, 1974.
6. American Association of Textile and Color Chemists Technical Manual, rev. ed. AATCC, 1981.
7. Deutsche Normen, Normenausschuss Textil und Textilmaschinen im DIN (German Standards, Standards Committee for Textiles and Textile Machinery within DIN). Berlin: Deut Verlag GmbH, Vol. 3, 1994.

8.   E Preininger. Plastverarbeiter 20(12):845, 1969.
9.   G Rosch. Textilbetrieb 91(10):53, 1973.
10. G Rosch. Textilbetrieb 91(6):59, 1973.
11. BJ Bichler. In International Textile Finishing Yearbook, (G Dierkes, ed.), Degussa, Frankfurt, 1983, p 51–53.
12. Bayer, AG. German Patent DOS 1,694,153, 1969.
13. K Eschle. Plastverarbeiter 21(7):631, 1970.

11
Plasticizers

Lewis B. Weisfeld
Consultant, Philadelphia, Pennsylvania

I. INTRODUCTION

Plasticizers are high-boiling organic liquids or low-melting organic solids which are added to tough or hard resins—chiefly poly(vinyl chloride) (PVC)—to impart flexibility, extensibility, and/or processability. They have varying degrees of solvating action on these resins. The main difference between ordinary solvents and plasticizers is volatility.

The softening action of plasticizers, *plasticization,* is usually attributed to their ability to reduce the intermolecular attractive forces between chains in the polymer (resin) system. Three general theories have been proposed to account for the mechanism of *plasticization:*

The lubricity theory, wherein the plasticizer is supposed to act by reducing intermolecular friction.

The gel theory, positing that the plasticizer destroys the intermolecular forces at specific active (polar) sites on the polymer chain.

The free-volume theory, which is the difference between the *total volume* and the *occupied volume.* Occupied volume includes the volume of the polymer molecules calculated from their van der Waals radii *plus* the volume associated with vibrational motions of individual bonds.

The *free-volume theory* is that most widely accepted today. The amount of free volume is essentially the same for all polymers at their respective glass transition temperatures, T_g. In essence, the free-volume theory incorporates elements of the gel and lubricity theories. Plasticizers increase the free volume available to the polymers chains and thus allows for greater internal chain rotation and unwinding.

It is characteristic of plasticizing substances that they lower the melting temperature, elastic modulus, and T_g of polymers but do not alter the chemical nature of the macromolecules. With plasticizers, we are able to vary the useful properties of an inexpensive commodity thermoplastic like PVC from a rigid, horny material to a soft, rubbery composition and all the hardnesses in between. It is important to note that very low plasticizer concentrations have the opposite effect: Some degree of embrittlement can occur as well as a loss in impact strength, and it is, therefore, essential not to use less than the minimum quantity appropriate for the particulate plasticizer used. At very low levels, some plasticizers may actually decease free volume and increase process viscosity.

II. PLASTICIZER REQUIREMENTS

The requirements for useful plasticizers include compatibility with the resin, permanence, and efficiency. Other desirable properties may include nonflammability, odorlessness, and nontoxicity. No one chemical compound can fulfill all these requirements, so blends are often used and property compromises are struck.

A. Compatibility

Compatibility is the term for the degree of mutual solubility of resin system and plasticizer. Although the polymer and the plasticizer may be mutually compatible, the formulation ingredients in the resin system may not be, and exudation could occur. Plasticizers which exhibit good compatibility are designated primary plasticizers, whereas those that show only partial compatibility are called secondary plasticizers. Branched-chain alkyl phthalates are invariably highly compatible with PVC, but epoxy adjuncts have only limited compatibility. Sometimes, hydrocarbons derived from petroleum residues are used as secondary plasticizers for cost considerations. Secondary plasticizers cannot be used alone.

Incompatibility is usually indicated by spewing or blooming on the surface of the plastics film or part, or by poor physical properties. Sometimes, incompatibility does not manifest itself immediately, but becomes apparent several months after fabrication.

Plasticizer compatibility can be determined via ASTM D3291. In this, *Standard Practice for Compatibility of Plasticizers in Poly(Vinyl Chloride) Plastics Under Compression,* the amount of plasticizer that spews due to the compressional stress set up inside a 180° loop bend is noted. Test specimens of the plasticized PVC sheet are bent through an arc of approximately 180°, with the inner radius of the bend equaling the thickness of the specimen. These bent specimens are secured in a jig designed to hold them in this conformation. At specified periods of time, a specimen is removed and bent 360° in the opposite direction, and the former inside of the loop (now the outside) is examined for evidence of spew.

B. Permanence

Resins must retain the plasticizer under various environmental circumstances for retention of useful properties. Major concerns are volatility, extractability, migration, and stability to heat and light. Volatility is usually a function of the molecular weight of the plasticizer and its vapor pressure.

A standard test method for determining plasticizer volatility is ASTM D2288. Plasticizers are simply heated in crystallizing dishes in a circulating air oven on a rotating turntable at either 105°C or 155°C. The specimens are removed after periods of 2, 4, and 24 h and the weight loss in these times is determined. However, perhaps more germane to actual usage, ASTM D1203 measures plasticizer loss from plastics using activated carbon to absorb the volatilized plasticizer, either in direct contact with the plastic material or using a wire cage, which prevents direct contact.

C. Efficiency

This is a specific plasticizer term and is defined by the concentration of plasticizer required to yield a certain degree of flexibility or modulus. Efficiency is expressed in terms of the relationship of one plasticizer to another. In general, monomeric plasticizers are more efficient than polymerics. When we compare one plasticizer to another, the comparison is usually made at equivalent efficiencies.

D. Other Properties

Other properties such as low-temperature properties, odor, taste (organoleptic effects), color, electrical properties, flux time, flammability, and weathering resistance may be important. Toxicity problems have become an increasing concern since several popular high-volume plasticizers have been shown to be carcinogenic in animal tests.

Fluxing time is determined by the number of seconds it takes for the plasticized formulation to band on a two-roll mill. Thermal properties such as brittleness temperature by impact testing (ASTM D746 and D1790) are important, and not simply for Arctic environments. For ASTM D746, *Brittleness Temperature,* the flexible plastic specimens are secured to a specimen holder which is immersed in a bath containing a heat-transfer medium that is cooled. The specimens are struck at a specified linear speed and then examined. The brittleness temperature is defined as the temperature at which 50% of the specimens fail.

Specific standards for flammability have been devised by Underwriters Laboratories Inc. and ASTM (through a wide variety of tests). For UL 94, the specimen is mounted vertically so that the lower end is placed 30.5 cm above a 0.6-cm-thick surgical cotton layer, and then ignited from a defined gas flame from below. The results of the test are classified as Class 94 V-0 if the afterflame time does not exceed 50 s and no specimen burns more than 10 and the cotton layer is not ignited by dripping material; Class 94 V-1 if the sum of the afterflame times does not exceed 250 s and no specimen burns longer than 30 s: Class V-2 if results are similar to V-1 but the cotton layer is ignited by the drippings.

III. STRUCTURE AND PROPERTIES OF PLASTICIZERS

Most plasticizers are used for the modification of poly(vinyl chloride) resins and are esters, the reaction products of acids or acid anhydrides with alcohols. Variations on the theme of plasticizers product types may be monobasic acids with polyhydric alcohols, polybasic acids with monohydric alcohols, and so forth. However, it must be remembered that plasticizers are useful with other resins as well, such as polyamides (nylons) and polyurethanes, for example.

A. Phthalate Esters

The most common commercial plasticizers are esters of orthophthalic acid; the isophthalates behave less favorably. Terephthalates, although more expensive, are becoming popular. The phthalate esters are usually derived from the reaction of monohydric alcohols with phthalic anhydride:

$$+ \ 2 \ C_8H_{17}OH \longrightarrow$$

$$- \ H_2O$$

phthalic anhydride   *dioctyl phthalate (DOP)*

The octanol used for the production of dioctyl phthalate is usually 2-ethylhexanol:

$$CH_3{-}CH_2{-}CH_2{-}CH_2{-}\overset{\overset{\displaystyle CH_2{-}CH_3}{\textstyle |}}{CH}{-}CH_2{-}OH$$

Di(2-ethylhexyl) phthalate, (DOP or DEHP), has, for decades, been the dominant plasticizer and the workhorse of the industry. Its is the standard of performance against which all other plasticizers are compared. As user demands have become more sophisticated, however, property requirements exceed the performance afforded by DOP alone. Also, DOP (DEHP) has come under increasing regulatory scrutiny because animal tests performed at the National Cancer Institute have shown it to be mildly oncogenic in rats and mice. Efforts are being made, under pressure from the Consumer Product Safety Commission, to remove DOP from products which may come into oral contact with children (pacifiers, teething rings). Another popular phthalate plasticizer, Santicizer 160 (butyl benzyl phthalate or BBP), is also under regulatory pressure. Although concern over the possible carcinogenic aspects of DEHP and dioctyl adipate (DOA) seems to have abated, new allegations over endocrine disrupter effects of phthalates in general (as well as bisphenol-A and alkylphenols) have erupted and legislative and regulatory pressure has grown.

A variety of monobasic alcohols may be used for commercial phthalate plasticizers. These are usually branched, such as isodecyl alcohol or 2-ethylhexanol, but may also be mixtures of linear alcohols such as C_7, C_9, and C_{11} alcohols. Pure linear alcohols are seldom used alone because blends provide a better balance of properties. Sometimes, two different alcohols may be used, as is the case with BBP mentioned previously.

The choice of alcohol has important effects on plasticizer properties. The following comparison among di(2-ethylhexyl) phthalate (DOP), diisodecyl phthalate (DIDP), and two linear blends illustrates the effects of branching within the phthalate esters (for comparisons, it is traditional to adjust plasticizer concentrations to achieve the same moduli efficiency rather than same concentration) (Table 1).

From this, we can conclude that phthalates based on branched alcohols appear to have better processability (faster fluxing time), but a higher brittle temperature and lower efficiency (more required to achieve the same modulus).

Table 1 Comparison of DOP, DIDP, and Two Linear Blends

	DOP	DIDP	Linear $C_6C_8C_{10}$	Linear $C_7C_9C_{11}$
Parts per hundred resin (phr)	54	58	49	50
Modulus at 100% elongation (psi)	1500	1500	1490	1500
Tensile strength (psi)	2660	2600	2700	2710
Elongation (%)	360	350	365	365
Brittle temperature, T_b (°C)	−27	−34	−36	−36
Activated carbon volatility (24 h at 90°C)	4.5	2.0	2.4	2.2
Fluxing time (s)	90	112	110	110
Compatibility (spew roll)	None	V.S.	V.S.	V.S.
Average MW[a] alcohol	131	158	137	145

[a] MW = molecular weight.

The volatility of the linear alcohols is lower than DOP but higher than DIDP; this function would seem to be more closely related to molecular weight.

Permanence is often a limiting factor in plasticizer selection, and high molecular weight seems to be key. For example, medium-molecular-weight (MW) phthalate esters such as DIDP will not pass the 90°C Underwriters Laboratories rating specification for building wire, which requires that the primary insulation be oven-aged for 7 days at 136°C for determining retained elongation values. The usual solution is the substitution of higher-MW phthalates such as ditridecyl phthalate (DTDP) for part of the plasticizer system, with the remainder being one of the trimellitates, such as tri(2-ethylhexyl) trimellitate. It is a necessary practice for a manufacturer of building wire to submit a sample of the product to Underwriters Laboratories (UL) for examination to establish compliance with UL 83, the standard for thermoplastic insulated wires, and the examination will include materials used in its manufacture. The most severe UL wire rating is the 105°C category that calls for oven-aging at 158°C. This segment of the PVC wire and cable market often uses trimellitates alone or in combination with DTDP or diundecyl phthalate (DUP). Other standards for wire and cable include ASTM D1047 (Specification for PVC Jacket for Wire and Cable), ASTM D2219 (PVC Insulation for Wire and Cable, 60°C Operation), and ASTM D2220 (same, for 75°C Operation). The National Electrical Manufacturers Association (NEMA) and the Insulated Cable Engineers Association (ICEA) have joined forces to create standards NEMA WC-5-1996 and ICEA S-61-403, both of which are similar to the ASTM specifications. For electrical wiring in automobiles, the Society of Automotive Engineers offers J1127 for battery cable and J1128 for low-tension primary cable.

Replacing branched-chain alcohols with normal or linear ones results in improved low-temperature properties, whereas other properties remain relatively unchanged. Improved low-temperature properties can also be achieved by substituting terephthalic acid for orthophthalic. High-solvating plasticizers like butyl octyl phthalate (BOP) and BBP are effective in adjusting melt viscosity and fusion time to produce high-quality PVC foams.

B. Trimellitate Esters

The trimellitic esters are distinguished by high thermal stability and low volatility, but the plasticizing action of the trimellitates in PVC is low compared to phthalates. They are usually not suitable where good low-temperature properties are required. Also, they are expensive, although recent concern over the possible carcinogenicity of phthalates and new trimellitic anhydride capacity augers well for lower prices. They are used mainly in PVC wire and cable, as mentioned earlier. In addition to trioctyl trimellitate (TOTM), the tris(isodecyl) and tris(iso-tridecyl) esters find use.

C. Aliphatic Dicarboxylic Esters (Monomeric)

These important plasticizers are comprised of the esters of adipic, azelaic, sebacic acids and glutaric acid:

$$\begin{array}{l} \text{COOH} \\ | \\ (\text{CH}_2)_n \\ | \\ \text{COOH} \end{array} \quad \begin{array}{l} n = 4, \text{ adipic acid} \\ n = 7, \text{ azelaic acid} \\ n = 8, \text{ sebacic acid} \\ n = 3, \text{ glutaric acid} \end{array}$$

They render unto PVC outstanding low-temperature properties and are distinguished by excellent light (UV) stability. In addition, their low viscosity (because of weak solvating action) is of great value in the manufacture of PVC plastisols. The most common esters are those of 2-ethylhexanol: dioctyl adipate (DOA), dioctyl azelate (DOZ), and dioctyl sebacate (DOS). To meet higher requirements, the oxo-process alcohols, isononyl and isodecyl alcohols, are also used. These yield somewhat more rigid plasticized materials with improved low-temperature elasticity and "drier"-feeling surfaces. The advantages and disadvantages of the 2-ethylhexyl diesters are compared to corresponding phthalate and trimellitate as presented in Table 2.

Table 2 Advantages and Disadvantages of 2-Ethylhexyl Diesters Compared to Corresponding Phthalate and Trimellitate

	DOP	TOTM	DOA	DOZ	DOS
Concentration (phr)	54	56	44	46	45
Modulus at 100% elongation	1500	1510	1490	1530	1520
Tensile strength (psi)	2680	2700	2700	2710	2700
Elongation (%)	360	350	360	365	370
Brittle temperature (°C)	−27	−24	−47	−52	−60
Volatility (24 h at 90°C)	4.5	0.85	7.2	5.6	2.3
Fluxing time (s)	90	130	280	300	310
Compatibility (roll spew)	None	v.s.	mod.	mod.	v.s.

The aromatic structures impart improved processability (flux time) but higher brittle temperatures. The aliphatics are more efficient (lower concentrations required to achieve equivalent moduli). Aliphatic esters are also less compatible because they are less polar and more linear.

One of the important uses of di(2-ethylhexyl) adipate or DOA has been for PVC meat and produce wrap. However, DOA (as well as DEHP or DOP) has had the finger of carcinogenicity pointed at it. Efforts are being made to replace it with plasticizers more immune to possible regulatory action.

D. Polymeric Plasticizers

The ultimate in nonvolatility is achieved with the polymeric plasticizers. These are polyesters for the most part, varying from low viscosity (100 mPa at room temperature) to very high (300,000 mPa). Viscosity is determined not only by molecular weight but also by the esterifiers used in the polyester structure. Condensation products of dibasic acids and glycols, they are usually terminated with monobasic acids and alcohols to control molecular weights within narrow ranges. Typical raw materials used for polymeric plasticizer production may include the following:

Dibasic Acids
  Adipic acid
  Azelaic acid
  Sebacic acid
  Phthalic acid
Glycols
  Ethylene glycol
  1,4-Butane diol
  Diethylene glycol
  Propylene glycol
  Dipropylene glycol
  Neopentyl glycol
Terminators
  Simple alcohols
  Fatty acids
  Phosphates
  Phosphonates

A typical polymeric plasticizer would be

Polybutylene adipate (carboxylic acid terminated)

Polymerics are more difficult to process than the monomeric plasticizers, but they impart greater permanence, better extraction resistance to oils and chlorinated solvents, and lower volatility. With these advantages go poorer low-temperature properties and lower efficiency. Some polymerics may have poor hydrolytic stability as well; sterically hindered neopentyl glycol offers (expensive) improvement in this regard. The composition of polymeric plasticizers is usually proprietary, and one is not able to assess structure/performance properties on the basis of the trade name.

E. Phosphate Esters

Triaryl phosphates have the important advantage of excellent flame retardancy, and they have enjoyed an expanding market especially because of the greater demand for this property. The phosphoric esters of alkylated phenols command increasing respect due to flame retardancy combined with improved light stability and lower volatility. Mixed esters (one alkyl group and two aromatic moieties) almost attain the level of DOP with respect to low-temperature resistance, but they have higher volatility. Phosphate plasticizers also require increased stabilizer levels. Some of the phosphates used for PVC plasticization are presented in Table 3.

Molecular weight is not the complete solution to all volatility problems. Although tri(2-ethylhexyl) phosphate (TOF) is usually considered the "plasticizer standard" in the phosphate series, it is seldom used today. Newer products include isodecyl diphenyl phosphate (Santicizer 148), which is claimed to lower smoke generation as well as reduce flammability. In addition to use in PVC, mostly for flame retardancy, phosphate esters are used to plasticize cellulose derivatives (nitrate and acetate–butyrate), polyurethane foam (also for flame retardancy), and others.

The combination of phosphate ester plasticizers and antimony oxide—itself an excellent flame retardant—is antisynergistic. The two should not be used together.

Table 3 Some Phosphates Used for PVC Plasticization

Plasticizer	Properties
Tri(2-ethylhexyl) phosphate $(C_8H_{17}O)_3P{=}O$	Light, stable, and flame retardant, gives plastisols of low viscosity but much more volatile than DOP
Octyl diphenyl Phosphate $(C_8H_{17}O)(C_6H_5O)_2P{=}O$	Plasticizing action similar to DOP, but more light (UV) sensitive
Tricresyl phosphate $(CH_3C_6H_4O)_3P{=}O$	Good flame retardant for articles subjected to high mechanical stress; poor low-temperature properties but excellent permanence
Trixylenyl phosphate $[(CH_3)_2C_6H_3O]_3P{=}O$	Exceptionally volatile, exceeding even tricresyl phosphate; poor light stability, but imparts good resistance to hydrocarbon extraction, low flammability, and good mechanicals

F. Other Plasticizers

1. Fatty Acid Esters

Fatty acid esters (pelargonates, laurates, myristates, palmitates, stearates, and ricinoleates) are included among plasticizers, but they are really a family of specialty products acting as extenders, lubricants, or secondary plasticizers. This class includes important lubricants for PVC, but they are really unsuitable as primary plasticizers because of poor compatibility. For nitrocellulose, mono-acid esters such as butyl acetylricinoleate and tetrahydofurfuryl oleate have found use.

tetrahydrofurfuryl oleate

2. Citric Esters

Citric esters, specifically tri(2-ethylhexyl) acetylcitrate, are on a par with DOP over the full range of plasticizer properties, but are nevertheless too expensive to occupy an important place with respect to volume consumption. Mostly, they are used for plasticization of cellulose esters where they compete successfully with sucrose esters. However, because of low toxicity, citric esters continue to make progress in medical applications such as PVC blood bags and tubing.

3. Sulfonic Esters and Sulfonamides

Sulfonic esters and sulfonamides are excellent PVC plasticizers whose properties match those of DOP in most every respect. An additional advantage is that they are difficult to saponify. Expense, however, is a liability. Toluenesulfonamide and toluenesulfonethylamide are special powder plasticizers for resins based on cellulose esters. Benzenesulfonbutylamide is compatible with polyamide 11 and 12 (nylons) and with copolyamides.

$$\underset{\underset{O}{\overset{O}{\|}}}{C_6H_5-S-NH-C_4H_9} \quad \text{benzenesulfonbutylamide}$$

$$\underset{\underset{O}{\overset{O}{\|}}}{R-S-O-C_6H_5} \quad \text{phenyl alkylsufonate (ASE)}$$

4. Glycolic Esters

Glycolic esters such as butyl phthalyl butyl glycolate are based on derivatives of hydroxyacetic acid, are similar in PVC compatibility to DOP, and are used sometimes in plastisols for resistance to petroleum hydrocarbons. However, the simple aliphatic glycolic esters are too volatile and therefore unsuitable as PVC plasticizers.

5. Benzoic Esters

neopentyl glycol dibenzoate

Benzoic esters such as the diglycol benzoates are liquid plasticizers rather like butyl benzyl phthalate (BBP) in plasticizing action. In the manufacture of PVC floor covering, benzoates offer the advantage of a highly soil-resistant surface. However relatively high cost precludes their general use.

Neopentyl glycol dibenzoate is especially hydrolysis resistant, a feature attributed to the steric effects of the neopentyl structure. Although rather expensive, it finds use in applications where the vinyl comes into repeated contact with aqueous substrates.

6. Secondary Plasticizers

Secondary plasticizers include chlorinated hydrocarbons (chloroparaffins), hydrocarbons, and epoxidized fatty acid esters. These are used for specific technical effects. PVC compatibility is limited in most cases. For the chloroparaffins, compatibility increases as chlorine content goes up, but plasticization is reduced. They are used mostly for their flame-retardant effect. As is the case with phosphate ester plasticizers, chloroparaffins require increased stabilizer concentrations.

Hydrocarbons are called ''extenders'' and can be useful in extending the shelf life of plastisols. Of limited compatibility, care must be taken to avoid exudation in the finished product. These are usually the least expensive ingredients that can be added to PVC, are sometimes used for such cost-reduction purposes, and are usually derived from petroleum refinery deep cuts and still bottoms. Triisopropyl biphenyl has been used extensively as a secondary plasticizer.

The epoxidized fatty acid esters likewise have limited compatibility Butyl and octyl epoxy stearate impart good low-temperature properties to PVC, equal to that of DOA, and epoxidized oleic acid, octyl ester, is less volatile and better in water extractability. Epoxidized soybean oil is added to PVC compounds because of its good stabilizing effect (and is, in fact, regarded as an important, FDA-sanctioned ''secondary stabilizer'') and low migration tendency, as is epoxidized linseed oil. Octyl epoxy tallate is more migratory. The best quality epoxy plasticizers for PVC are those with the highest oxirane values and lowest iodine numbers, so care must be taken in specification.

IV. APPLICATIONS

The major markets for flexible (plasticized) PVC are (1) wire and cable, (2) film and sheeting, (3) molding and extrusion, (4) flooring, and (5) textile and paper coatings. We have already seen how complex the wire and cable area can be in considering plasticizer selection in view of UL requirements. Although phthalate esters are predominantly used in wire and cable applications, the trimellitates and polymerics come into their own at the upper end of the specifications, especially for high temperature and oil resistance.

Vinyl flooring is an important market requiring large quantities of plasticizers such as butyl benzyl phthalate (BBP) and 2,2,4-trimethyl-1,3-pentanediol diisobutyrate (TXIB), and particularly the benzoic esters, as mentioned earlier. These plasticizers resist staining to a high degree, and BBP has little tendency to migrate into and soften flooring adhesives. Linear phthalates, adipates, and azelates are not appropriate because of poor stain resistance and high fluxing temperatures, as well as cost.

The superior cleanability of today's vinyl flooring can be related to the flooring's stain resistance. The best plasticizers for stain resistance combine benzoate esters with a volatile solvent component. The solvent component of the plasticizer volatilizes during processing, evaporating at the surface. It leaves a hardened "plasticizer-starved" layer on top of the flooring sheet which is very resistant to stainants. The benzoate esters provide enough plasticization to add some flexibility to the flooring. However, they do not migrate readily, leaving the surface hard for some time. When these aromatic structures finally get to the surface, they have low mutual solubility with common stainants, which tend to be oily and aliphatic. These plasticizers are generally used only in the plastisol or organosol wear layer of the flooring (for cost reasons) and overlay a decorative layer and a foam layer plasticized with more common phthalates. Other polymers can be used for the wear layer, notably urethanes, but they are not found all that frequently.

Another major market for plasticizers is *food packaging*. Plasticizers considered safe for use according to Title 21, Code of Federal Regulations (Sections 178.3740, 175.300, and 177.1210), include epoxidized soya and linseed oils, various adipates, azelates, and sebacates, and a number of polymeric and polyester plasticizers. DOP (DEHP) is permitted in contact with foods of high water content by Prior Sanction (21 CFR 181.27), but we do not know how long this provision will last.

Although the vinyl industry is by far the largest consumer of plasticizers, they do find their way into other polymers as well. The plasticizers used for vinyls generally work well with cellulose nitrate, cellulose acetate propionate, cellulose acetate butyrate, and ethyl cellulose. Cellulose acetate requires lower alkyl phthalates (dimethyl or diethyl) for plasticization. Some grades of nylon can be softened with toluene ethyl sulfonamides, and these may sometimes be added during the polymerization reaction. Polyvinyl butyrals are most often plasticized with glycol derivatives, adipates, and phosphates. Polyvinyl acetate responds favorably to a wide variety of PVC plasticizers, but of the more polar types. Polyurethanes are generally compatible with all those commonly used for PVC. For that matter, plasticizers which migrate from PVC to urethanes must be avoided in such composites (e.g., vinyl-film-coated urethane foams) by limiting PVC plasticizer selection to fairly high-molecular-weight polyester types, or even to nitrile rubber.

V. HEALTH CONCERNS

Plasticizers for flexible vinyl, particularly the phthalates, became a health issue in the early 1980s. Studies performed at the National Cancer Institute showed that di(2-ethylhexyl) phthalate (DOP or DEHP) caused hepatocellular carcinomas and adenomas in both rats and mice at high feeding levels. Similar studies with di(2-ethylhexyl) adipate (DOA) disclosed that it, too, was carcinogenic for female mice (and probably male mice as well) but not for rats of either gender.

This plasticizer toxicity issue has abated somewhat since it was subsequently found that the rodent metabolism of the esters was quite different from that of humans, and even "switched" to benign at slightly lower feeding levels. Nevertheless, all phthalates are under review by the Environmental Protection Agency and test data must be submitted regularly by industry groups under the Negotiated Test Rule provisions of the Toxic Substance Control Act (TSCA). DEHP must still be labeled as a "potential human carcinogen."

Most recently, all phthalate plasticizers have come under suspicion as endocrine disrupters or estrogen mimics, even at extremely low levels of exposure. Such effects may be of more importance to the environment (amphibians, birds, and reptiles) than to humans, as no human epidemiology has yet been found.

On August 3, 1996, President William Jefferson Clinton signed the Food Quality Protection Act into law. Among other things, the measure repeals the infamous Delaney Clause, the 40-year-old amendment to the Food, Drug, and Cosmetics Act that effectively banned carcinogens in food or food packaging, regardless of concentration or risk. This may well signal a shift away from the "cancer paradigm" of the past several decades and a refocusing of health, safety, and environmental concerns on the effects of manufactured chemicals on endocrine systems. Although recognition of the role that chemicals play on reproductive responses is not new, the public is becoming newly aware and will be demanding more legislative and regulatory control of manufactured chemicals and industries that generate them. We can anticipate that plastics additives, both as indirect food additives for food-contact purposes (regulated by the Food and Drug

Administration) and as nonfood adjuvants which may migrate into the environment (regulated by the Environmental Protection Agency), will be profoundly affected by new regulations targeting endocrine disruption effects.

The Safe Drinking Water Act Amendments of 1996 and the Food Quality Protection Act decree that pesticide manufacturers and some producers of other chemicals will have to test their products' ability to cause endocrine disruption. The water law requires the EPA to implement an endocrine screening and testing program for chemicals found in drinking water to which a substantial population is exposed; the food law directs the same for all pesticides, new and old. At this point, no one really knows for sure how many chemical or chemical manufacturers will be affected by the endocrine screening program.

Displacing our previous preoccupation with chemical carcinogenesis, estrogen mimics/endocrine disrupters will be the focus of intense environmental concern for the next decade. These ubiquitous and mostly synthetic chemicals masquerade as sex hormones, and extensive epidemiological evidence points to human fertility problems. The young are especially at risk. Suspects include the common phthalate plasticizers, alkylphenol-based surfactants, bisphenol-A-based resin systems, and certain persistent organochlorine compounds. A particularly good article on the subject can be found in Ref. 2. Near-birth exposure of laboratory rats to small amounts of two endocrine-disrupting chemicals caused a 5–21% reduction in testicle size and daily sperm production on maturity, according to a new study by the Reproductive Biology Unit of Britain's Medical Research Council. Rats were exposed to one of the compounds, the plasticizer butyl benzyl phthalate, at a level 300 times below the one previously thought necessary for testicular toxicity, and yet the response was invoked. The other chemical was p-octylphenol. Toxicologists have only recently discovered the possible estrogenicity of other widely used chemicals such as bisphenol-A, phthalates, and some alkylphenolic compounds [3].

In addition to the above-cited British study [4], French scientists have reported a decline in the sperm count of Parisian men between 1977 and 1992 [5], although a control group in Toulouse showed no such regression. They attribute this to the supposition that industrial pollution is greater in the urban environment. Although not conclusive, evidence is strong that estrogenic properties of phthalates, alkylphenols, and bisphenol-A (from polycarbonates and food can linings) may be responsible. Chemical firms are now developing reliable and repeatable assays for environmental estrogens and endocrine disrupters [6].

Phthalate plasticizers are currently under attack by environmental groups in the wake of a U.K. report by the Ministry of Agriculture, Fisheries and Food that found traces of the materials in baby formula milk, coupled with concern about reproductive (estrogenic) effects. It was speculated that flexible PVC tubing plasticized with phthalates may have been responsible for the baby formula contamination, but the source is yet unknown [7–10].

Finally, the EPA has announced the availability of the "Special Report on Environmental Endocrine Disruption: An Effects Assessment and Analysis," which states that no conclusive evidence has emerged that industrial chemicals disrupt hormonal systems in humans. The report provides an overview of the current state of the science for endocrine disruption. Its major components are an introduction to the endocrine system and the endocrine-disruption hypothesis, a review of potential human health and ecological risks, and an analysis section, including an overview of research needs. The agency found that, except for cases involving workplace exposure, there has been no causal relationship shown between these chemicals and hormonal disruption leading to adverse health effects. Nevertheless, the EPA stated that such chemicals have been shown to disrupt hormonal activities in animals and pose a potential risk to humans, so research into so-called endocrine disrupters should be intensified. The report represents an interim assessment pending a more extensive review expected to be issued by the National Academy of Sciences later in 1997. An electronic version is accessible on EPA's Office of Research and Development home page on the Internet at http://www.epa.gov/ORD. It can also be obtained by contacting ORD Publications Office, Technology Transfer Division, National Risk Management Research Laboratory, U.S. Environmental Protection Agency, 26 W. Martin Luther King Drive, Cincinnati, OH 45268; telephone (513) 569-7562; facsimile (513) 569-7566 (62 FR 12185-12186 (3/14/97) [11–14].

VI. SOME STARTING FORMULATIONS

A. Basic, General-Purpose Flexible PVC Formulation

PVC resin, high MW (IV > 0.95)	100.0 phr
Plasticizer (DOP)	30.0–80.0
Plasticizers/processing aid (ESO)	5.0
Stabilizer (barium/cadmiun or barium/zinc)	3.0
Filler (calcium carbonate)	0–30.0
Lubricant (stearic acid)	0.5
Pigment	0–3.0

B. Wire and Cable Starting Formulation

PVC resin, high MW (IV > 0.95)	100.0 phr
Plasticizer (diisodecyl phthalate)	50.0
Stabilizer (tribasic lead sulfate)	5.0
Filler (calcined clay)	75.0–15.0
Filler (calcium carbonate)	5.0–20.0
Lubricant (petroleum wax)	0.5

C. Automotive-Trim Starting Formulation

PVC resin, high MW (IV > 0.90)	100.0 phr
Plasticizer (diisodecyl phthalate)	50.0
Stabilizer (Ba/Zn)	3.0
Fungicide (Vinyzene BP-5)	0.5
Filler (CaCO$_3$)	10.0–50.0
Lubricant (stearic acid)	0.5
Pigment	As needed

D. Refrigeration-Gasketing Starting Formulation

PVC resin, high MW (IV > 0.95)	100.0 phr
Plasticizer (polymeric type)	50.0–100.0
Stabilizer (Ba/Zn or Ca/Zn)	3.0
Filler (CaCO$_3$)	20.0–60.0
Lubricant (stearic acid)	0.5
Pigment	As needed

E. Transparent Beverage Tubing Starting Formulation

PVC resin, high MW (IV > 0.90)	100.0 phr
Plasticizer (DOP or DOA)	50.0
FDA stabilizer (Ca/Zn)	3.0
Lubricant (stearic acid)	0.5

F. Calendered Film/Sheet Starting Formulation

PVC resin, high MW (IV > 0.90)	100.00 phr
Plasticizer (DOP)	40.0–60.0
Plasticizer/stabilizer (ESO)	0–5.0
Stabilizer (Ba/Cd or Ba/Zn)	2.0–4.0
Filler (CaCO$_3$)	0–20.0
Lubricant (stearic acid)	0.1–0.5
Pigment	0–5.0

VII. SUPPLIERS OF PLASTICIZERS

Aristech Chemical Corporation
600 Grant Street
Pittsburgh, PA 15219
(412) 433-7700

C. P. Hall Company
7300 S. Central Avenue
Chicago, IL 60638
(312) 767-4600

Exxon Chemical Company
Intermediates Department
P.O. Box 3272
Houston, TX 77253-3272
(713) 870-6998

Hatco Chemical Corporation
King George Post Road
Fords, NJ 08863
(973) 738-1000

Monsanto Chemicals
Specialties Division
800 N. Lindbergh Boulevard
St. Louis, MO 63167
(314) 694-1000

BASF Corporation
Chemical Division
100 Cherry Hill Road
Parsippany, NJ 07054
(973) 316-4792

Eastman Chemical Products Co.
P.O. Box 431
Kingsport, TN 37662-5280
(800) 327-8626

FMC Corporation
2000 Market Street
Philadelphia, PA 19103
(215) 299-6000

Huls America Inc.
P.O. Box 456
Piscataway, NJ 08536
(732) 980-6931

Argus Chemical Division
Witco Corporation
520 Madison Avenue
New York, NY 10022
(212) 605-3600

REFERENCES

1. D Dieckmann, W Eldridge. Annual Technical Conference of the Society of Plastics Engineers, Dallas, TX, 1990.
2. D Lutz. The Sciences 12, January/February 1996.
3. Chem Week 24, 1 January 1996.
4. Br Med J 312:467, 1996.
5. Br Med J 312:471, 1996.
6. Chem Ind (London) 156, 4 March 1996.
7. Chem Market Reporter 5 and 9, 3 June 1996.
8. Chem Eng News 12, 3 June 1996.
9. Chem Ind (London) 397, 3 June 1996.
10. Chem Week 29, 12 June 1996.
11. Chem Eng News 19, 17 March 1997.
12. Chem Week 17, 19 March 1997.

13. Chem Market Reporter 7, 24 March 1997.
14. Plast News 4, 24 March 1997.

BIBLIOGRAPHY

Carcinogenesis Bioassay of Butyl Benzyl Phthalate in F344/N Rats and B6C3F$_1$ Mice (Feed Study). NTP Technical Report Series No. 213, 1982; NIH Pub. No. 82-1769.
Carcinogenesis Bioassay of Di(2-ethylhexyl)adipate. NTP Technical Report Series No. 212; NIH Pub. No. 81-1768.
Carcinogenesis Bioassay of Di(2-ethylhexyl)phthalate in F344 Rats and B6C3F$_1$ Mice (Feed Study). NTP Technical Report Series No. 217, 1982; NIH Pub. No. 82-1773.
Colborn, T, Dumanoski, D, Myers, JP. Our Stolen Future: Are We Threatening Our Fertility, Intelligence, and Survival? A Scientific Detective Story. New York: Dutton, 1996. See also Natural History (American Museum of Natural History), 42–49, March 1996.
Connor, M. Additives. In: Modern Plastics Encyclopedia '97. New York: McGraw-Hill, 1997, pp C-9 and C-99.
Federal Food, Drug, and Cosmetics Act of 1938, 52 Stat. Section 1040 (1938); 21 USC Section 301 et seq.
Federal Register (daily) and Code of Federal Regulations. http://www.access.gpo.gov/su—docs/
Food Additives Amendments of 1958, Pub. L. 85-929, 72 Stat. Section 1784 (1958).
Food Quality Protection Act of 1996, Pub. L. 104-170, 110 Stat Section 1486 (1996) et seq.
Krauskopf, LG. Plasticizers: Types, properties, and performance. In: LI Nass, ed. Encyclopedia of PVC, 2nd ed. New York: Marcel Dekker, 1988, Vol 2, p 143.
Regulatory Update for the Plastics Industry (monthly). http://www.regup.plastics.com/
The Safe Drinking Water Act Amendments of 1996, Pub. L. 104-182, 110 Stat. S1613 et seq. (August 6, 1996); 42 USC S201 et seq.
Sears, JK, Darby, JR. The Technology of Plasticizers. New York: Wiley–Interscience, 1982.
Sears, JK, Darby, JR. Solvation and plastization. In: LI Nass, ed. Encyclopedia of PVC, 2nd ed. New York: Marcel Dekker, 1986, Vol 1, p 435.
Toxicology and Carcinogenesis Studies of Tris(2-ethylhexyl) Phosphate in F344/N Rats and B6C3F$_1$ Mice (Gavage Study). NTP Technical Report Series No. 274, 1984; NIH Pub. No. 84-2530.
Weisfeld, LB. Environmental and health concerns in formulating vinyl compounds. In: EJ Wickson, ed. Handbook of PVC Formulating. New York: Wiley, 1993.
Wickson, EJ. Handbook of PVC Formulating. New York: Wiley, 1993.

12
Processing Aids

Robert P. Petrich
Rohm and Haas Company, Philadelphia, Pennsylvania

John T. Lutz, Jr.
JL Enterprises, Bensalem, Pennsylvania

I. INTRODUCTION

Except for additives that enhance the physical properties of a polymer, virtually all others could be classified as "processing aids." Stabilizers and antioxidants prevent or minimize thermal degradation during the melt processing of a polymer. Therefore, stabilizers and lubricants could be classified as processing aids, because without them, melt processing would not be practical. In a similar manner, plasticizers, lubricants, and, possibly, fillers also enhance the processing of many polymers. These additives are covered in detail in their individual chapters in this volume.

Other methods of enhancing processing by reducing melt viscosities and temperatures include the addition of low molecular weight species of the same polymers. In many ways, these lower-molecular-weight chains act as plasticizers by interrupting the regularity and entanglement of the higher-molecular-weight polymer chains.

For most commodity polymers, "processing aid" additives are already incorporated to yield, in reality, proprietary compounds that have carefully designed and balanced processing and performance properties. End users rarely need to add other components to achieve good processing and part quality, and seldom do they have the opportunity to do so.

Unlike these other plastics that come to the fabricator in a ready-to-use condition, poly(vinyl chloride) (PVC) resin is supplied in neat, or pristine, form for the bulk of the applications. True, ready-to-use pellets or powder blended formulations do comprise a large portion of the "PVC" market; but by far the largest usage of PVC, worldwide, is in the in-house formulations made by individual processors/fabricators. All formulators of PVC need to know how to select and use processing aids. They must understand the interaction between processing aids and the PVC particulate structure. Modifying PVC is unlike modifying other polymers that display a rather specific and relatively sharp melting point. The user of processing aids for PVC must constantly be aware of the changing character of the PVC "melt" to the compound. The remainder of this chapter is devoted to providing a better understanding of the function of processing aids in PVC.

II. PROCESSING AIDS FOR POLY(VINYL CHLORIDE)

A. Mechanism of Operation

Although processing aids have been used commercially in PVC compounds for more than 40 years, the mechanism of their action has not been elucidated clearly. This is not too surprising, considering the lack of understanding of PVC rheology that prevailed until recently. Some of the early proposals for processing aid mechanisms range from simple additivity of the melt viscosity of the high-molecular-weight processing aid with that of the PVC resin, to a complex proposal that the processing aid influences the melting and recrystallization kinetics of the PVC crystallites. Ryan [1] discussed some of these early hypotheses, but could not find evidence to prove that any of them were correct. Today, much more is known about the morphology of PVC. We now know that PVC has a well-defined particulate structure which is very difficult to destroy at temperatures below the degradation point [2–6]. The rheological characteristics of PVC melts are determined by this particulate structure. With this understanding of PVC morphology, we can say today that the mechanism of action of processing aids is related to their effects on these particulate flow processes.

Processing aids are used in PVC formulations for many reasons, but they can all be grouped into two general categories:

1. They accelerate and control the fusion process in PVC compounds.
2. They strongly affect the rheological characteristics of the fully fused PVC melt.

Any proposed mechanism for the action of processing aids must consider these two categories of effect.

If one considers first the mechanism by which processing aids improve the fusion characteristics of PVC powder blends, it is generally accepted that they accelerate the process by increasing the interactions between the PVC grains. Gould and Player [7] state that processing aids improve the fusion process by increasing particle-to-particle friction, hastening the breakdown of the PVC grains and the ultimate exchange of chains between particles. They also suggest that the processing-aid molecules improve the heat transfer during this process, by virtue of increasing contacts between grains. Menges et al. [4,5] argue that the PVC grains are bound together during the fluxing process by a "putty," consisting of PVC polymerization auxiliaries and minor additives in the PVC formulation. These authors consider the processing aid to be one important component in this "putty." Gould and Player attribute the particular effectiveness of acrylic processing aids in increasing interparticle friction to their high surface hardness. Bohme [8], on the other hand, bases his explanation of acrylic-processing-aid performance on the capability of such agents to increase the coefficient of friction between the PVC melt and the metal surfaces of processing machinery. He argues that the acrylic processing aid causes the compound to adhere better to the metal surface, thus improving the heat transfer and the shear between the compound and the machinery surfaces. Krzewki and Collins [9] also expressed the opinion that processing aids accelerate the fusion process due to increased interparticle and particle–metal friction.

1. Improvement of Fusion

The first effect of a processing aid to be observed during the processing of a rigid PVC powder blend would be during the densification/plastification/fusion process. Processing aids greatly shorten the time and the heat/shear history, necessary to melt and homogenize the compound. In the early days of the development of rigid PVC, this effect was first noted in laboratory experiments on two-roll mills. The addition of a processing aid shortened considerably the time required to flux the powder blend, form a homogeneous band with smooth edges, and develop a uniform rolling bank. With the more modern laboratory test equipment available today, especially instrumented kneader mixers such as the Brabender or Haake machines, one can more accurately determine the effects of processing aids on the fusion characteristics of PVC compounds. When a predetermined quantity of PVC powder blend is run under a controlled temperature/speed condition, the recorded evolution of the torque needed to turn the blades as a function of time gives a direct measure of the fusion process. For example, the two curves shown in Fig. 1 compare a clear PVC compound having a high K value,* run

---

* "K value" is a shorthand definition of the molecular weight of PVC. Its use is intended to minimize confusion among inherent and specific viscosity values obtained by different methods among suppliers of PVC. Higher "K values" yield a higher molecular weight. The K value is related to relative viscosity as follows:

$$\log \eta_{\text{cyclohexanone}} \text{ rel } = \left(\frac{75K^2 \times 1^{-6}}{1 + 1.5KC \times 10^{-3}} + (K \times 10^{-3}) \right) C$$

where C is the constant concentration of PVC in the solvent [10].

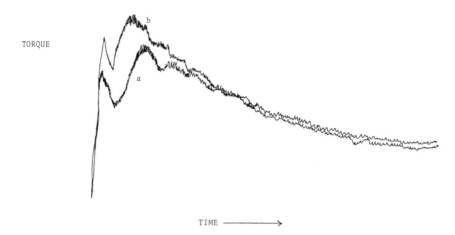

Figure 1 Gottfert kneader chamber fluxing curves for PVC powder compounds with 0 (curve a) and 1.5 phr (curve b) of the acrylic processing aid.

under the same conditions, without processing aid and with 1.5 phr of the acrylic processing aid added. The point of maximum torque is generally accepted as representing the beginning of plastification. Comparing the times to maximum torque peak, we can see that the addition of the processing aid reduces the plastification time from 50 to 27 s. We can also note that the maximum torque is considerably increased, reflecting the increased interaction of the PVC particles with each other and the chamber surfaces.

These instrumented kneader mixers can also be run in a programmed temperature mode, where the bowl temperature rises at a preset rate. Although the curves obtained from these experiments differ somewhat in overall shape, one can again note in this case a substantial shortening of the time required for fusion when a processing aid is included in the formulation (Fig. 2).

Processing studies on extruders fitted with instruments for continuous measurement of pressures and temperatures also can demonstrate the effect of processing aids in improving the fusion of PVC powder blends. Table 1 contains data on the melt pressure distribution along the barrel of a single-screw laboratory extruder, for the same compound run with and without an acrylic processing aid. The compound with a processing aid shows higher melt pressure readings in the first zones, illustrating that the fusion point for this compound is moved closer to the feed port than for the control without a processing aid.

Pazur and Uitenham [11] have confirmed this effect with direct visual observations of the melting process along a screw that was quickly removed from the extruder. Their data, shown in Fig. 3, indicate that the melt pool in the screw

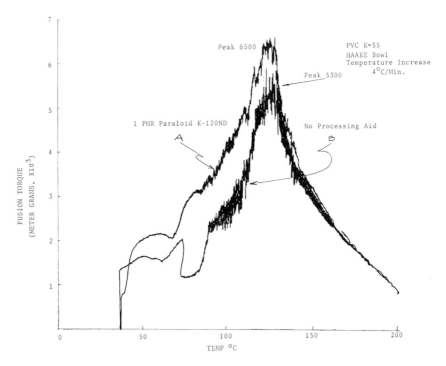

Figure 2   Haake programmed temperature rise kneader chamber fusion curves for PVC powder compounds with 0 (curve A) and 1 phr (curve B) of the acrylic processing aid.

Table 1   Effect of Processing Aid on Melt Pressure Distribution—Single-Screw Extrusion

Processing aid and use level, phr	Melt pressure (bar)		Back pressure on screw (kN)
	10.5D	Die entry	
None	290	275	25.5
Standard acrylic			
2 phr	290	275	26.5
4 phr	325	305	28.5
"High-efficiency" acrylic,			
3 phr	320	315	29.5

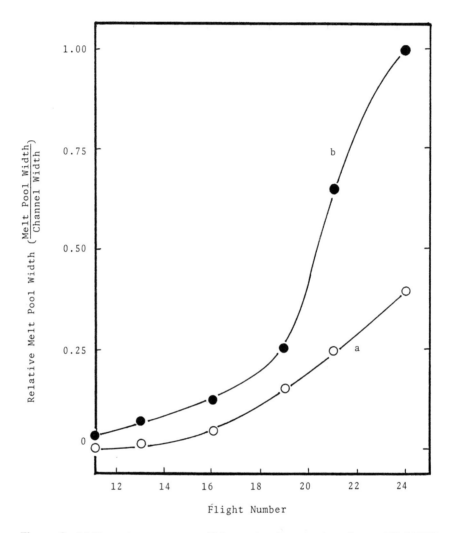

Figure 3 Melting rate versus screw flight number for extrusion of unmodified PVC powder blend (curve a) and the same compound with a processing aid and impact modifier added (curve b). (From Ref. 11.)

channel is at least twice as wide for a compound containing a processing aid and impact modifier as for the unmodified control compound.

B. Effects of Processing Aids on the Rheology of PVC Melts

The most important function of processing aids in rigid PVC is the improvement of the rheological characteristics—homogeneity, strength, and elasticity—of the melt after fusion and during fabrication. Unmodified rigid PVC compounds are notably poor in this respect, in particular when processed at low to moderate temperatures. Under such conditions, the crystalline regions in the primary PVC particles do not melt, and the degree of molecular interdiffusion between particles is very low, so the interparticle strength is very poor. In this circumstance, the molecules of the processing aid function to bind together the particles and improve the melt strength. The ability of the most widely used acrylic processing aids to perform this function is based on three inherent characteristics:

1. Excellent compatibility with PVC, even on the molecular scale
2. Flexible, elastic characteristics of the acrylic polymer chains
3. Long, high-molecular-weight polymer chains

Even when processing temperatures are raised sufficiently to allow greater particle breakdown and interpenetration of PVC molecules from one particle to another, these processing aids continue to provide a strong enhancement of melt strength, because they mix intimately with the shorter, stiffer PVC chains and provide a longer range viscoelastic network. This network greatly assists in binding together the PVC domains, equalizing stress concentrations within the melt during fabrication, and, thus, greatly reducing processing defects [7]. Evidence for these effects is presented in Fig. 4.

1. Homogenization

The homogenizing action of processing aids, tying together the semicrystalline PVC domains into a more uniform viscoelastic structure, can be observed quite easily even on laboratory equipment. Figure 4 shows the results of a laboratory roll-milling experiment. Figures 4a and 4b compare an unmodified, high K-value PVC compound with an identical one containing 2 phr of an acrylic processing aid. Each compound was milled for 5 min at 175°C; thus, both are well beyond the point of fusion. It is evident that the compound containing the processing aid is much more homogeneous: The rolling bank is smooth, the edges are well knit, and the surface of the mill band is uniform and glossy. This effect is the key to successful production of a high-quality surface in high-speed, rigid PVC calendering. In addition, the homogenization effect is very important in extrusion

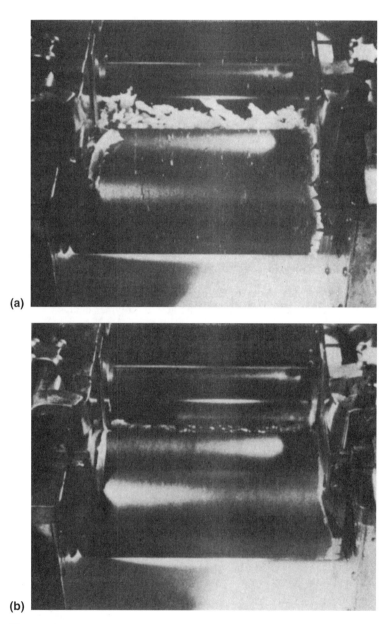

Figure 4 Milling characteristics of rigid PVC compound with no processing aid (a) compared to the same formulations with 2 phr of a acrylic processing aid (b).

processes, for uniform delivery of constant quality and quantity of melt to the dies. Likewise, in injection molding, this homogenization effect is important in obtaining good quality surfaces and avoiding weakness at weld lines. In all types of processes, the processing aids assist in binding together the myriad formulation ingredients to avoid plateout, splay, and so forth.

2. Elasticity and Extensibility

Notwithstanding the fusion and homogenization effects discussed earlier, the most important reason for adding processing aids to rigid PVC compounds is to improve the elasticity and extensibility of the melt. This effect can be observed even in simple laboratory experiments such as milling. An experienced operator can easily sense by hand the improvement in tear strength and extensibility that occurs when a processing aid is added to the formulation. The earliest processing aids were developed using just this simple kind of test. However, today, more sophisticated instrumentation is available to allow more accurate measurement of the effects. Several laboratory tests have been developed to measure the improved elasticity in extrusion processes. One approach is to simply increase extrusion haul-off speed until the extrudate begins to tear or crack [7,11]. The data of Gould and Player, given in Table 2, demonstrate the dramatic improvement processing aids can give in the maximum drawdown ratio of extrudate emerging from the die.

If the takeoff equipment is instrumented to allow direct measurement of force and speed ratio, additional information can be gained. The Gottfert Rheotens, which allows such measurements, has been used to characterize the effects of various processing aids on PVC melt elasticity. Petrich [12] used the Rheotens to compare the melt strength and elongation effects of a standard acrylic processing aid, a "high-efficiency" (specialized composition and higher molecu-

Table 2 Limiting Extruder Takeoff Rate as a Function of Processing Aid Content

Processing aid content (%)	Maximum extrudate draw ratio without tearing
0	1.32
3.0	2.5

Source: Ref. 7.

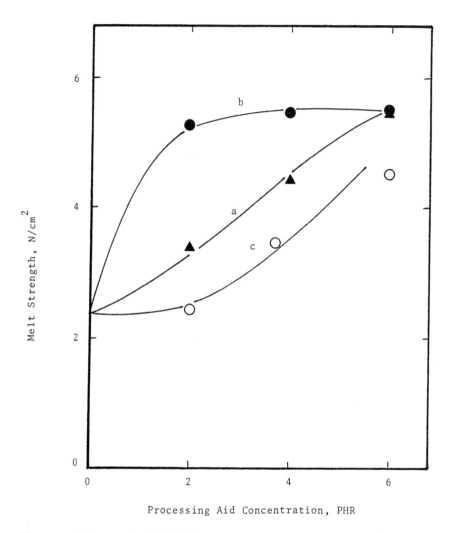

Figure 5 Melt strength of rigid PVC extrudates as a function of processing aid content: curve a = a standard acrylic processing aid, curve b = a high-efficiency acrylic processing aid, and curve c = a lubricating processing aid. (From Ref. 12.)

lar weight) acrylic processing aid, and a lubricating processing aid. Figure 5 illustrates a substantial, essentially linear improvement in the breaking stress of the molten extrudate as the concentration of standard acrylic processing aid is increased (curve a). The "high-efficiency" processing aid (curve b) exhibits a much more dramatic effect, reaching a plateau value at only 2 phr, which is

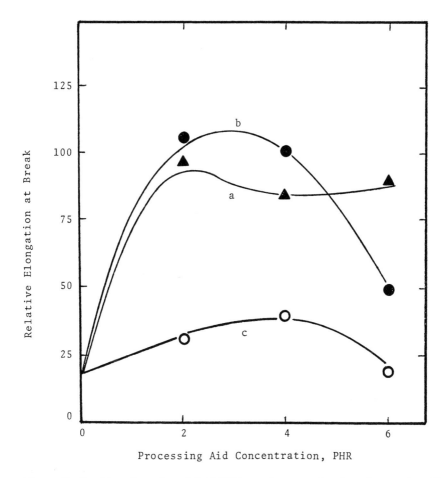

Figure 6 Breaking elongation of rigid PVC extrudates as a function of processing aid content: curve a = a standard acrylic processing aid, curve b = a high-efficiency acrylic processing aid, and curve c = a lubricating processing aid. (From Ref. 12.)

equivalent to that reached by the other product at 6 phr. The lubricating process aid (curve c), because it is a dual-purpose product, exhibits somewhat lower efficiency than the standard. The corresponding data on the elongation at break for these melt extrudates show some interesting effects (Fig. 6). Even at 2 phr, the standard processing aid causes a sharp improvement in melt elongation, which remains relatively constant at higher concentrations. On the other hand, the "high-efficiency" product is essentially equal to the standard at 2 and 4 phr, but it shows a reduction in melt elongation at the high concentration of 6 phr. It was

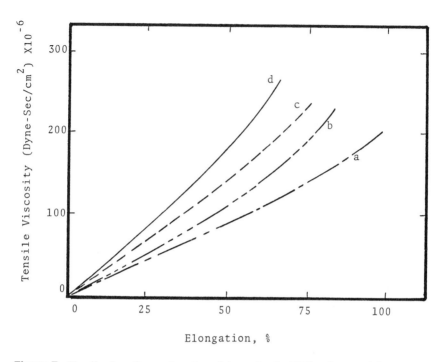

Figure 7   Tensile viscosity as a function of elongation for PVC melts containing varying concentrations of processing aid: curve a = no processing aid, curve b = 1.0 phr of a standard processing aid, curve c = 1.5 phr of a standard processing aid, and curve d = 0.5 phr of a high-efficiency processing aid. (From Ref. 11.)

theorized that this very high-molecular-weight product requires higher melt temperatures for optimum elongation at such high concentrations. (In effect, this is not of great practical concern because very few PVC compounds utilize such high processing aid levels.) Pazur and Uitenham used the Gottfert Rheotens to study the effect of processing aid on the melt extension properties of a twin-screw PVC pipe compound [11]. They showed (Fig. 7) that the addition of a processing aid progressively increases the tensile viscosity of the compound at any given elongation. In addition, their data indicate that a high-efficiency processing aid can outperform the standard type in this respect, even at much lower concentrations. They also observed that a certain minimum melt temperature and processing aid concentration are necessary for effective action. The data shown in Fig. 8 indicate that the standard processing aid begins to improve the tensile viscosity of the base stock only at temperatures above about 190°C. These results, like those of Petrich, also show some tendency for higher optimum processing temperatures for the "high-efficiency" processing aids.

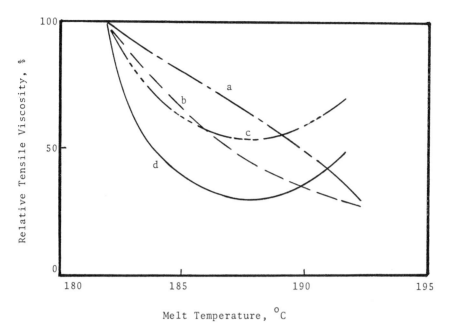

Figure 8 Relative tensile viscosity as a function of temperature for PVC melts containing varying concentrations of processing aid: curve a = no processing aid, curve b = 1.0 phr of a standard processing aid, curve c = of a 1.5 phr standard processing aid, and curve d = 0.5 phr of a high-efficiency processing aid.

Gould and Player [7] have developed a more sophisticated test procedure, measuring the postextrusion swell on the extrudate from a converging zero-length die on a ram extruder. Measurements of swelling ratio and flow pressure allow the calculation of the recoverable strain ϵ_r and the average extensional stress σ_ϵ. Plots of the reversible strain as a function of average extension stress, as shown in Fig. 9, indicate a substantial improvement when a processing aid is added to the formulation. The processing aid has no effect on the initial, Hookean region of the curve. However, it increases the value of ϵ_{rc}, the critical, recoverable strain point at which the curve departs from linearity and gives considerably higher recoverable strain values over a wide range of stress in the nonlinear region. This increase in the critical strain ϵ_{rc} is a key feature of processing-aid action because melt fracture is thought to begin to occur in the dies at ϵ_{rc}. The general term "melt fracture" covers a wide variety of defects that occur during polymer processing, especially extrusion. The semicrystalline, particulate nature of PVC melts causes them to have poor extensibility and, thus, to be especially susceptible to melt fracture problems. Addition of minor amounts of processing aid is an effective

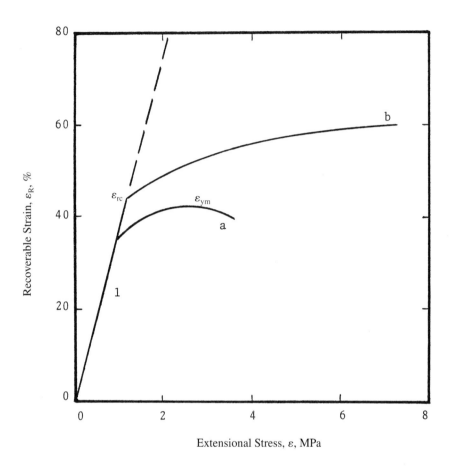

Figure 9 Relationship between recoverable strain and extensional stress for PVC melts extruded through converging knife edge die: curve a = no processing aid, curve b = 3 phr of a processing aid, and curve 1 = linear hookean region; critical and maximum strains given by ϵ_{rc} and ϵ_{rm}, respectively. (From Ref. 7.)

solution to these problems. This is illustrated in Fig. 10. The photomicrographs show the surface of extrudates at varying speeds and with different concentrations of processing aids. The control material, with no processing aid, shows a very rough surface finish when extruded at a screw speed of 40 rpm (Fig. 10a). Compound b, containing 1.5 phr of a standard acrylic processing aid, shows much less melt fracture at 40 rpm. Compound c, containing 1.2 phr of the ''high-efficiency'' processing aid, gives a completely smooth surface finish under the

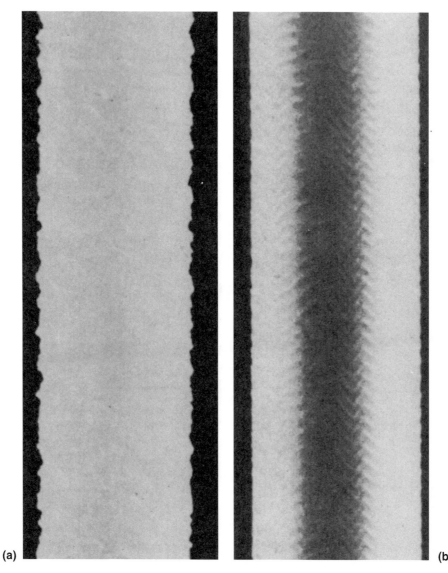

(a)   (b)

Figure 10   Effect of processing aids on surface roughness of rigid PVC extrudates: (a) no processing aid, 40 rpm screw speed, (b) 1.5 phr of a standard acrylic processing aid, 40 rpm, (c) 1.2 phr of a high-efficiency acrylic processing aid, 40 rpm, and (d) 1.2 phr of a high-efficiency acrylic processing aid, 60 rpm.

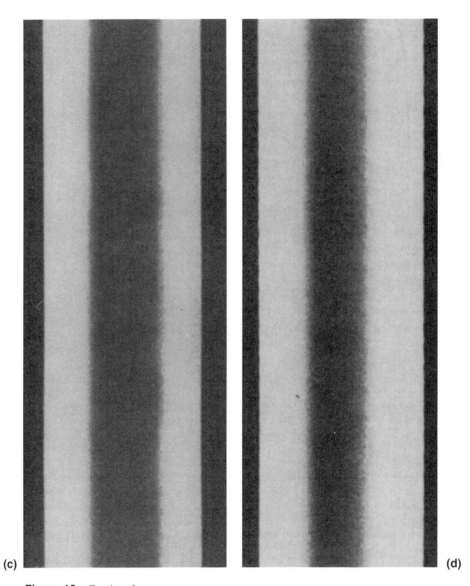

(c)   (d)

Figure 10 Continued.

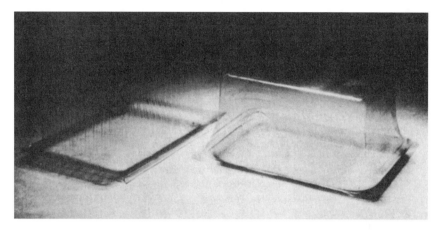

Figure 11   Improvement of thermoformability of rigid PVC film through the use of processing aids: left, package containing 2 phr of an acrylic processing aid; right, package containing no processing aid.

same conditions. Figure 10d shows that this compound, with 1.2 phr of a "high-efficiency" acrylic processing aid, gives a satisfactory surface even at 60 rpm screw speed.

The improvement of melt strength and extensibility given by processing aids can also be illustrated by thermoforming experiments. Figure 11 shows a good example of the thermoforming benefits of a processing aid in commercial PVC film. The container on the left was made with a commercially calendered film containing 2 phr of an acrylic processing aid. This film easily thermoformed into a good package, with full corners, relatively uniform wall thickness, and good definition of decorative/functional details. The package on the right was made from the same film formulation, but without a processing aid. Under the same thermoforming conditions, the corners are not fully filled, and the film tore before drawing down to meet the sides of the mold. Because of these dramatic improvements in thermoformability, PVC calenderers frequently use more of the processing aid in their formulation than is necessary strictly for good processing on the calender, to facilitate the forming of the films by the end user. Likewise, the improved melt strength, extensibility contributed by processing aids, can also be useful in other postfabrication operations such as belling of pipe and welding of profiles.

3.   Melt Viscosity

Although processing aids normally raise the melt viscosity of PVC compounds slightly, a key factor in their success is that this effect is far smaller than their beneficial effect on melt elasticity/extensibility. The increase in viscosity actually depends on the viscosity of the base formulation and on the test being used. Ryan [1] found that the melt viscosity of a high-molecular-weight PVC compound, measured in a capillary rheometer, increased only slightly over a broad range of temperatures and shear rates when an acrylic processing aid was added. Capillary rheometer measurements on lower K-value PVC resins have shown a significant increase in melt viscosity. However, in practical torque rheometer tests at shear rates similar to those used in commercial processing equipment, the increase in melt viscosity associated with processing aid addition is usually no greater than 5–10%. These melt viscosity increases usually have no significant detrimental effect on commercial PVC fabrication processes.

C. Effect of Processing Aids on Other Properties of PVC Products

Because modern processing aids exert their beneficial effects on PVC processing characteristics at levels as low as 1–3 phr, they have very little effect on the physical properties of rigid PVC materials. The polymers used as processing aids generally have fairly good physical properties themselves, related to their high-molecular-weight and high-T_g compositions. In addition, the compatibility of most processing aids with PVC avoids problems of inhomogeneities in the product, which could serve as initiation sites for failure.

Physical properties of compounds containing normal use levels of processing aids normally are within experimental error of the properties of control samples without a processing aid. In fact, in many cases, the physical properties of fabricated PVC products containing processing aids are *better* than unmodified controls, due to the better fusion in compounds containing a processing aid. Likewise, the optical properties of PVC compounds are not greatly changed by the addition of processing aids, and the improvements in surface quality, gloss, and so on usually outweigh any changes in the actual clarity of the product.

The heat stability of rigid PVC compounds containing processing aids, especially the most popular commercial acrylic processing aids, is at least as good as that of unmodified control compounds.

Overall, the effect of commercial processing aids on properties other than processibility can generally be assumed to be neutral.

D. Types of Processing Aids

The use of plasticizers was the earliest attempted solution to the problems associated with the inherently poor processing of PVC. A variety of low-molecular-weight chemical compounds were found to be sufficiently soluble and mobile to penetrate the PVC structure, allowing it to soften at considerably lower temperatures. Thus, the minimum acceptable processing temperature was lowered considerably. However, these plasticizers also caused a sharp reduction of the modulus and strength of PVC, even at low concentrations. Nevertheless, the resulting material had a useful balance of properties and found extensive commercial use.

The original commercial processing aids for rigid PVC were discovered as a result of a search for a ''high-temperature plasticizer''—a material that would provide improved fluidity at processing temperatures but remain inert at use temperatures.

During the 1950s, it was found that some high-molecular-weight PVC-compatible polymers could function in this way. By the late 1950s, the first commercial processing aid for rigid PVC, Paraloid K-120, was made available to PVC compounders by Rohm and Haas. Processing aids quickly became accepted as an essential component of rigid PVC formulations, and combined with developments in improved stabilizers and processing machinery, this allowed a very rapid growth of rigid PVC applications. In particular, processing aids provided great assistance in allowing the use of simpler, more economical fabrication techniques directly from powder blends. Thus, it could be said that the development of processing aids was one of the important factors in the establishment of PVC as a leading material for rigid pipe, siding, window profiles, and other extrusions, as well as clear films and bottles.

In those early days of development of processing aids, the first compositions reported in the literature as having processing-aid functionality were copolymers of styrene with acrylonitrile [13] or methyl methacrylate [14].

Although such compositions have been made and used captively by PVC producers, they were not produced and sold on the open market purposely as processing aids for PVC. The first commercially promoted processing aids, introduced by Rohm and Haas in the late 1950s, were of the acrylic type (i.e., methacrylate/acrylate copolymers), with methyl methacrylate predominating. Today, more than 35 years later, this type of composition remains dominant. Its success is based on the inherent compatibility of the methacrylate polymers with PVC, their specific rheological properties, and their intrinsic stability to heat and light. Many acrylic monomers are available, and they can be combined in various ratios and structures, as well as in polymers of different molecular weight. Thus, a wide variety of acrylic processing aids can be tailored to any application desired. These various combinations have been the subject of many patents [15–20].

Outside the acrylic field, poly(α-methyl styrene) has been used somewhat in commercial PVC applications [21]. Newer compositions suggested, but not yet used in significant volume, including polyneopentylene terephthalate and polyalkylene carbonate [22,23].

In addition to conventional processing aids, many multipurpose materials have been conceived. The most successful of these has been a class of lubricating processing aids, which combine very strong external lubrication effects with the conventional rheological improvements of acrylic processing aids [24–27]. These materials are typified by Paraloid K-175. Besides such commercial additives, similar principles have been applied to internal modification of PVC resins during their production to give "easy processing," lubricated resins [28,29]. Other experimental multipurpose processing aids have been tested, but without extensive commercial use to date. These include processing aids that reduce the viscosity of the PVC compound. Both acrylic [30,31] and polyester [32] compositions have been described. Other compositions have been designed to improve the thermal stability of the PVC compound, through incorporation of glycidyl methacrylate [33] or oxirane [34] functionality. Finally, the possibility of raising the heat-distortion temperature of PVC compositions has been suggested through the use of processing aids incorporating methacrylate copolymers with bicyclic side chains [35] or styrene–acrylonitrile–acrylamide terpolymers [36].

E. Applications for Processing Aids

All PVC compounds benefit from the accelerated fusion (time or temperature), more complete homogeneity, improved melt strength/extensibility, and correction of plateout contributed by the processing aids. In this subsection, we summarize separately (Tables 3–13) the specific benefits imparted in each processing appli-

Table 3 Extrusion: Benefits of Processing Aids

Fast, controlled fusion of PVC dry blend
    Shorter residence time; better thermal stability; faster output
    Uniform delivery of melt to die; reduced surging; better control of wall thickness; no
        wasted "overweight parts"
    Less sensitivity to temperature variations
Increased hot-melt strength
    Faster startups; fewer line tears
    Higher output rates
    Multiple dimension profiles from same die due to greater pulldown potential
    Good retention of gas for foam formation
Improved appearance
    Excellent gloss
    Excellent interior and exterior profile surfaces
    No melt fracture
    Minimum hangup in barrel
Prevention of plateout
    Consistent appearance
    Reduced cleanup downtime
Improved homogeneity
    Increased finished part ductility, minimized brittle failures
    Enhanced cold formability

Table 4 Extrusion: Typical Extrusion Formulations (Parts per 100 Parts of PVC)

Pressure pipe (twin screw)	Telephone duct (twin screw)
100 PVC ($K = 67-68$)	100 PVC ($K = 67-68$)
0.2–0.4 Tin stabilizer	0.2–0.4 Tin stabilizer
0.8 Calcium stearate	0.8 Calcium stearate
1.2 Paraffin wax (165°F)	1.2 Paraffin wax (165°F)
1.0 TiO_2	1.0 TiO_2
0.5–0.75 Acrylic processing aid	0.7 Acrylic lubricant processing aid
0.5 $CaCO_3$	3–5 Acrylic or CPE impact modifier
PVC siding (twin screw)	20–30 $CaCO_3$
100 PVC ($K = 67$)	PVC window profile (twin screw)
5–6 Acrylic or CPE impact modifier	100 PVC ($K = 67$)
0.5–1 Acrylic processing aid	5–6 Acrylic or CPE impact modifier
0.5 Acrylic lubricant processing aid	1.0 Acrylic processing aid
1.6 Tin stabilizer	0.5–1.0 Acrylic lubricant processing aid
1.3 Calcium stearate	1.5 Tin stabilizer
1.0 Paraffin wax (165°F)	1.2 Calcium stearate
10–14 TiO_2	1.0 Paraffin wax (165°F)
Interior profile	8–12 TiO_2
100 PVC ($K = 67$)	Opaque sheet and film
6–10 Impact modifier	100 PVC ($K = 60-62$)
1.0 Acrylic processing aid	8–12 Impact modifier
1.0 Acrylic lubricant processing aid	2.0 Acrylic processing aid
1.5 Tin stabilizer	1.0 Acrylic lubricant processing aid
0.8 Calcium stearate	2.0 Tin stabilizer
1.2 Paraffin wax (165°F)	2.0 Calcium stearate
3–5 $CaCO_3$	0.7 Paraffin wax (165°F)
Pigments as needed	TiO_2 as needed

cation (what they mean to the "customer") and give typical formulations for each application.

The illustrations presented are for rigid PVC, but processing aids provide similar advantages for plasticized PVC as well. Given the multitude of "plasticized PVC" applications—from 20 phr plasticizer "semirigids" to 90$^+$ phr plasticizer "ultrasoft" compounds—we will not attempt to provide additional formulations. A rule of thumb would be that "semirigids" respond well to processing aids in the 1–5-phr range; "ultrasoft" compounds frequently require 5–10 phr of a processing aid to perform the desired function.

Table 5 Blow-Molded Bottles/Containers: Benefits of Processing Aids

---

Fast, controlled fusion of PVC dry blend
  Faster rates, uniform melt from powder blends
  Consistent melt delivery from pellets
  Uniform parisons, less overweight
Increased hot-melt strength
  Improved melt strength for high blow ratios
  Controllable parison length, thickness, uniformity
Improved appearance
  Excellent gloss and clarity/sparkle[a]
  Elimination of melt fracture, surface defects, flow lines
Prevention of plateout
  Longer runs, more production
  Elimination/correction of ''black speck'' problems
Improved homogeneity
  Improved ductility
  Less scrap due to trim breakage
  More uniform impact resistance

---

[a] Especially with acrylic lubricant processing aid Paraloid K-175.

Table 6 Blow-Molded Bottles and Containers: Typical Formulations (Parts per 100 Parts of PVC)

General purpose	Food-grade type
100 PVC ($K = 58$)	100 PVC ($K = 58$)
2.0 Butyltin stabilizer	2.0 Ca/Zn stabilizer
0.5 Glycerol monostearate	4.0 Epoxy soybean oil
0.2 Ester wax (OP)	2.0 Acrylic processing aid
2.0 Acrylic processing aid	0.5 Acrylic lubricant processing aid
0.5 Acrylic lubricant processing aid	0.3 Rhodiastab 50
12–14 Clear MBS or ABS impact modifier[a]	0.5 Glycerol monostearate
	0.2 Ester wax (OP)
Blue toner as needed (e.g., 0.06 of 1% concentrate)	0.5 Approved phosphite (e.g., TNPP) (optional)
	12–13 Clear MBS impact modifier[a]
	Blue toner as needed (e.g., 0.06 of 1% concentrate)

[a] Grades include Paraloid BTA III N-2 and Paraloid BTA-733.

Table 7 Cellular (Foam) PVC Profiles: Benefits of Processing Aids

Fast, controlled fusion of PVC dry blend
   Rapid fusion; lower temperature fusion allows better utilization of blowing agents
   Uniform melt delivery to die
Increased hot-melt strength
   Retention of gas
   Low/uniform density
   Fine cell structure
   Integral skin
   Fast rates through sizer
Improved appearance
   Smooth, uniform skin
   Uniform dimension in free-blown extrudates
Prevention of plateout
   Long, productive runs between cleanouts
   Minimum postdecoration surface problems
Improved homogeneity
   Better ductility
   Improved impact; reduced splitting and brittleness—especially when impact modifiers
      are used

Table 8 Cellular (Foam) PVC Profiles: Typical Formulations (Parts per 100 Parts of PVC Resin)

	Twin screw	Single screw
PVC resin ($K = 62$)	100	100
Paraloid KM-318F[a]	8–9	8–9
Azodicarbonamide	0.5	0.5
Tin stabilizer	1.5	1.5
Calcium stearate	0.8	0.8
Paraffin wax (165°F)	1.0	1.0
Oxidized polyethylene wax	0.1	—
CaCO$_3$	5.0	5.0
Pigments		As needed

[a] Specially designed to give processing aid/impact-modifier/lubricant package. Alternate systems include 6 parts of high-molecular-weight acrylic processing aid; 3 or 5 parts of acrylic impact modifier or CPE; and 1 part of acrylic lubricant processing aid.

Table 9   Calendering: Benefits of Processing Aids

Fast, controlled fusion of PVC dry blend
    Improved dispersion of ingredients
    Maximum homogeneity of melt delivered to calender rolls
    Maximum color intensity; possibility of lower pigment levels
Increased hot-melt strength
    Smooth, air-free rolling bank; smooth uniform sheet
    Elimination of flow lines
    Higher takeoff speeds; faster production rate; less scrap
    Improved gage control
Improved appearance
    Improved gloss (and clarity/sparkle for clears)[a]
    Smooth edges; wider usable sheet
Prevention of plateout
    Improved production rates and machine utilization due to less downtime for cleaning
    Fewer rejects due to post decoration failures[a]
Improved homogeneity
    Better ductility; less brittle sheet
    Less scrap due to edge-trim brittle failures
    Improved thermoformability and cold forming

---

[a] Especially with acrylic lubricant processing aid Paraloid K-175 substituted for part of external lubricant.

Table 10 Calendering: Typical Formulations (Parts per 100 Parts of PVC)

Clear sheet film
   100 PVC ($K = 60$)
   10–15 MBS or ABS impact modifier[a]
   1.6–2.0 Acrylic processing aid
   0.5–1.0 Acrylic lubricant processing aid
   2.0 Tin stabilizer
   0.5–0.8 Glycerol monostearate
   0.2–0.4 Ester wax (OP)
   0.05–0.2 Ester wax (E)
   Blue toner as needed (e.g., 0.06 of 1% concentrate)
Opaque sheet film
   100 PVC ($K = 60$)
   6–12 General-purpose or high-efficiency MBS, acrylic, or ABS impact modifier
   1.6–2.0 Acrylic processing aid
   0.5–1.0 Acrylic lubricant processing aid
   1.6–2.0 Tin stabilizer
   0.5–0.8 Glycerol monostearate
   0.2–0.4 Ester wax (OP)
   0.05–0.2 Ester (E)
   TiO_2 as needed
Nonstress whitening clear (packaging) sheet film
   100 PVC ($K = 60$)
   10–15 Crease-whitening-resistant MBS[b] impact modifier
   1.6–2.0 Acrylic processing aid
   0.5–1.0 Acrylic lubricant processing aid
   2.0 Tin stabilizer
   0.5–0.8 Glycerol monostearate
   0.2–0.4 Ester wax (OP)
   0.05–0.2 Ester wax (E)
   Blue toner as needed (e.g., 0.06 of 1% concentrate)

[a] Special grades include Paraloid BTA III N-2.
[b] Specific grades include Paraloid BTA-702.

Table 11 Injection-Molded Parts: Benefits of Processing Aids

Fast, controlled fusion of PVC dry blend
 Rapid fusion of powder blends
 Thorough fusion of pellets
 Short residence times
 Improved weld line strength
Increased hot-melt strength
 Uniform melt/flow in mold[a]
 Faster cycles; faster mold filling[a]
 Reduced shear burning; faster rates[a]
Improved appearance
 Elimination of splay (splash), craters, and other surface defects
 High, uniform gloss
 No shear-burn streaks[a]
 No gate-end blemishes
Prevention of plateout
 Increased molding condition ranges (time, temperature, speed)
 Longer runs between cleanouts
Improved homogeneity
 Improved ductility and impact strength, especially at weld line
 Nonbrittle flash/sprue removal

[a] Especially true for the acrylic lubricant processing aid Paraloid K-175.

Table 12 Injection-Molded Parts: Typical Formulations (Parts per 100 Parts of PVC)

Pipe fittings
 100 PVC ($K = 57$)
 1–5 MBS, ABS, acrylic[a] or CPE impact modifier
 0.75–1.0 Acrylic processing aid
 0.5 Acrylic lubricant processing aid
 1.5–2.0 Tin stabilizer
 0.5 Calcium stearate
 0.5 Paraffin wax (165°F)
 1.5 TiO_2
High-flow applications
 100 PVC ($K = 52$)
 10 High-efficiency MBS[b] or ABS impact modifier (10–12 high-efficiency acrylic impact modifier[a] for exterior applications)
 1.2–2 Acrylic processing aid
 1.0–1.5 Acrylic lubricant processing aid
 2.0–2.2 Tin stabilizer
 2.7 Glycerol monostearate
 0.3 Polyethylene wax, oxidized
 TiO_2 as needed (10–12 for exterior applications)

[a] Grades include Paraloid KM-334.
[b] Grades include Paraloid BTA-753.

464

Table 13 Thermoforming Operations: Benefits of Processing Aids

Fast, controlled fusion of PVC dry blend
  Provides uniform sheets and films
Increased hot-melt strength
  Deep, uniform draws without tearing
  Minimum thin spots and blowouts
  Fast forming cycles; less preheating
Improved appearance
  Improved gloss and uniformity
  Minimized swirls and surface blemishes
Prevention of plateout
  Long cycles between mold cleanups
Improved homogeneity
  Good ductility
  Good draws at lower temperatures, less cool-down time (= faster rates)

Note: For typical extrusion formulations, see entries for film and sheet in Tables 4 and 10.

III. APPENDIX

The major U.S. suppliers/producers are provided (see Tables A1–A4). The currently available grades of processing aids (primarily for PVC) are described as to general composition. Some of these products have small amounts of additional monomers added to give slightly different compositions and specialized performance. It is not our purpose to provide analytical results on composition or specific methods of manufacture. Those interested in these details should refer to the patent literature.

Table A1   Major Suppliers of Polymeric Processing Aids

Company	Tradename	Grade	Type[a]
Amoco Chemicals	Resin 18		α-Methyl styrene
Atochem, USA	Celukavit	N, S	Styrene–acrylonitrile based copolymers
Kaneka, Texas	Kane Ace	PA-20	Acrylic type II
		PA-100	Acrylic type III
Metco, USA	Metablen	P-501	Acrylic type I
		P-530	Acrylic type II
		P-551	Acrylic type II
		P-700	Acrylic type III
		L-1000	Acrylic type III modified
Richardson Polymer	—	P-210D	Styrene/methylmethacrylate copolymer
Rohm and Haas	Paraloid	K-120N	Acrylic type I
		K-120ND	Acrylic type I
		K-125	Acrylic type II modified
		K-147	Acrylic type I
		K-175	Acrylic type III
		KM-318F	Acrylic type II, modified: Combined impact/processing aid/lubricant; high die-swell for rigid PVC foams

[a] The following polymer types are referenced:
Acrylic type I: methylmethacrylate/ethylacrylate copolymer
Acrylic type II: methylmethacrylate/butylacrylate copolymer
Acrylic type III: methylmethacrylate/butylacrylate/styrene/copolymer (lubricating processing aid)

Table A2   Processing Aid Product Designations/Recommended Applications

Processing aid type	Supplier	Trade names®/grades
ABS	GE Plastics	Blendex 590, 862, 863, 864
Acrylics, modified acrylics	Barlocher	Barorapid 3F, 10F
	BASF	Vinuran KR3833, KR3835
	Elf Atochem	Metablen P-501, P-530 P-550, P-551, P-552 P-700,[c] VP-701,[c] P-710[c]
	Huls	Vestiform R210, R315, R450
	Kaneka Texas	Kane Ace PA-10, PA-20, PA-30, PA-40, PA-50, PA-100,[c] PA-101[c]
	Protex	Modarez APVC8, APVC100
	Rohm and Haas	Paraloid K-120N, K-120ND, K-125, K-130B, K-400,[b] K-415[b] K-175[c]

Note: Generally, these processing aids can be used in clear PVC.
[a] Recommended for opaque PVC.
[b] Recommended for cellular PVC.
[c] Lubricating processing aid.

Table A3 Typical Processing Aids and Their Applications

Processing aid	Siding	Profiles	Cellular	Pipe	Extruded film/sheet		Calendered film/sheet		Pipe fittings
					Clear	Opaque	Clear	Opaque	
Paraloid									
K-120N	X	X		X		X		X	X
K-120ND	X	X			X		X	X	X
K-125		X				X			X
K-130B					X		X		
K-175	X	X	X	X					X
K-400			X						
K-415			X						
Metablen									
P-501	X			X					
P-530			X						
P-550	X								
P-551					X	X	X	X	
P-552					X		X		
P-700	X								
P-710	X								
Kane Ace									
PA-10	X	X							
PA-20	X	X			X		X		X
PA-30			X						
PA-40			X						
PA-50					X		X		
PA-100	X	X							
PA-101	X	X							

Table A4 Representative Suppliers for Processing Aids

North America	
Elf Atochem N.A.	3 Parkway, Philadelphia PA 19102
General Electric Company	5th and Avery Streets, Parkersburg, WV 26102
Kaneka Texas Corp.	17 S. Briar Hollow, Houston, TX 77027
Rohm and Haas Company	Independence Mall West, Philadelphia, PA 19105
Europe	
Barlocher	Riesstrasse 16, D-8000 Munchen 50, Germany
Huls	Postfach 1320, D-4370, Marl, Germany
Kaneka Belgium	Westraat 34, B-1040 Brussels, Belgium
Protex	B.P. 177, 6 Rue Barbes, 92305 Levallois, France
Rohm and Haas Company	La Tour de Lyon, 185 Rue de Bercy, 75579 Paris-Cedex, France
Asia	
Kanegafuchi Chemical Co.	2-4, 3-chome, Nakanoshima, Kita-ku, Osaka, Japan
Kureha Chemical Industry	9-11, 1-chome, Nihonbashi Horidome-Cho, Chuo Ku, Tokyo, Japan

ACKNOWLEDGMENT

We thank Marcel Dekker, Inc., and Leonard I. Nass, editor of *The Encyclopedia of PVC,* for permission to reprint a portion of Chapter 16, ''Polymeric Modifiers; Types, Properties and Performance,'' by V. E. Malpass, R. P. Petrich, and J. T. Lutz, Jr.

REFERENCES

1. C Ryan. SPE J 24:89, 1968.
2. A Berens, V Folt. Polym Eng Sci. 8:5, 1968.
3. P Faulkner. J Macromol Sci Phys. B11:251, 1975.
4. G Menges, N Berndsten. Kunststoffe 66:735, 1975.
5. G Menges et al. Kunststoffe, 69, 562 (1979).
6. E Rabinovitch, J Summers. J Vinyl Technol. 2(3):165, 1980.
7. R Gould, J Player. Kunststoffe 69:7, 1979.
8. K Bohme. Angew Makromol Chem 47:243, 1975.
9. R Krzewki, E Collins. SPE ANTEC 1981, Preprint 570, 1981.
10. Plastics (Int J) 98, May 1963.
11. A Pazur, L Uitenham. SPE ANTEC 1981, Preprint 573, 1981.
12. R Petrich. Reprints PRI Conference on Processing of PVC, London, 1978.
13. B. F. Goodrich. U.S. Patent 2,646,417 (1953).

14. B. F. Goodrich. U.S. Patent 2,791,600 (1957).
15. ICI. U.S. Patent 3,373,229 (1965).
16. Stauffer. U.S. Patent 3,764,638 (1973).
17. Rohm and Haas. U.S. Patent 3,833,686 (1974).
18. Kanegafuchi. British Patent 1,378,434 (1974).
19. Japanese Geon. U.S. Patent 3,673,283 (1972).
20. Tenneco. U.S. Patent 3,874,740 (1974).
21. A Wilson, V Raimondi. Reprints PRI Converence on Processing of PVC, London, 1978.
22. U.S. Patent 4,105,624 (1978).
23. Air Products. U.S. Patent 4,137,280 (1979).
24. Rohm and Haas. U.S. Patents 3,859,384 and 3,859,389 (1975).
25. R Petrich. Mod Plast 49(8):74, 1972.
26. R Graham. Am Chem Soc Div Org Coat Plast 34:172, 1974.
27. Protex. French Patent 2,324,660 (1974).
28. BASF. British Patent 848,153 (1960).
29. BASF. U.S. Patent 4,051,200 (1976).
30. Mitsubishi Rayon. German Patent 2,163,986 (1972).
31. Rohm and Haas. U.S. Patent 3,867,481 (1975).
32. Rohm and Haas. German Patent 2,017,398 (1970).
33. du Pont. U.S. Patent 3,096,313 (1967).
34. Rohm and Haas. U.S. Patent 3,284,545 (1966).
35. Rohm and Haas. U.S. Patent 3,485,775 (1969).
36. Monsanto. U.S. Patent 3,584,079 (1971).

13
Specialty Additives

Saul Gobstein
Sa-Go Associates, Inc., Shaker Heights, Ohio

I. COMPLEXITY OF ADDITIVE TERMINOLOGY

A. Additives Defined

It is known and documented that in the early development of plastics, useful products could not be manufactured unless certain additives were incorporated in the polymer matrix. The incorporation of these additives, forming a compounded plastic, overcame the limitations of the raw polymer. For example, the use of heat stabilizers allowed the polymer to be processed and gave the plastic a useful service life while plasticizers also assisted in processing and gave the product elasticity.

Mascia [1] defines plastic additives as "those materials that are physically dispersed in a polymer matrix without affecting significantly the molecular structure of the polymer." The definition should be expanded to include "but they do affect the processing of the polymer, as well as the performance in its service life." Based on this definition, all materials added to a polymer are additives.

B. Importance of Additives

In the plastics industry, the words *resin, polymer,* and *plastic* are used interchangeably. They are not synonomous. The *American Heritage Collegiate Dictionary* [2] defines the term resin as "any of numerous physically similar polymerized synthetic or chemical modified natural resins including thermoplastic materials such as polyvinyl chloride, polystyrene, and polyethylene; and thermosetting materials such as polyesters, epoxies, and silicones that are used with fillers, stabilizers, pigments and other components to form a plastic." Plastic [3] is defined as "capable of being shaped or formed" and comes from the Greek word "plasti-kos"—fit for molding. Therefore, it is the additives which transform the resin or polymer into a useful plastic.

The first generation of additives was mainly concerned with overcoming the limitations of the polymers. The second generation of additive technology is directed to the building of desired properties into a plastic part by the use of additives.

C. Classification of Additives

There is no clear-cut classification system for plastics additives. Mascia [4] classifies additives according to function. This system is not adequate because an additive may perform more that one function. Another problem is that one additive with a specific function may interfere with another additive's function with regard to compatibility, interaction, dispersion, and/or distribution. Therefore, the final choice is a compromise based on the need for and the requirements of a specific function at the lowest cost.

II. ANTIBLOCKING AGENTS AND SLIP ADDITIVES

Antiblocking and slip agents have been used extensively in packaging film since the beginning of the cellophane era. The major market is in polyolefins [5]. However, there is also a market in poly(vinyl chloride) film and plastisols used in bottle caps, as well as in other films.

The problem occurs between adjacent surfaces which are very smooth and have a high gloss. The contact between these types of surfaces is nearly perfect, eliminating air between the surfaces, and therefore the surfaces adhere to one another. Because there is no air layer between the surfaces, it is difficult to separate them. Blocking can also occur due to pressure or temperature effects that induce fusion of the surfaces in contact.

A. Blocking and Slip Defined

The term "block" [6] is defined as the ability of a layer of film to adhere to another layer of film. In many cases, this phenomenon may be minimized by processing conditions (line speed) and by avoiding high storage temperatures. It also can be controlled by antiblocking additives.

The term "slip" means the ability of a layer of film to slide past another layer of film. A high coefficient of friction [7] denotes low slip, and a low coefficient of friction denotes high slip.

B. How They Function

Slip and antiblocking agents are microscopic-sized particles that are incorporated into a plastic formulation. These particles are completely or partially incompatible in the polymer matrix. The additive will exude or bloom to the surface, forming a microscopic bump, thus forming a surface irregularity that will give slip and antiblocking characteristics.

It is important to maintain a balance between slip and antiblocking agents. Blocking will usually decrease with the addition of an antiblocking agent up to a maximum point. Beyond that point, there is no further blocking. Excessive use of the antiblocking agent therefore only decreases the optical properties. The same is true for slip additives.

Antiblocking agents and slip additives do interact with one another as well as exerting an effect on the lubrication system, the film gauge, line speed, type of film, type of resin, and other additives in the formulation. It is possible to get "wet blocking" when a slip additive is used alone or in excess [8]. It is also possible to get "heavy blocking" when antiblocking agents are used alone or in excess due to the pressure, which flattens out the microscopic bumpy surface.

These additives have one characteristic in common: They are incompatible in the polymer. In the case of slip additives, they also migrate to the surface during or after processing and exhibit lubricating features. Increasing the slip additive concentration will decrease the coefficient of friction very quickly until it reaches the critical level. An increase in concentration after this point will not decrease the coefficient further.

C. Types of Materials Used

Traditionally the most popular antiblocking agents are diatomaceous earth or synthetic silicon dioxide. These products have the least interference with clarity and are usually approved by the U.S. Food and Drug Administration (FDA). Fatty acid amides are also used, in very small quantities—usually less than 0.25 phr. When the amides are used, the lubrication system should be checked to make sure it also is balanced.

The types of materials that are used as slip or antiblocking agents are given in Table 1. A partial list of manufacturers of antiblocking agents and slip additives is presented in Table 2. New developments include replacement of silica with less abrasive synthetic zeolite and the possible use of a graphitic (but colorless) material, boron nitride [9].

III. ANTISTATIC AGENTS

A. Need for Antistatic Agents

Static electricity can cause several problems in the manufacture and fabrication of plastic parts as well as in the service life of the plastic. The charge on the

Table 1 Products Used as Slip and Antiblocking Agents

	Antiblock	Slip
Saturated fatty acid amides	X	X
Unsaturated fatty acid amides	X	X
Metal salts of fatty acids	X	X
Diatomaceous earth	X	
Synthetic silica and zeolites	X	
Waxes	X	X
Calcium carbonate	X	
Talc	X	
Boron nitride	X	?

plastic part or plastic particle, while in process, is unpredictable. A static electric discharge in a dust-filled or flammable atmosphere can cause an explosion. A static charge on a resin particle in a low-humidity atmosphere can cause agglomeration. Static charges on videotapes or disks can cause audio or video distortion. Static charges on computer components can cause loss of memory. The buildup of a static charge on a film or sheet may cause difficulty in separation.

B. Theoretical Considerations

A static-charged particle or a static-charged plastic surface occurs when two surfaces are parted after close contact. The charge buildup is determined by the

Table 2 U.S. Manufacturers of Antiblock and Slip Additive

Company	City, State
American Hoechst Corp., Ind. Chem. Div. (amides)	Somerville, NJ
Armak Co., Div. of Akzo Nobel (amides)	Chicago, IL
Cab-O-Sil Div., Cabot Corp. (silica)	Tuscola, IL
Davidson Chem. Div., W.R. Grace & Co. (amides)	Baltimore, MD
Degussa Corp. (silica)	Teterboro, NJ
Witco Chem. Corp. (amides)	Memphis, TN
PQ Corp. (zeolite)	Philadelphia, PA
Carborundum Corp. (boron nitride)	Amherst, NY
Ferro Corp. (metal salts)	Cleveland, OH
Lonza Inc. (amides)	Fairlawn, NJ

rate of generation and, simultaneously, by the rate of decay. Frictional forces on the polymer particles while in processing can build up a static charge on the particles.

The static electricity buildup on plastic parts and polymer particles is complex and not well understood. ASTM D257 states that ''surface resistance or conductance cannot be measured accurately, only approximated, because more or less volume resistance or conductance is nearly always involved in the measurement. The measurement value is largely a property of the contamination that happens to be on the specimen at the time.'' This explains why DC resistance and conductance test methods may not be truly correlated with actual service life conditions.

Static electricity is a surface phenomenon. It is either a buildup or a deficiency of electrons. Electrostic charges on poor conductive surfaces such as plastics can be generated by frictional forces or ionized air [10].

To dissipate the charges, Mascia [11] teaches us that the rate of surface generation can be decreased to some extents by reducing the intimacy of contact, whereas the rate of charge decay can be increased substantially by rendering the surface conductive by the formation of a conductive layer. This conductive layer can be achieved by the addition of an antistatic agent which migrates to the surface and forms conductive paths through absorption of atmospheric moisture.

Schmidt [12] claims that charges can be dissipated into the air by use of an antistatic agent. The mechanism involves water molecules in the air becoming attracted to the charged end of the antistatic molecule. The charge is transferred from the surface of the plastic to the water molecules which then can diffuse the charge into the air environment. The antistatic agent is then free to pick up another charge from the plastic and transfer it to the water molecules in the air. This can go on and on unless the antistatic agent is removed from the surface of the plastic.

C. Practical Aspects

The buildup of electrical charges may be dissipated in several ways:

1. *Grounding:* If the problem occurs in processing, ground the equipment. This will undoubtedly solve the process problem but may not solve the problem of static buildup of the plastic part after it is manufactured and in service.

2. *Increasing Moisture Content:* Increasing the moisture content of the air can dissipate the static charge in processing. For example, in the early days of dry-blending rigid PVC, there were minor problems of balling and particle agglomeration while blending. With the new regulations for vinyl chloride monomer (VCM), starting in 1974, blending problems became almost catastrophic, especially when humidity was

low. By introducing water while the resin was being charged into the blend, the problem was overcome. This required higher discharge temperatures from the blender (above 212°F), which, in turn, resulted in better dispersion and distribution of additives.

3. *Ionization:* The rate of static dissipation in a plastic part while in process can be increased by increasing the electrical conductance of the surrounding air by ionization. This method usually charges and discharges surface static electricity. In practice, this method has been found lacking, as there may be static buildup farther down the process line or in the finished part.

4. *External Antistatic Agents:* An antistatic agent is applied to a plastic part, film, sheet, or monofilment surface in solution form. The solvent evaporates, leaving the antistatic agent on the plastic surface. As long as the antistatic agent remains on the surface, protection is provided. Once the agent is removed by washing, rubbing, or solvent, the part will no longer be protected.

5. *Internal Antistatic Agents:* Internal antistatic agents are added to the polymer matrix. After processing, these additives exude to the surface, forming a conductive layer and thereby making the plastic itself more conductive. If the surface is washed, rubbed, or attacked by solvent, exudation will occur again when the surface is dried and will reestablish the conductive layer.

Care must be excercised in incorporating internal antistatic agents into the formulation. The lubrication system should be checked to determine whether it is out of balance (internal versus external lubrication). The performance of the stabilization system as well as other ingredients in the formulation should be checked for possible interaction with the antistatic agent.

D. Testing

1. Ash Pickup

Testing antistatic properties of a plastic material is done by many standard quantitative testing procedures as well as homegrown tests. One very popular homegrown test for rigid PVC phonograph records is the dust or ash pickup test. This is a simple test in which cigarette ashes are dumped on a phonograph record and the record is checked to see if the ashes are rejected from its surface.

2. Corona Discharge Electrometer [13]

In this test, a 5.5-in. × 5.5-in. × 103.0-in. pressed polished specimen is placed on a grounded aluminum plate for 5 s. The corona discharge is 1 in. from the specimen. A charge of 10 kV ionizes the air without arcing, charges the panel, and completes the circuit through the aluminum plate. The charged panel is then placed under a 3-in. static detector. The detector is in an insulated glove box 36 in. × 16 in. × 24 in. The humidity is controlled by specific saturated salt solutions. Readings for the voltage decay can be taken from 1 to 5 min or up to 60 min, depending on the type of material used.

3. Surface Resistivity Measurements

These measurements have been standardized under ASTM D257-66. As pointed out in this and the procedures above measuring surface resistivity has many drawbacks. Surface resistivity is a measure of charge lost from the surface and not lost into the air. It also does not take into account contamination or impurities that may improve or retard static dissipation. Therefore, it is possible to have a material with poor surface resistivity and good antistatic properties.

This test is an excellent one for measuring conductivity, but it does not actually measure antistatic behavior. Equipment for this test is available from Kiethley Instrument Co. of Cleveland, OH.

4. Soot Chamber Test [14]

This is a subjective test which attempts to measure dust buildup on a plastic part. A charge is induced on a plastic part, which is then placed in a chamber. The chamber contains a filter paper soaked with an aromatic solvent. The paper is ignited and allowed to burn completely. The chamber is then opened and the part examined for soot buildup. The degree of soot buildup is rated from 1 to 10.

5. Charge Induction and Decay Between Electrodes

This method is used to test plastic sheet for antistatic properties. The method which is adapted from method 4046 of Federal Test Method Standard 101B [15], essentially tests for sparking in an explosive atmosphere. It measures the time required to charge a plastic surface and the time it takes to dissipate that charge.

E. Materials Used

Materials used to impart antistatic properties to a plastic are ionic materials and hydrophilic materials which will attract moisture to the surface of the plastic [16]. The ionic group consists of quaternary ammonium compounds or amines. The hydrophilic group consists of polyglycols (derivatives of ethylene oxide). Derivatives of sulfonic acids, phosphoric acid, and polyhydric alcohols are also used

Table 3   U.S. Manufacturers of Antistats

Manufacturer	Trade name	City, State
Alframine	Electrosol 325	Paterson, NJ
Cytec	Cyastat	Wayne, NJ
American Hoechst	Hostastat	Somerville, NJ
Witco	Markstat	Brooklyn, NY
Buckman Laboratories	Bubond	Memphis, TN
Lonza	Aldo	Greenwich, CT
Michel	Michel	New York, NY

as antistatic agents [16]. Most products are proprietary. A partial list of manufacturers is presented in Table 3.

F.   Selection of an Antistat

Antistat consumption for all plastics grew 45% in a 5-year period from 2150 metric tons in 1980 to 3125 metric tons in 1985 [17]. Most of the growth was due to packaging of food and electronic parts. However, acceptance of antistats in halogenated polymers is still limited because, in most cases, they interfere with the function of other ingredients in the formulation.

For example, quaternary ammonium compounds give a rigid PVC compound good antistatic properties but hurt the heat stability. Polyether-type antistats have a smaller effect on heat stability but only render marginal antistatic properties.

In plasticized compounds, the type and amount of plasticizers that are used play an important role in antistatic properties [18].

IV.   FOAMING AGENTS, FOAM PROMOTERS, AND FOAM ACCELERATORS

A.   Introduction to Cellular Foam

There are two types of cellular foam: open cell and closed cell. In the closed-cell type, each cell is individual, usually spherical in shape, and completely enclosed by plastic walls. This type of cell structure has good insulating properties as well as a high degree of buoyancy. In the open-cell type, all the cells are interconnected. This type of cell structure is known for its absorbency and capillary action.

A foam is a solid–gas composition consisting of a continuous polymer phase and a gas phase, either continuous or discrete, created by a foaming agent. The cellular structure depends on the nature of the cell-forming process, that is, whether it is a physical change of state, a chemical decomposition, or some other chemical reaction [19].

Polymers can be expanded in one of the following ways:

1. Whipping air into a plastisol or emulsion, followed by rapid curing of fusion
2. Mixing in a liquid component which vaporizes when heated
3. Dissolving a gas into the compound when it is liquid or in the plastic state
4. Adding components that chemically react and generate a gas within the plastic
5. Incorporation of hollow glass beads
6. Thermal decomposition of a foaming agent that liberates a gas

B. Promoters for Mechanical Foaming

Foaming promoters are adjuncts that facilitate the formation of uniform cells and can increase the stability of the foam. For example, they can change the surface tension of a foamable plastisol, modify the melt viscosity of an expandable compound, or provide nucleation sites for cell formation in gas-saturated polymer melts. Adjuncts that can decrease the decomposition temperature of the blowing agent are called foaming agent (or blowing agent) promoters.

Foam can be prepared by vigorous mechanical agitation of plastisols. In a typical mechanical foaming process, a compounded plastisol that contains a suitable foam promoter and compressed air is fed into a mixing head of a continuous mixer [20], where a flowable froth is produced. The froth can be continuously cast on fabrics or release paper by a coating knife [21] or, if desired, it can be discharged into a mold. The foam is gelled and fused in a hot-air oven with radiant or high-frequency heating. Unless properly stabilized, the entrapped air diffuses rapidly and the froth collapses before or during the fusion. Foam promoters are used to prevent this occurrence.

Foam promoters are surface-active agents of the nonionic or anionic type. Usually, a combination of several oil-soluble surfactants affords the best balance of desired foaming properties [22]. Aqueous solutions of surfactants have also been suggested for this purpose [23], but the most successful process [24] uses a water-in-oil (plasticizer) emulsion prepared with anionic wetting agents [25].

The currently marketed foam promoters [26] are blends of surface-active materials such as potassium oleate or morpholine linoleate [27] in a water-in-oil emulsion, combined with heat stabilizers and dissolved in a plasticizer.

Satisfactory foaming in the mixing device and trouble-free conveying of the froth require plastisols with either Newtonian or thixotropic flow properties. For good air retention properties, the foamable plastisol should be designed around highly solvating plasticizers.

Plastisol foam obtained by this process displays an open-cell structure and can be made in a range of apparent densities from 12 to 60 lb/ft^3.

C. Physical Blowing Agents

Physical blowing agents undergo a phase change during foaming. For example, compressed gases dissolved under pressure in a plastisol can develop a cellular structure on release of the external pressure. Volatile liquids incorporated in a mix are capable of producing a foam by passing from a liquid to a gaseous state.

The preferred agents for physical foaming processes are usually odorless, nontoxic, chemically stable compounds that will not adversely effect the physical properties or the thermal stability of polymers. Inert gases and volatile liquids are the most commonly used agents of this class. Occasionally, leachable solids have also been used in preparing microporous compositions [28].

Because physical foaming agents leave no solid residues, nucleating additives are frequently needed to create sites for cell formation [29]. Although physical foaming agents are generally inexpensive, their efficient use ordinarily requires specialized equipment designed to produce one specific cellular product.

1. Gaseous Blowing Agents

Nitrogen, air, and carbon dioxide are prime examples of gaseous blowing agents. The first two are used in the high-pressure ''gassing'' process; the latter lends itself to low-pressure adsorption methods.

The high-pressure gas expansion method was commercially developed in the mid-1930s for cellular materials based on cross-linkable elastomers [30]. Later, several modifications of the original process permitted the manufacture of expanded thermoplastics, primarily PVC [31]. The basic process operates as follows: The compound is placed in a mold or autoclave and is saturated under high pressure with nitrogen or air while heat is applied to fuse the resin. Following the fusion step, the gas-saturated melt is cooled under pressure and, when cold, removed from the pressurized vessel. At this point, the uniformly distributed gas bubbles contain gas at higher than atmospheric pressure, but, essentially, no expansion occurs because the gas bubbles are encapsulated within the fused resin. When the preexpanded body is heated to temperatures above the heat distortion point, the gas in the cells expands the soft matrix. At the same time, a pressure equilibrium of the gas in the cell and the surrounding environment is brought about.

Because the solubility of nitrogen in polymers is rather low, extremely high pressures, on the order of 3000–10,000 psi, are required to saturate the melt. An improved absorption device facilitates the gassing step [32]. Other improvements of the process include saturating plasticizers with gas prior to blending with the polymer [33] and using special plasticizers that can readily be saturated with nitrogen [34]. In spite of all these improvements, gas expansion requires heavy-duty equipment capable of handling gas-saturated melts at high temperatures.

2. Liquid Blowing Agents

Many useful physical blowing agents are found among volatile liquids with boiling points not exceeding 110°C. These compounds are usually selected from odorless, nontoxic, noncorrosive, and nonflammable liquids having good thermal stability in the gaseous state.

Because the efficiency of liquid blowing agents is directly related to the ratio of the specific volume of vapor to the volume of liquid, products with high specific gravity combined with low molecular weight are most effective. Fluorinated aliphatic hydrocarbons have all these desirable properties and, therefore, are ideal physical blowing agents [35] (Table 4). The less expensive chlorinated hydrocarbons, such as methylene chloride, also have been suggested for making cellular compositions [36].

For many years, chlorofluorocarbons (CFCs) appeared unique in their ability to dissolve in many polymers and then, with higher temperature and decreased pressure, to expand while assisting the polymer to form thin walls, yielding a low-density foam with small cell size. In many cases, CFCs then remained in

Table 4 Properties of Physical Blowing Agents

Blowing agent	Molecular weight	Density (g/cm^3 at 25°C)	Boiling point or range (°C)
Cyclohexane	84.17	0.774	80.8
Toluene	92.13	0.862	110.6
Trichloroethylene (HCFC[a])	131.40	1.466	87.2
1,2-Dichloroethane (HCFC)	98.97	1.245	83.5
Trichlorofluormethane (CFC[b])	137.88	1.476	23.8
1,1,2-Trichlorotrifluoroethane (CFC)	187.39	1.565	47.6
Acetone	58.08	0.810	56.2
Methyl ethyl ketone	72.10	0.810	79.6

[a] HCFC: hydrochlorofluorocarbon.
[b] CFC: chlorofluorocarbon.
Source: Ref. 37.

the foam permanently, leading to excellent thermal insulation. This technology enabled thin-walled refrigerators and other insulated containers. In the 1980s, the progressive decrease in the atmospheric ozone layer was linked to escape of CFCs, a finding now generally accepted. In polyesters, nitrogen or carbon dioxide can replace CFCs. This is more difficult in polystyrene, polyurethanes, and polyethylene foam. CFC usage has now been almost entirely discontinued.

A short-range solution has been replacement with hydrochlorofluorocarbons (HCFCs). The latter are still ozone reactive, but much less so than CFCs. It is likely that their usage will eventually be also banned. In general, replacement of CFCs with HCFCs tends to lead to lower cell uniformity, particularly at low densities. At times, the resultant poorer appearance can be compensated for by pigmenting the foam.

A long-range solution may lie in the use of supercritical fluids, in particular, supercritical carbon dioxide, to replace CFCs and HCFCs. Supercritical carbon dioxide dissolves in a wide variety of polymers but requires high-pressure equipment to maintain the supercritical state.

One of the main advantages of liquid blowing agents is the endotherm that accompanies the change from the liquid to the gaseous state. Hence, low-boiling liquids can be used in conjunction with chemical blowing agents to lower the decomposition exotherm of these agents. Tetrachloroethanes [38], methyl chloride, ketones, and aromatic hydrocarbons [39] are most widely used for this purpose.

D. Chemical Blowing Agents

Inorganic and organic compounds that liberate large volumes of gas, as a result of thermal decomposition at elevated temperatures, constitute a group of additives known as chemical blowing agents. These agents are usually solid products with good thermal stability at ambient temperatures. At elevated temperatures, the agents undergo rapid decomposition in well-defined temperature intervals which are determined by the chemical nature of the agents and the environment in which they are being decomposed. The most commonly used foaming agents liberate, in addition to nitrogen, other noncondensable gases such as carbon dioxide, carbon monoxide, and hydrogen.

The ideal blowing agent has the following characteristics:

Be easily dispersed
Not have an adverse effect on process properties
Not plate out
Not have a deleterious effect on heat or light stability
Leave an odorless, colorless, nonstaining, nonmigrating, nontoxic residue
Not leave a corrosive or toxic decomposition product

Release gas over a short, well-defined temperature range
Release gas at a controllable rate
Not affect the gelation or fusion rate of a plastisol
Function equally well in open or closed cellular production conditions
Be stable in storage
Leave a residue that could form nucleation sites for cells
Leave a residue that would not detract from electrical properties
Not have detrimental dermatological properties in applications requiring
    contact with skin
Allow the formulator to develop foams for a wide range of applications
Not be expensive

The major advantage of using a chemical blowing agent is that these additives can produce a cellular structure from a compound which is being processed in conventional equipment.

1. Inorganic Blowing Agents

Inorganic blowing agents, mostly alkali salts of weak acids, have been of limited use. The most important of these agents are listed in Table 5. They can liberate gas either by thermal dissociation or in the presence of activators by chemical decomposition. The thermal dissociation of an inorganic salt is an endothermic reversible reaction, and the reaction rate and equilibrium point depend on external

Table 5   Inorganic Blowing Agents

Chemical description[a]	Supplier	Decomposition temp. (°C)	Gas yield [mL (STP)/g]	Main decomposition product
1. Ammonium bicarbonate	Many	60°C	850	NH_3, CO_2, H_2O
2. Sodium bicarbonate	Many			CO_2, H_2O
A		100–140°C	534	CO_2, H_2O
B		Ambient		
3. Sodium borohydride	Ventron Div. Morton-Thiokol			
A	Co.	300°C	2370	H_2
B		100°C		

[a] Key: A, thermal dissociation; B, chemical decomposition in acidic medium.

pressure. For this reason, the use of inorganic blowing agents is limited to foaming at atmospheric conditions.

As foaming agents, inorganic salts have the advantage of low cost, but this is counterbalanced by poor storage stability and the difficult dispersion properties of these products. Because gases generated by inorganic foaming agents either are readily condensable (e.g., water vapor) or have high diffusion rates (e.g., hydrogen), foams made with these agents are dimensionally unstable and require long annealing periods.

Among the inorganic salts, ammonium bicarbonate is of some interest because it leaves no residue on decomposing:

$$(NH)_4) HCO_3 \rightarrow NH_3 + CO_2 + H_2O$$

Assuming water vapor as the expandable gas, the gas yield of ammonium bicarbonate [850 mL (STP/g)] is one of the highest among all chemical blowing agents. The salt has been suggested as an additive to dinitrosopentamethylenetetramine in the expansion of vinyl plastisol [40].

Sodium bicarbonate liberates a smaller volume of gas than the ammonium salt but its thermal stability is better. In storage, it has a tendency to pick up moisture and to cake. For use in plastics sodium bicarbonate must be carefully ground by repeated passing through a colloid mill, three-roll mill, or similar high-shear device.

The thermal decomposition of sodium bicarbonate takes place in a wide temperature range ($100-140°C$) with the attendant evolution of carbon dioxide and water vapor:

$$2NaHCO_3 \rightarrow Na_2CO_3 \rightarrow CO_2 + H_2O$$

The alkaline residue of sodium carbonate is objectionable in any application where it can damage coatings and substrates. Because the thermal dissociation of sodium bicarbonate decreases rapidly with increased pressure, application of the salt is limited to expansion under atmospheric pressure [41].

To increase the efficiency of sodium bicarbonate, acidic additives such as stearic acid are used [42]. Total decomposition of the bicarbonate can be brought about by chemical reaction of the salt with water-soluble acids:

$$2NaHCO_3 + 2H^+ \rightarrow 2Na^+ + 2H_2O + 2CO_2$$

This system has been suggested for frothing of plastisols. The reaction may proceed at room temperature; therefore, the reactive ingredients must be predispersed in plastisol and stored separately. When the two components are blended, carbon dioxide foams the compound. So that the evolved gas in the plastisol can be entrapped and breakdown of the foam prevented, the compound should contain a fatty acid soap as a thickening agent.

A similar technique can be applied to sodium borohydride, which generates a much larger volume of gas than sodium bicarbonate:

$$NaBH_4 + 2H_2O \rightarrow 4H_2 + NaBO_2$$

A typical plastisol foaming system consists of a plastisol containing a buffered solution of sodium borohydride and a second plastisol with an acidic activator [43].

With strong organic acids, such as acetic and oxalic, the gas generation is extremely rapid, and it becomes difficult to control the frothing of plastisol. Using buffered acid solutions offers a convenient means of controlling the reaction. Best results have been reported with glycine [44], glycerol [45], and isocyanates [46].

2. Organic Blowing Agents [47]

Organic compounds that release nitrogen as the main component of the liberated gas are the most important foaming agents. The thermal decomposition of organic blowing agents is an irreversible exothermic reaction independent of external pressure. The decomposition rates of these products are governed by temperature, or time, or both, and are independent of concentration. Interaction with other ingredients in the formulation or impurities may change the course and the rate of the blowing agent decomposition.

Chemically, organic foaming agents are characterized by the functional groups shown in Table 6. Blowing agents with the same functional groups may display widely different decomposition temperatures, depending on their molecular structure, but their decomposition exotherms are essentially equal. For example, azides generate very high exotherms, whereas sulfonhydrazides have relatively low caloric effects. A large exotherm generated during rapid decomposition of the blowing agent may raise the internal temperature of the foaming system

Table 6 Characteristic Functional Groups of Organic Blowing Agents

—N=N—	Azo
$>$N—NO	N-Nitroso
—SO$_2$—NH—NH—	Sulfohydrazo
—N$\langle$ (N, N ring)	Azido

Table 7   Commercial Blowing Agents

Chemical composition	Melting point, or decomposition point in air	Decomposition range in compound	Gas yield [mL (STP)/g]	Manufacturer[a]
Azobisisobutyronitrile (AZDN), $(CH_3)_2(CN)C—N{=}N—C(CN)(CH_3)_2$	105°C	90–105°C	136	2
4,4′-Oxybis (benzenesulfon-hydrazide) (OBSH), $O(C_6H_4—SO_2—NH—NH_2)_2$	164°C	130–160°C	125	4,5,6
1,1-Azobisformamide (ABFA) (azodicarbon amide), $H_2N—CO—N{=}N—CO—NH_2$	195–200°C	150–200°C	220	1,3,4 ,5,6
p-Toluenesulfonyl semicarbazide (TSC), $C_6H_4(CH_3)SO_2NNNHSONH_2$	227°C	190–200°C		5,6

[a] Manufacturer	Address	Trade name
1. Astra Zeneca	Wilmington, DE	Ficel
2. Dupont	Wilmington, DE	Vaso
3. Mobay Chemicals	Pittsburgh, PA	Porofor
4. Olin Chemicals	Stamford, CT	Nitropore, Kempore
5. Uniroyal	Naugatuck, CT	Celogen
6. Dongjin Chemical	Akron, OH	Unicell

beyond the degradation point of the polymer. Charring of the compound may occur if the heat is not dissipated quickly.

With the exception of azides, all functional types of organic blowing agents are represented among commercial products. Table 7 gives the properties of commercially available blowing agents for vinyls [47].

a.   Azo Compounds.   Aromatic azo compounds, such as diazoaminobenzene, were the first useful nitrogen-releasing blowing agents for high polymers. Because of their toxic and staining properties, they became obsolete when aliphatic azo derivatives were made commercially available. Of these, azobisisobutyronitrile and azobisformamide are particularly significant in cellular technology.

b.   2,2′-Azobisisobutyronitrile (AZDN).   This compound has found utility in the production of white, nonstaining, ordorless cellular vinyl products by the plastisol-casting method. The low decomposition temperature of the blowing agent, combined with a moderate exotherm, permits the production of large cellular bodies without the danger of polymer degradation. Also helpful in this respect is the fact that the decomposition residue, tetramethylsuccinonitrile (TMSN), is a hydrogen chloride acceptor, which facilitates the stabilization of PVC compounds.

Unlike many other organic blowing agents, AZDN is insensitive to activation by compounding ingredients. The decomposition rate of AZDN depends only on the temperature of the environment (Table 7). At 30°C and below, the storage stability of the compound is excellent. AZDN is a flammable solid that ignites easily from an open flame and continues to burn when the flame is removed.

Simultaneously with the evolution of nitrogen, AZDN generates free radicals useful in the polymerization initiation or grafting of monomers with ethylenic linkages ($-C{=}C-$) [48]. Because of this dual functionality, AZDN has found application in a novel process [49] for cellular cross-linked PVC having excellent dimensional and thermal stability. In the process, maleic anhydride and styrene (or acrylonitrile) are copolymerized and grafted onto the PVC chain. Maleic anhydride is then hydrolyzed to the acid, which is later caused to react with a diisocyanate to cross-link the polymer. As a free-radical donor, AZDN induces the copolymerization and grafting. At the same time, it forms minute gas cells that act as nucleation sites for the formation of gas cells when the isocynate-terminated polymer is caused to react with water.

The main disadvantage of AZDN as a blowing agent is the toxicity of the residual tetramethylsuccinonitrile. It can be driven off in annealing ovens or removed from the surface by washing with warm water. The health hazards connected with handling and processing AZDN therefore require special safety measures. This limits the utility of AZDN as a foaming agent. Today, free-radical polymerization of monomers accounts for most of the AZDN consumption.

c. Sulfonylhydrazides. Monosubstituted sulfonylhydrazides, such as benzene or toluenesulfonylhydrazide, have been of limited use in cellular technology [50]. They leave odoriferous thio compounds as residues:

$$4R-SO_2-NH-NH_2 \rightarrow 4N_2 + 6H_2 + R-S-S-R + R-S-SO_2-R$$

Symmetric, disubstituted sulfonylhydrazides offer a significant improvement in this respect because of the polymeric nature of the residue [51]. Decomposition of the sulfonhydrazides follows an internal redox reaction with a low level of heat evolution; the exotherm generated during the oxidation of the hydrazide groups is partially compensated by the endotherm of the sulfonyl reduction.

Among the symmetric sulfonylhydrazides, 4,4'-oxybis (benzenesulfonylhydrazide) (OBSH) [52] is the most widely used in the United States, whereas in Europe, benzenesulfonylhydrazide (BSH) [53] is more popular. Commercial OBSH is a finely ground powder that decomposes slowly at 130°C and more rapidly at 150°C. The generated noncondensable gas consists predominantly of nitrogen (98%) and the residue is a white, nontoxic solid. The blowing agent ignites easily from a spark or flame and continues to decompose with the evolution of copious smoke, even after the source of ignition has been removed. Under normal storage conditions, however, the compound is stable for a prolonged period. Like other sulfonylhydrazides, OBSH can be activated to decompose at lower temperatures in the presence of oxidizing agents such as ferric chloride, peroxides, and organic bases such as triethanolamine.

Cellular polymers produced with OBSH display a uniform, medium to fine pore structure and are nontoxic, odorless, and nonstaining, although occasionally they discolor slightly under UV exposure. Because the decomposition temperature of OBSH is too high for chemical frothing of most plastisols and too low for processing in calendered compositions, this substance finds use only in expansion of plastisols based on low-molecular-weight PVC homopolymers or copolymers [54]. Another application of OBSH makes use of its bifunctionality, which is capable of cross-linking elastomeric blends containing diene rubbers, such as NBR, with PVC. In these compounds (useful as pipe insulation), OBSH acts as a blowing and vulcanizing agent [55].

d. Sulfonyl Semicarbazides. These compounds represent a novel group of chemical blowing agents [56] with higher decomposition temperatures than the hydrazides from which they have been derived. Of particular interest is toluene sulfonylsemicarbazide (TSC) [57], a compound that has a decomposition range of 210–225°C and a gas yield of 143 mL (STP)/g. Because of its high decomposition point, TSC has been suggested as a foaming agent for rigid vinyl compositions and other structural plastics.

e. 1,1'-Azobisformamide (ABFA). Frequently called azodicarbonamide. ABFA is the most versatile blowing agent for vinyls. It has a unique combination of valuable properties that fulfill the requirements of an ideal chemical foaming agent [58].

The commercially available azobisformamide is an orange yellow to pale yellow powder which decomposes in air above 195°C (the color depends on the particle size distribution). The agent and its residue are odorless, nonstaining, and, when properly compounded, nondiscoloring. Because ABFA and its residue are nontoxic, use of the blowing agent in nontoxic applications has been sanctioned under several FDA regulations [59].

Although ABFA is insoluble in common solvents and plasticizers, it disperses readily in most compounds. The storage stability of the agent, both dry and in a dispersed form, is excellent. Significantly, ABFA is the only organic foaming agent that does not support combustion and is self-extinguishing.

When azobisformamide is heated in an ester plasticizer above its decomposition point, it delivers gas at 200 mL (STP)/g, corresponding to one-third of its molecular weight. The gas phase consists of 65% N_2, 32% CO, and 3% CO_2. Under alkaline conditions, ammonia is also generated. The composition of the residue depends on the medium in which the blowing agents is decomposed. The residue is usually white and consists mainly of hydrazobisformamide (biurea), cyanuric acid, and urazole. In the presence of alkalis, ABFA hydrolyzes rapidly at 100°C with the liberation of nitrogen, carbon dioxide, and ammonia. Polyhydric alcohols [60] and alcoholamines are equally capable of hydrolyzing ABFA at elevated temperatures.

Because of its high decomposition temperature in air, ABFA was originally considered unsuitable as a foaming agent [61]. However, a later observation that many additives, including vinyl stabilizers, can lower the decomposition temperature of ABFA led to successful commercial application [62].

E. Chemical Blowing Agent Activators

Vinyl stabilizers containing lead, cadmium, and zinc are the most efficient activators. These additives depress the decomposition temperature of ABFA and increase its rate of decomposition. The activating effect of the basic metal salts is derived from the instability of the corresponding azodicarboxylates that are formed as the first step in the decomposition of ABFA. Because the azodicarboxylates of lead, cadmium, and zinc decompose at lower temperatures than the diamide, they initiate the thermal breakdown of ABFA [63]. The activation of ABFA by metal salts is concentration dependent [64], and the rate of ABFA decomposition increases as the amount of metal salt in the compound increases. Some proprietary compounds offered in the trade are said to activate the ABFA catalytically [65].

Because of the high-temperature stability of barium azodicarboxylate, barium salts exert no activating effect on ABFA, and in their presence, the decomposition rate of ABFA is as low as it is in air. However, barium compounds show synergistic activity with lead and zinc in reducing the decomposition temperature of ABFA [66]. The activation of ABFA is promoted by alkaline conditions and inhibited in acidic media. For example, adding fumaric acid to lead-stabilized, ABFA-containing plastisol increases the decomposition temperature of the blowing agent to 190°C, compared with 165°C in the absence of the acid.

The particle size of ABFA is another factor that governs degree of activation by stabilizers [67]. Because ABFA is insoluble, activation of the blowing agent takes place in a heterogeneous system and, therefore, is directly related to the active surface area. ABFA of large particle size and small surface area is more difficult to activate than material of fine particle size having a large area of activation. Recognition of this fact has led to the commercial acceptance of an azobisformamide series of blowing agents with particle size distribution tailored to specific applications [68].

1,1'-Azobisformamide lends itself to being compounded with dry blends [69] as well as plastisols. Successful processing techniques applicable to these foamable compounds have been used in the manufacture of expanded vinyl fabrics [70] and floor coverings [71]. Other uses include low-density open-cell foam [72], extruded low-density gaskets [73], high-density profiles [74], expanded jackets for electrical wires and cables, slush-molded products [75], crown cap liners [76], injection-molded soles, and expandable inks [77].

F. Methods of Incorporation

In the compounding operation, chemical blowing agents and promoters are handled like other additives. Because of the heat sensitivity of the foaming agents, careful temperature control is essential to avoid premature gas generation.

For homogeneous distribution of the blowing agent in the compounds, the agents are predispersed. High-shear equipment is usually used in this operation. Occasionally, wetting agents are added to facilitate the dispersion of the solids. For the convenience of processors, ready-to-use blowing agent dispersions are commercially available.

Blending of the blowing agent dispersion depends on the type of compound involved in the processing. Low-shear churns and blenders are used for adding blowing agent dispersions to plastisols, and ribbon blenders or high-intensity mixers are used in processing of blends and calendering compounds. When only small quantities of blowing agent are to be added to an extrusion or injection-molding compound, resin pellets are coated with the blowing agent by tumbling although this leads to reduced consistency compared to the metered addition of a dispersion.

G. Test Methods

The apparent density of a cellular product is obtained through the trial-and-error approach at tailoring the formulation to blow under optimum conditions. Unfortunately, there is no single definition of what constitutes optimum conditions. The optimum condition will depend on the type of process, the type and amount of blowing agent, and the formulation. The factors of chemical composition of the blowing agent that delineate its efficiency are its molecular weight, number of moles of gas that can be split off, and decomposition mechanism. The efficiency of chemical blowing agent is measured by the quantity of gas given off on decomposition. This quantity is called gas yield and is defined as the gas volume at normal temperature (0°C) and pressure (760 mm Hg) per unit of weight. The actual gas yield of a blowing age in a composition may differ appreciably from the theoretical or stoichiometric gas yield because of side reactions induced by other additives present; for example, stabilizers, pigments, or resin impurities (e.g., residual emulsifiers). The gas yield values of most of the common chemical blowing agent are given in Table 7.

Determination of the actual gas yield from a chemical blowing agent heated beyond its decomposition point is the subject of ASTM method D1715 [78] issued jointly by ASTM and SPI. In this test, because the blowing agent undergoes decomposition in an inert environment and the heating rate is low, the test has little relevance to blowing agent effectiveness, nor is this method applicable to measurements of the rate of decomposition—for example, the gas yield as a function of temperature or time or both.

Differential thermal analysis (DTA) combined with thermogravimetric analysis (TGA) can be used advantageously to follow the decomposition behavior of chemical blowing agents. For example, an organic foaming agent generates an exotherm during the decomposition, and the evolved heat as well as the volume of the liberated gas can be plotted as a function of temperature. The interval between the onset and the peak of the exotherm is considered the decomposition temperature. From the corresponding TGA graph, the gas yield in this temperature range can be read [79].

The torque rheometer test method [80] allows the determination of blowing agent stability and gas evolution rate under conditions bearing a close relation to actual processing. Significant information on the thermal behavior of blowing agents in polymers also can be obtained when their actual performance (i.e., their efficiency in lowering the apparent density of the polymer) is determined as a function of time or temperature or both [81].

V. FUNGICIDES, BACTERICIDES, AND OTHER BIOCIDES

Biodeterioration of plastics can occur due to microorganisms and macroorganisms. Deterioration by microorganisms, fungi, and bacteria is probably the most commonly recognized form and is primarily a chemical destruction.

Damage by macro-organisms is chiefly physical rather than chemical. It involves gross destruction, such as the gnawing or boring of holes, instead of molecular changes and the rupturing of chemical bonds. Birds, insects, and rodents can penetrate plastic, when it represents a barrier to a source of food or water. Among the macro-organisms, those broadly classified as foulers, such as barnacles, do no particular damage to coatings. Those known as borers, including mollusks and crustaceans, are capable of penetrating the jackets of underwater cables.

End products known to have been damaged by microbes include upholstery, wall covering, floor covering, wire and cable, tarpaulins, tents, baby pants, outerwear, and shower curtains.

A problem caused by the indirect action of microorganisms is the phenomenon called pink staining. Certain organisms produce colored metabolic by-products that are soluble and readily migrate through amorphous polymers. It is not necessary that they grow on the plastic itself; they can cause the staining by growing on some other substrate in contact with it. A typical example is vinyl-coated fabric in which the vinyl is frequently stained due to microorganisms that

Table 8 U.S. Producers of Microbicides

Elf Atochem	Rahway, NJ
Huls America	Piscataway, NJ
Ferro Chemical	Toledo, OH
Ventron Div. Morton Int'l.	Beverly, MA
R. T. Vanderbilt Company	East Norwalk, CT
Akros America	New Brunswick, NJ

are flourishing in the supporting textile, even though the vinyl layer shows no evidence of growth. The name pink staining derives from the color of the stain and represents the first case in which a microbiological cause was proved. The pink-stain-producing organism is *Streptoverticillum rubrireticule* [82]. Stains of different colors have been found to have a similar origin but to involve other microorganisms.

Biodeterioration of plastics has been comprehensively reviewed by Wessel [83] and the use of biocides for its control by Scullin et al. [84]. A textbook by Greathouse and Wessel [85] is a good source of general information on deterioration of materials.

Additive chemicals used to prevent degradation by microorganisms are collectively called microbicides; they included fungicides, which kill fungi, and bactericides, which kill bacteria. Although not scientifically correct, the term *microbicide* is often used in industry to include chemicals that inhibit growth but do not kill. The proper term for these kinds of agent are *fungistat* and *bacteristat.* Table 8 contains a partial listing of producers of microbicides for PVC. The chemicals used to prevent destruction by the higher forms of life, chiefly members of the animal kingdom, include both lethal agents, or poisons, and repellents.

Although halogebated polymers are not susceptible to microbial attack, as far as is known, some of the modifying agents provide a nutrient source for both bacteria and fungi. The modifiers known to be microbial nutrients include some lubricants, and plasticizers in particular. Even when only nonsusceptible ingredients are used, microbial growth (commonly called mildew) may occur on the surface because of contamination with nutrients from an external source.

A. Mechanism of Biodeterioration

Deterioration by fungi and bacteria is chemical, involving changes in composition and breaking of chemical bonds. It appears to be caused by the action of enzymes, which the microorganisms produce, on additives in the formulation, especially plasticizer. Of the two, fungal deterioration seems to be the most prevalent and has received the most attention. This may be because fungi frequently cause discoloration, whereas bacterial damage is less readily apparent. Other manifestations of microbial degradation include changes in odor or weight, embrittlement, exudation, loss of tensile strength, loss of elongation, and changes in electrical properties.

The enzymes furnish energy, provide a mechanism for moving nutrients through the cell wall, allow the organism to extract nutrients from the substrate, and stabilize the internal chemistry of the cell to variable external environments [86].

The large amount of investigation of fungal and bacterial deterioration has been largely empirical [87]. The little that is known is not well understood and is controversial [88]. A major concern has been in the field of electronics. It was thought that solving the moisture problem would also solve the biodeterioration problem, based on the idea that the fungi and bacteria cannot live without water. In many cases, this has solved the deterioration problem; however, there have been exceptions.

The adverse effects of fungi detract from esthetic characteristics by discoloration and spotting. In electrical or electronic applications, fungal organisms may destroy the dielectric properties of the system.

Bacterial deterioration is more insidious than fungal deterioration. The bacteria usually attack the plasticizer, causing stiffening as well as creating foul odors.

Stahl and Pessen [89], in 1953, studied the growth of 1 fungus and 1 bacterium on 47 plasticizers, including an 18-member homologous series of sebacates. In 1957, this work was expanded and extended by Berk et al. [90], who presented results of an in vitro study of the ability of 24 species of fungi to use 127 plasticizers as the sole source of carbon. Hueck and van der Plas [91], in 1960, made a further contribution to the understanding of the biological deterioration of plasticizers.

Table 9 summarizes the relative microbial susceptibility of some important types of plasticizers. It is essential to realize that these are only generalizations; they are useful guidelines but should not be interpreted narrowly. For example, although phthalates and phosphates tend to be resistant and to be superior to saturated diesters of aliphatic dibasic acids, and branched-chain esters tend to be more resistant than straight chains, there are exceptions as well as varying degrees of microbial degradation. Resistance also varies according to the organism present, which ordinarily cannot be predicted or controlled during the service life of a plastic article, and according to the environmental conditions prevailing during exposure. Even for studies under carefully controlled laboratory test conditions and with pure cultures of microorganisms, the literature contains conflicting data reported by different investigators on a given plasticizer. Table 10 lists the microorganisms that are used to study the deterioration and degradation.

Table 9 Microbial
Susceptibility of Classes of
Plasticizers

Susceptible
  Sebacates
  Epoxidized oils
  Epoxidized tallate esters
  Polyesters
  Glycolates
Moderately susceptible
  Adipates
  Azelates
  Pentaerythritol esters
Resistant
  Phthalates
  Phosphates
  Chlorinated hydrocarbons
  Citrates

Table 10 Microorganisms Used to Study Degradation

Fungi	*Mucor* sp.
Alternaria tenuis	*Curvularia geniculata*
Aspergillus flavus	*Stemphylium consortiale*
Cladosporium herbarum	*Glomerella cingulata*
Paecilomyces varioti	*Myrothecium verrucaria*
Penicillium funiculosum	*Stachybotrys atra*
Trichoderma viride	Bacteria
Pullulavia pullulans	*Pseudomonas aeruginosa*
Aspergillus niger	*Serratia marcescens*
Aspergillus versicolor	*Bacillus subtilis*
Penicillium piscarium	*Escherichia coli*
Penicillium luteum	*Staphylococcus aureus*
Aspergillus oryzae	*Streptomyces rubrireticuli*
Fusarium sp.	

Where resistance to microbiological degradation is important for successful application, the compounder should determine its suitability by in vitro testing with the appropriate organism or organisms, or preferably by exposure to conditions which duplicate or closely simulate those of actual use, for instance, soil burial, tropical climate, or submersion under water.

B. Biocidal Chemicals

Many thousands of chemicals of widely differing structures are known to have biocidal activity. Of these, several hundred are used commercially as economic poisons for protection against insects, birds, rodents, marine organisms, weeds, algae, and microbes [92]. These have been chosen on the basis of a suitable combination of efficiency, cost, lack of danger to humans, and lack of side effects on the material being protected. The number of pesticidal chemicals suitable for polymers and plastics is still smaller, due to limitations of solubility and compatibility, among other factors [93]. The relatively few microbicides from this group that are useful are listed in Table 11.

Three factors are chiefly responsible for the small number of acceptable compounds: compatibility, heat stability, and processing temperature. Compatibility is inherent in the chemical structure of the biocide, but problems can sometimes be overcome by formulation with other additives, as in the case of copper 8-quinolinolate. Many potentially useful biocides are eliminated because of adverse effects on heat stability. These effects can sometimes be alleviated by using other additives or by developing a carefully balanced stabilizer system. High-temperature processing cycles rule out other biocides. Some may decompose, others may volatilize completely from the compound, and still others may volatilize sufficiently to present a health hazard to workers during processing.

Of the microbicides shown in Table 11, those most widely used are N-(trichloromethylthio)phthalimide and N-(trichloromethylthio)-4-cyclohexe-1,2-dicarboximide. They have high activity against a spectrum of organisms, can be

Table 11 Microbicides

Quaternary ammonium carboxylates
N-(Trichloromethylthio) phthalimide
Bis(tri-n-butyltin) oxide
N-(Trichloromethylthio)-4-cyclohexene-1,2-dicarboximide
Bis(8-quinolinolato) copper
p-Hydroxybenzoic acid esters
Condensate of 10,10'-oxybisphenoxarsine and epoxidized soybean oil

---

used to make transparent as well as opaque compounds, and are useful in calendering and extruding as well as in plastisols. The various 10,10^1-oxybisphenoxarsine derivatives are also widely employed [94], as is 2-n-octyl-4-isothiazoline. The quaternary ammonium carboxylates are chiefly limited to polyolefins because of an adverse effect on PVC heat stability that precludes use in the more rigorous conditions found in calendering and extrusion. They can be used in both clear and pigmented compounds.

Bis(8-quinolinolato)copper, commonly called copper-8-quinolinolate, is a highly effective fungicide, but its use is primarily restricted to military applications because of the color it imparts. By itself, the chemical has difficult and limited compatibility, but it can be compounded to overcome this problem. Means for doing this have been described in several patents [95–99].

Other materials that are occasionally used in special applications are zinc borate and modified barium metaborate. These two chemicals can also be used as part of a flame-retardant system. Tri-n-butyltin compounds are mainly used in marine applications such as boat bumpers, life preservers, marine antifouling coatings, and so forth. Trialkyltin compounds, which are toxic to a variety of aquatic life, have come under regulatory attack, particularly in Europe, and will probably disappear from the market. Only organic arsenicals have comparable effect against marine fouling organisms. These are also under attack as environmentally undesirable.

The usual concentration of microbicide ranges from 0.1% to 5%, based on the weight of the plasticizer. The optimum level varies with the specific microbicide used, the susceptibility of the plasticizer or other biodegradable additive, and the degree of protection required for the end use. The recommendations of manufacturers of microbicides should be followed in this regard.

The ideal biocide would have the following characteristics:

Be highly effective as both a fungicide and a biocide
Be effective against a wide range of organisms
Not interact with other ingredients in the formulation
Be dispersed and distributed easily
Be essentially nontoxic
Be safe to handle
Not create any environmental problems
Not contribute color or odor to the formulation
Not detract from the processing properties of the compound

The previous discussion has been concerned with the use of biocides to prevent microbial attack and degradation of compounds themselves. There has been some interest in incorporating microbicides into plastics to keep the finished articles free from microbes in order to protect the user. Examples of such articles are toys, toothbrushes, hairbrushes, telephones, and dishes. Two of the chemicals claimed to be useful for this purpose are 2-hydroxy-5-chlorobenzoic-3′,4′-dichlor-oanilide and 3,5,3′4′-tetrachlorosalicylanilide [100].

C. Insecticides

Published information on the attack of insects is scanty and conflicting. Gay and Wetherly [101] reported that semirigid PVC and rigid PVC are virtually resistant to attack by termites. However, plasticized PVC used in wire and cable is very susceptible to attack. The degree of termite penetration varied with plasticizer concentration and structure. Susceptibility increased with increasing level of plasticizer and decreased when tricresyl phosphate was used. Other effective means of control were incorporation of the insecticides aldrin and dieldrin and incorporation of certain inorganic fillers [102].

D. Rodent Repellants

Poly(vinyl chloride) compounds such as cable jackets and food packages have been attacked by rodents, especially rats and mice. Damage to cables is apt to occur if they are placed in locations that block the passage of rodents to food sources. Welch and Duggan [103] studied effects of adding rodent-repellent chemicals to calendered film and found a trinitrobenzene complex at a concentration of 0.025 g/in.^{2} to be effective against roof rats, Norway rats, and house mice. Two others had some activity but were less efficient: tetramethylthiuram disulfide and a zinc dimethyldithiocarbamate/cyclohexylamine complex, both rubber accelerators of pronounced odor.

E. Marine Microbiocides

The literature on damage to plastics by marine organisms is scanty, and that on its control by chemical agents is even more sparse. Connolly [104] found that during a 7-year exposure in the sea, there was little damage by micro-organisms, but extensive penetration by boring mollusks. Carlston and Whitting [105] found a copolymer of vinyl chloride and vinyl acetate to be a suitable binder for a ship bottom paint to prevent the attachment of fouling organisms such as barnacles. This polymer provided controlled leaching of toxicants to inhibit growth of foulers but was not itself degraded. Suitable toxicants for the coating were copper, cuprous oxide, and mercuric oxide. As stated previously, various tri-n-butyltin derivatives been found useful in such coatings, although perhaps even more objectionable.

F.   Test Methods

Many different approaches have been made toward evaluating resistance to microbial degradation and the effectiveness of microbicides. These range from laboratory Petri dish tests using pure cultures of organisms to actual environmental exposure under use conditions, and from a study of individual compounding ingredients to the testing of complete compositions.

A realistic approach is to prepare samples of the total composition, with and without a microbicide, in some convenient form such as film, sheeting, or coated fabric. These can then be cut into small squares and placed in Petri dishes containing a medium of mineral salts agar or malt agar; the medium is inoculated with spore suspensions as earlier and the dishes are incubated and observed for growth on the specimens as a function of time. A disadvantage of this procedure is that it gives only a qualitative picture, the visual observation of growth, but it is valuable as a rapid screening method. To make it more informative, larger containers for the agar medium, such as Pyrex baking dishes, can be used so that larger specimens can be exposed. After inoculation and incubation, the samples can be removed and subjected to some quantitative measurements. Such properties as stiffness, tensile strength, ultimate elongation, electrical resistance, color, surface tack, or change in weight can be measured by standard procedures.

An interesting technique is the respirometric method that Siu and Mandels [106,107] developed and Burgess and Darby [108,109] modified. The latter correlated results with a direct weight-loss method. This procedure uses differential manometers to measure oxygen absorbed by microorganisms growing specimens in inoculated nutrient agar. Quantitative estimations of microbial susceptibility in as little as 4 days are indicated.

The microorganisms associated with degradation belong to different groups of fungi and bacteria. Although tens of thousands of microorganisms are known, relatively few have been isolated from compositions that have deteriorated in actual service. There is little agreement regarding the organisms to use in laboratory testing. Table 10 presents some that have been used by investigators in this field. Both pure cultures and mixed cultures are used.

Several test methods are used: exposing compounds in such various finished forms as film, coated wire and cable, coated fabrics, and fibers in a tropical room, on an outdoor test fence, immersed in the sea, or buried in soil beds. These procedures give valuable information on the fate of various compounds under actual use conditions. In addition, they readily permit the exposure of entire finished products or of samples large enough to permit quantitative measurement of a variety of chemical and physical properties.

Along with visual observations, such data are helpful in characterizing a formulation, estimating its service life, and studying the mechanism of degradation. As an example, Baskin and Kaplan [110] used a 14-day soil burial test to study the mildew resistance of coated fabrics. They used changes in breaking (tensile) strength and stiffness as criteria of resistance. The stiffness test gave the more reliable results. A typical soil burial procedure is method 5762, U.S. Federal Specification CCC-T-191b.

VI. REODORANTS AND ODOR PREVENTION

A frequent problem is the presence of undesirable odor. These vary in intensity and in type, depending on many factors, including quality and composition of raw materials, thermal history during processing, and conditions of use. The importance of the problem is largely a function of the nature of the end product; a malodor might be tolerated in an industrial or agricultural tarpaulin but would be quite unacceptable in upholstery or clothing. Among the odors detectable in various compounds are those that have been described as musty, rancid, butyric, phenolic, and mercaptalike.

Although plasticizers or their impurities or degradation products are frequently responsible for odor formation, other compounding ingredients may also contribute and should be considered when a solution to an odor problem is sought. These other ingredients include stabilizers, lubricants, fillers, and colorants. Trace residues of solvents from inks used to print and decorate films and sheeting have also been a source of odor. Finally, microbial degradation may cause malodor formation.

A. Chemicals for Odor Control

Products used to neutralize a malodor or to replace it with a pleasant odor are known variously as odorants, deodorants, reodorants, masking agents, perfumes, and fragrances. The development of a suitable material is an art, and the specific chemicals employed are closely guarded proprietary secrets. Blends of components are frequently employed from a variety of chemical classes, including esters, ethers, alcohols, aldehydes, ketones, acids, terpenes, and phenols.

Factors to be considered in compounding a suitable product include (1) the nature of the malodor and its intensity, (2) the requisite lasting power of the deodorant or odorant, which is dependent on its evaporation rate, which may in turn be affected by other additives, (3) any special requirements, such as FDA sanction if the end use is in food packaging, and (4) whether the objective is merely to neutralize an unpleasant odor or to replace it with a detectable, pleasant scent. Pantaleoni [111] has presented an excellent review of the subject of industrial perfumery.

A reodorant for a specific compound is best selected through cooperation with a producer of fragrances, due to the sophisticated nature of the art. A partial

Table 12   U.S. Producers of Reodorants

Alpine Aromatics, Inc., Metuchen, NJ
Dodge & Olcott, Inc., New York, NY
Fritzsche Brothers, Inc., New York, NY
Givaudan Corporation, New York, NY
Harwick Standard Chemical Company, Akron, OH
Mermix Chemical Company, Chicago, IL
Noville Essential Oil Company, Inc., North Bergen, NJ
Polak's Frutal Works, Inc., Middletown, NY
R. T. Vanderbilt, Norwalk, CT

---

list of these manufacturers can be found in Table 12. To achieve the best results, the reodorant specialist requires information on the identity of all compounding ingredients, the time and temperature of processing, the nature of the finished product, and its intended use and life expectancy. To supply a detectable odor when one is desired in the finished product for increased sales appeal or special effect, a wide variety of scents are available. These include leather, cedar, citrus, floral, mint, and talcum powder. It is also possible merely to neutralize an unwanted odor. Concentrations of reodorants range from 0.01% to as much as 1%.

Instead of using deodorants or perfumes, one can approach the problem in another way, preventing the occurrence of the odor by chemical means. One source of odor in ester plasticizers involves trace quantities of aldehydes and ketones in the alcohols used in preparation of the plasticizers. It is possible to minimize or eliminate this source by chemically treating the alcohols. Sodium borohydride has been used for this purpose, to reduce the carbonyl compounds to alcohols [112,113]. Lithium borohydride has also been suggested [114].

Attack by microorganisms is another cause of odor in PVC compounds during service. The use of chemicals for controlling microbial growth was discussed in a previous section of this chapter.

B.   Test Methods

No satisfactory instrumental method has been devised for measuring and evaluating odor. It is usually judged by the human olfactory sense, by an organoleptic panel using an previously agreed upon arbitrary scale. Such a panel generally consists of five or more persons, preferably trained in odor detection and evaluation.

It is interesting that there are people, known as anosmics, whose sense of smell is seriously impaired or totally lacking. It is also important to recognize that odor is quite subjective and that a given scent may be regarded as acceptable or even pleasant by one individual and unpleasant by another.

A common technique in odor evaluation is to accelerate or increase formation of the odor by exposing the test specimens in closed containers to an elevated temperature or a high relative humidity or both before the organoleptic panel begins the judging.

A problem related to odor is the taste imparted to food in contact with an article or very near it. An example of such an article is a refrigerator door gasket. One test method is to place a specimen in a closed jar containing butter, which does not touch the specimen. After a suitable aging period, an organoleptic panel tastes the butter and evaluates it for any off-flavor. These subjective methods may be replaced shortly with instrumental methods based on head-space gas chromatography with mass spectral (GC/MS) analysis of the components. An area of interest is the odors in automotive interiors.

REFERENCES

1. L Mascia. The Role of Additives in Plastics. London: Edward Arnold, 1974, p 1.
2. American Heritage Collegiate Dictionary. New York: American Heritage Publishing Co., 1969, p 1106.
3. American Heritage Collegiate Dictionary. New York: American Heritage Publishing Co., 1969, p 1003.
4. L Mascia. The Role of Additives in Plastics. London: Edward Arnold, 1974, p 2.
5. NM Molnar. J Am Oil Chem Soc 51:84–87, 1974.
6. ASTM D1893. Method of Test for Blocking of Plastic Film, 1972.
7. ASTM D1894. Method of Test for Coefficients of Friction of Plastic Film, 1975.
8. AM Birks. Plast Technol 131, July 1977.
9. SG Hatzikiriakos et al. SPE Conference Polyolefins 2000, Houston, TX, Feb. 2000.
10. JK Rogers. SPE J 29:28–34, January 1973.
11. L Mascia. The Role of Additives in Plastics. London: Edward Arnold, 1974, p 106.
12. WI Schmidt. Modern Plastics Encyclopedia. New York: McGraw-Hill, 1984, p 103.
13. JL Roger. SPE J 29:28–34, January 1973.
14. M Cubera. Plast Compound 6(2):29, 1983.
15. Federal Test Method Standard No. 101B, Method 4046, Electrostatic Properties of Material, 1969.
16. JL Roger. SPE J. 29:28–34, January 1973; L Mascia. The Role of Plastics in Additives. London: Edward Arnold, 1974, p 107.
17. Chemicals and Additives Special Report '83. Mod Plast 60:75, September 1983; Sa-Go Associates Estimates, private communication, 1986.
18. JL Rogers. Plast Eng 29:52–56, February 1973.
19. ASTM Standards, D-883-75a, Part 35, Nomenclature Relating to Plastic.

20. Oakes Mixer, E. T. Oakes Corp., Islip, NY; Euromatic, Stork America Corp., New Canaan, CT.
21. P Schmidt, A Polte. Kunststoffe 57:25, 1967.
22. PV Batsch. U.S. Patent 2.861,963 (1958) (to U.S. Rubber Co.)
23. MS Maltenfort. U.S. Patent 2, 966, 470 (1960) (to Chemical Research Association).
24. KM Deal, DC Morris, RR Waterman, Ind Eng Chem Prod Res Dev 3:209, 1964; Chem Eng News 63, September 23, 1963.
25. RR Waterman, KM Deal, PA Whitman, (U.S. Patent 3,288,729 (1966) (to R. T. Vanderbilt Co.).
26. RT Fomade. Vanderbilt Co., Inc, Norwalk, CT.
27. RR Waterman, DC Morris. U.S. Patent 3,301,798 (1967). (to R. T. Vanderbilt Co.).
28. Chem Eng News 37(36):42, 1959.
29. RH Hansen. SPE J 18:77, 1962.
30. EV Osberg. India Rubber World 97(37–39):48, 1937.
31. H Lindemann. U.S. Patent 2,751,627 (1956) (to Lonza Elect. & Chem. Works).
32. H Lindemann. U.S. Patent 2,829,117 (1958) (to Lonza Elect. & Chem. Works).
33. O Fuchs. German Patent 1,065,169 (1959) (to Dynamit A. G.).
34. O Fuch. German Patent 1,060,508 (1959) (to Dynamit A. G.).
35. DD Lineberry. U.S. Patent 3,052,643 (1962) (to Union Carbide Corp.); Farbwerke Hoechst A.G. British Patent 890,398 (1962).
36. RH Peterson, WK Asbeck. U.S. Patent 3,122,515 (1964). (to Union Carbide Corp.).
37. Encyclopedia of Polymer Science and Technology, Vol. 2. New York: Wiley–Interscience, 1965, p 534.
38. JG Hawkins. British Patent 818,224 (1959) (to Whiffen & Sons).
39. GR Sprague, FM Scantlebury. U.S. Patent 2,737,503 (1956) (to B.F. Goodrich Co.).
40. Sorbo Ltd. British Patent 728,666 (1955).
41. TW Sarge, FH Justin. U.S. Patent 2,695,427 (1954) (to Dow Chemical Co.).
42. WT Ten Broech, Jr. U.S. Patent 2,478,879 (1949) (to Wingfoot Corp.).
43. TF Bush, U.S. Patent 2,909,493 (1959) (to B.F. Goodrich Co.).
44. DR Jones. Br Plast 248–250, May 1962.
45. Farbenfabriken Bayer, A. G. German Patent 1,184,951 (1962).
46. WJ Vokousky. U.S. Patent 3,084,127 (1963) (to B.F. Goodrich Co.); see also Tech. Service Bull. FF1202F/TF 915. Cleveland, OH: B.F. Goodrich Chemical Co.
47. The following papers deal in comprehensive manner with the chemistry and technology of organic blowing agents: RA Reed. Plast Prog 1955:51, 1956; RA Reed. Br Plast 33(10):468, 1960; HA Scheurlen. Kunststoffe 47:446, 1957; HR Lasman. Encycl Polym Sci Technol 2:532, 1965; RL Heck, III. Plast Compound 1(4):52, 1978; Plast Compound 3(6):64, 1980.
48. M Hunt. U.S. Patent 2,471,959 (1949) (to E.I. duPont de Nemours & Co.); RE Burk, U.S. Patent 2,500,023 (1950) (to E.I. duPont de Nemours & Co.).
49. Y Landler. U.S. Patent 3,200,089 (1965) (to Kleber Colombes); J Cell Plast 3(9): 404, 1967; Mod Plast 45(2):94, 1967.
50. F Lober, M Bogemann, R Wegler. U.S. Patent 2,626,933 (1953) (to Farbenfabriken Bayer, A. G.).
51. BA Hunter, DL Schoene. Ind Eng Chem 44:119, 1952.

52. DL Schoene. U.S. Patent 2,552,065 (1951) (to U.S. Rubber Co.).
53. G Wick, D Homann, P Schmidt. Kunststoffe 49:383, 1959.
54. Celogen OT. Bulletin 200-B90. Naugatuck, CT: Uniroyal Chemical: Nitropore OVSH, Technical Bulletin ONE-01. Wilmington, MA: National Polychemicals, Inc.
55. L Clark, T Grabowsky, A Poshkus. U.S. Patent 2,849,028 (1958) (to Armstrong Cork Co.); L Clark. U.S. Patent 2,873,259 (1959) (to Armstrong Cork Co.).
56. BA Hunter. U.S. Patent 3,152,176 (1964) (to U.S. Rubber Co.).
57. BA Hunter. U.S. Patent 3,235,519 (1966) (to U.S. Rubber Co.); Celogen RA. Bulletin 200-B67. Naugatuck, CT: Uniroyal Chemical.
58. RA Reed. Plast Prog 1955:51, 1956; HR Lasman. Mod Plast Encycl 41-1A:369, 1967.
59. Subpart F, 121.2550, 121.1085, 121.2562.
60. WB Curtis, BA Hunter. Patent 2,806,073 (1957) (to U.S. Rubber Co.).
61. BIOS Report 1150, No. 22, 21–23.
62. Bulletin PKB-2. Wilmington, MA: National Polychemicals, Inc., 1958; RR Barnhart. Comp Res Rep. 38, Celogen AZ, 2–3, 1958; RA Reed. Br Plast. 33(10):471, 1960.
63. Tech News, 12, 1963.
64. Bulletin OKE-44. Wilmington, MA: National Polychemicals, Inc. EB Harris. 1(2): 296, 1965.
65. Advance Div., Carlisle Chem. Works, Inc. (now Akzo/Interstab Chemicals), New Brunswick, NJ, ABC Blowing Agent Catalyst.
66. RE Lally, LM Alter. SPE J. 23(11):69, 1967.
67. HR Lasman, JC Blackwood. Plast Technol 9(9):37, 1963.
68. Bulletin OKE-42. Wilmington, MA: National Polychemicals, Inc.
69. TV Hartmen, RR Kozlowski, T Podnar. Foams from Vinyl Dryblend Powders. Akron, OH: SPE Retec, 1965.
70. D Cram, M Lavender, RA Reed, A Schoffield. Br Plast 24, January 1961; Mass., Bulletins OKE-02, OKE-46, OKE-47, and OKE-48 Wilmington, MA: National Polychemicals, Inc.; A Werner, Mod Plast 39(2):135, 1961; A Hackert. Kunststoffe, 52(10):624, 1962; Imperial Chemical Industries, Ltd., Plastics Div., IS 881; M Lavender. Plastics 27(299):65, 1962; RJ Meyer, DI Esarove. Plast Technol 5(3): 27, 1959; General Tire & Rubber Co. British Patent 833,416 (1960); PE Roggi, Chartier,
71. RA. U.S. Patent 2,964,799 (1960) (to U.S. Rubber Co.); P Smith. Melliand Textile Rep., Int. Ed. (3), 1963.
72. PV Lanthier, HR Lasman. Preparation of low density cellular thermoplastics. SPE Retec, Palisades Section, 1964.
73. D Esarove, RJ Meyer. Plast Technol 5(3):27, 1959.
74. RJ Meyer, D Esarove. Plast Technol 5(4):32, 1959; ER Dilley. Plast Inst Trans. 34(109):17, 1966.
75. TJ Rhodes. U.S. Patent 2,907,074 (1959) (to U. S. Rubber Co.); DDM Streed, EA Luxenberger. U.S. Patent 2,974,373 (1961) (to U. S. Rubber Co.); WR Hickler, JL Powell. U.S. Patent 2,939,180 (1960) (to B. F. Goodrich Co.); SL Strickhouser, RJ Van Twisk. U.S. Patent 2,917,749 (1959) (to U.S, Rubber Co.).

76. TH Brillinger. U.S. Patent 3,032,826 (1962) (to W. R. Grace & Co.).
77. RF Nairn. U.S. Patent 2,961,332 (1960) (to Congoleum-Nairn).
78. ASTM D1715. Method of Test for Gas Evolved from Chemical Blowing Agents for Cellular Plastics.
79. RH Hansen. SPE J. 181:80–82, 1962.
80. C Benning. Plast Technol. 13(4):56, 1967.
81. HR Lasman, JC Blackwood. Plast Technol. 9(9):37, 1963.
82. CC Yeager. Environmental effects on Flexible vinyl systems. SPI Vinyl Film & Sheeting Meeting, 1969.
83. CJ Wessel. SPE Trans. 4(3):193, 1964.
84. JP Scullin, MD Dudarevitch, AI Lowell. Encyclopedia of Polymer Science and Technology. Vol. 2. New York: Wiley Interscience, 1965, p 379.
85. GA Greathouse, CJ Wessel, eds. Deterioration of Materials: Causes and Preventative Techniques. New York: Reinhold, 1954.
86. EL Cadmus. Additives For Plastics. Vol. 1. New York: Academic Press, 1978, p 219.
87. DV Rosato, RT Schwartz. Environmental Effects on Polymeric Materials. Vol. I. New York: Wiley, 1968, p 990.
88. CJ Wessel. Time Dependent Effect in Plastics Materials: Biodeterioration. Plastics Institute of America, 1963.
89. Stahl, W. H., and Pessen, H., Appl. Microbiol., 1(1), 30 (1953).
90. S Berk, H Ebert, L Teitell. Ind Eng Chem. 49(7):1115, 1957.
91. EH Hueck-van der Plas. Plastica 13:1216, 1960.
92. O Johnson, N Krog, JL Poland. Chem Week 92(21):117, 1963; 92(22):55, 1963.
93. JP Scullin, MD Dudarevitch, AI Lowell. Encyclopedia of Polymer Science and Technology. Vol. 2. New York: Wiley–Interscience, 1965, p 379.
94. P Baker. Plast Compound 35, October 1978.
95. WE Field. U.S. Patent 2,567,905 (1951) (to Monsanto Chemical Co.).
96. RW Malone. U.S. Patent 2,567,910 (to Monsanto Chemical Co.).
97. JR Darby. U.S. Patent 2,632,746 (1953) (to Monsanto Chemical Co.).
98. JR Darby. U.S. Patent 2,632,747 (1953) (to Monsanto Chemical Co.).
99. JR Darby. U.S. Patent 2,689,837 (1954) (to Monsanto Chemical Co.).
100. WK Teller. U.S. Patent 3,005,720 (1961) (to Weco Products Co.).
101. FJ Gay, AH Wetherly. Australia Commonwealth Scientific and Industrial Research Organization, Div. of Entomology, Technical Paper No. 5, 1962, p 31.
102. CJ Wessel. SPE Trans. 4(3):193, 1964.
103. JF Welch, EW Duggan. Mod Packag 25(6):130, 1952.
104. RA Connolly. Mater Res Stand. 3:193, 1963.
105. EF Carlston, LR Whiting. U.S. Patent 2,592,655 (1952) (to the United States of America).
106. RGH Siu, GR Mandels. Text Res J 20:516, 1950.
107. GR Mandels, RGH Siu, J Bacteriol 60:249, 1950.
108. R Burgess, AE Darby. Br Plast 37(1):32, 1964.
109. R Burgess, AE Darby. Br Plast 38(3):165, 1965.
110. AD Baskin, AM Kaplan. Appl Microbiol 4:288, 1956.

111. R Pantaleoni. Chem Eng News 31(17):1730, 1953.
112. Process Stream Purification Through Hydride Chemistry. Technical Booklet. Ventron Corporation, Metal Hydrides Division.
113. RH Wise. U.S. Patent 2,867,651 (1959) (to Standard Oil Co.).
114. WA Dimler, AA Schetelich. U.S. Patent 2,957,023 (to Esso Research and Engineering Co.).

Index